Mechanics of Particle- and Fiber-Reinforced Polymer Nanocomposites

Mechanics of Particle- and Fiber-Reinforced Polymer Nanocomposites

From Nanoscale to Continuum Simulations

Sumit Sharma
Dr B R Ambedkar National Institute of Technology Jalandhar
Jalandhar 144 011, Punjab
INDIA

WILEY

This edition first published 2021

Published by Jon Wiley & Sons, Inc., Hoboken, New Jersey
Published simultaneously in Canada.

Registered Office
John Wiley & Sons, Inc., 111 River Street, Hoboken, NJ 07030, USA

Editorial Office
111 River Street, Hoboken, NJ 07030, USA

For details of our global editorial offices, customer services, and more information about Wiley products visit us at www.wiley.com.

Wiley also publishes its books in a variety of electronic formats and by print-on-demand. Some content that appears in standard print versions of this book may not be available in other formats.

Library of Congress Cataloging-in-Publication Data

Names: Sharma, Sumit, 1985- author.
Title: Mechanics of particle- and fiber-reinforced polymer nanocomposites : from nanoscale to continuum simulations / Sumit Sharma.
Description: First edition. | Hoboken, New Jersey : Wiley, 2021. | Includes index.
Identifiers: LCCN 2021002139 (print) | LCCN 2021002140 (ebook) | ISBN 9781119653622 (cloth) | ISBN 9781119653639 (adobe pdf) | ISBN 9781119653646 (epub)
Subjects: LCSH: Nanocomposites (Materials) | Polymers.
Classification: LCC TA418.9.N35 S477 2021 (print) | LCC TA418.9.N35 (ebook) | DDC 620.1/92–dc23
LC record available at https://lccn.loc.gov/2021002139
LC ebook record available at https://lccn.loc.gov/2021002140

Cover Design: Wiley
Cover Image: © metamorworks/Getty Images

Set in 9.5/12.5pt STIXTwoText by SPi Global, Chennai, India

Dedicated to *my parents Mr. Ashok Sharma and Mrs. Nirmal Sharma; my daughter Dhriti Sharma and wife Mrs. Rajni Sharma; and my teachers Dr. Rakesh Chandra and Dr. Pramod Kumar.*

Contents

Preface

This book has resulted due to the guidance of my supervisors, **Dr. Rakesh Chandra** and **Dr. Pramod Kumar** at Dr. BR Ambedkar National Institute of Technology, Jalandhar. The author is highly grateful to **Dr. S.K. Tiwari** and **Mrs. Deepa Bedi** for their invaluable contribution. Without their esteemed guidance, support, and motivation, this book might not have materialized. I am very grateful to my parents, my wife, and my little daughter, who have always motivated me during my life. There was an urgent need for such a book because it caters to the need of students and researchers working in the field of multi-scale modeling of polymer composites. This book will provide the readers with an overview of the most commonly used tools for multi-scale modeling of particle and fiber reinforced polymer composites. The Editor is working in the area of molecular dynamics (MD) simulation of composites since the last nine years. Before that the Editor has worked in the area of "Finite Element Modeling" of composites using NISA and MATLAB.

Biovia Materials Studio, formerly known as Accelrys Materials Studio, is a versatile tool for atomistic modeling. It is being widely used in industry for simulation of proteins, DNA, and various other biological substances. It is widely used for predicting the properties of nanomaterials such as carbon nanotubes, polymers, metals, ceramics, and various types of nanocomposites. In this book, the readers will find all the basic steps necessary for multi-scale modeling of any material. After reading this book, the readers will be able to model their own problems for predicting the properties of polymer-based composites.

In Chapter 1, different types of fibers and matrix materials have been explained in detail in order to give the readers an introduction about the constituents of a composite material. The fibers discussed are natural and advanced fibers such as glass fiber, boron fiber, carbon fiber, Kevlar fiber, etc. The polymer matrix, metal matrix, and ceramic matrix composites have been explained in detail along with the processing techniques. In addition to the above, the basic terminology associated with the composites has been defined.

Chapter 2 is an attempt to identify the topics that are most relevant to nanocomposite materials and structures and review representative journal publications that are related to those topics. Articles covering developments in modeling techniques, from continuum to atomistic methods, are emphasized. Publications, in which mechanical properties such as stiffness and damping and thermal properties such as thermal conductivity have been predicted, are reviewed. Chapter 3 reviews and evaluates models that predict the stiffness of short-fiber composites. These include the dilute model based on Eshelby's equivalent inclusion, the self-consistent model for finite-length fibers, Mori–Tanaka type models, bounding models, the Halpin–Tsai equation, and shear lag models. Several models are found to be equivalent to the Mori–Tanaka approach, which is also equivalent to the generalization of the Hashin–Shtrikman–Walpole lower bound. The essential ideas and numerical implementation of fast Fourier transform-based homogenization and asymptotic homogenization have been discussed.

In Chapter 4, the properties predicted by the mathematical models are compared with finite element method (FEM) models, and the results given by Finegan et al. [1]. Cox shear lag model is used to find the longitudinal storage and loss moduli and the corresponding loss factors. Eshelby's model is also used to predict different storage moduli such as, E'_{11}, E'_{22}, G'_{23}, and G'_{12} and corresponding loss factors, that is, η_{11}, η_{22}, η_{23}, and η_{12}. Similarly, Mori–Tanaka model, Chow model, Modified Halpin–Tsai model/Finegan model, Hashin–Shtrikman model, Lielens model, and Self-consistent model can be used to predict the storage moduli and damping of the composite. Also, different models have been compared for Type 1 RVE. The strain energy values are used to find the storage moduli and the corresponding loss factors.

In Chapter 5, the principle of Dynamic Mechanical Analyzer (DMA) and its working is explained. Specimen of vapor grown carbon fiber (VGCF)/polypropylene (PP) composite is tested on DMA. The storage modulus of the nanocomposite in fiber direction and damping loss factor are then evaluated, and the results are compared with FEM and Mori–Tanaka model. Chapter 6 provides the reader an initial overview of molecular dynamics. Various molecular modeling methods, such as molecular dynamics, Monte Carlo, Brownian dynamics, dissipative particle dynamics, lattice Boltzmann methods have been discussed in order to highlight how these are different from each other. Types of force-field, potential, ensemble, thermostat, boundary conditions, and molecular dynamics methodology have been discussed in detail for better understanding of the readers. After reading Chapter 6, the audience will be able to understand the basic terminology of molecular modeling.

Chapter 7 provides the details of molecular dynamics (MD) simulation procedure for nanofiber- and nanoparticle-reinforced polymer composites. MD simulation has been used to study the effect of carbon nanofiber (CNF) volume fraction (V_f) and aspect ratio (l/d) on mechanical properties of CNF reinforced PP composites. MD simulations have also been used to investigate the effects of organic/inorganic components on the properties of silica nanoparticle/hydroxyapatite fiber reinforced bis-GMA/TEGDMA dental composite materials.

Chapter 8 presents a structural mechanics approach to modeling the deformation of carbon nanotubes. Fundamental to the proposed concept is the concept that a carbon nanotube is a geometrical frame-like structure, and the primary bonds between two nearest-neighboring atoms act like load-bearing beam members, whereas an individual atom acts as the joint of the related load-bearing beam members. By establishing a linkage between structural mechanics and molecular mechanics, the sectional property parameters of these beam members are obtained. The accuracy and stability of the present method is verified by its application to graphite. Chapter 9 summarizes the results of Chapters 1–8 and outlines the future scope of work.

An attempt has been made here to cover thoroughly all the tools related to multi-scale modeling of polymer-based composites so that the users working in this field can use this book as the text for modeling their own problems as per their requirement. The Editor is highly thankful to all the contributors of this book for their excellent work. The Editor will also be highly grateful to the potential readers for sending their valuable suggestions, if any, so that this book can be improved further.

April 2020

Sumit Sharma
Dr. BR Ambedkar National Institute of Technology
Jalandhar-144011, Punjab
India

Reference

1 Finegan, I.C., Tibbetts, G.G., and Gibson, R.F. (2003). Modeling and characterization of damping in carbon nanofiber/polypropylene composites. *Composites Science and Technology* 63 (11): 1629–1635.

Biography

Dr. Sumit Sharma is working as an Assistant Professor in the Department of Mechanical Engineering in Dr. BR Ambedkar National Institute of Technology (NIT) Jalandhar, Punjab. Before joining this institute, he worked as an Assistant Professor in the School of Mechanical Engineering in Lovely Professional University, Phagwara. He completed his PhD in Composite Materials from NIT Jalandhar in 2015. He did his MTech in Mechanical Engineering, also from NIT Jalandhar in 2010. He graduated in Mechanical Engineering (with honors) from Kurukshetra University in 2007.

Dr. Sumit Sharma is working in the area of Molecular Dynamics simulation of composites since the last eight years. He has more than 30 research articles in reputed journals such as Computational Materials Science, Composites Part B, Composite Science and Technology, Journal of Composite Materials, JOM, IEEE, etc. He is also the reviewer of various journals such as Computational Materials Science, Composites Part B, Composite Science and Technology, Computational Condensed Matter, Carbon, etc. He is the member of ASTM, MRS, and ISME societies. He has published the following books:

1. **Sharma S.** (2020), "*An Introduction to Molecular Dynamics Simulation of Polymer Composites*", NOVA Publishers, ISBN: 9781536174083.
2. Han B., **Sharma S.**, Nguyen T.A., Longbiao L., Bhat K.S. (2020), "*Fiber-Reinforced Nanocomposites: Fundamentals and Applications*", Elsevier Publishers, ISBN: 9780128199046.
3. **Sharma S.** (2019), "*Molecular Dynamics Simulation of NanoComposites Using Biovia Materials Studio, LAMMPS and Gromacs*", Elsevier Publishers, ISBN: 9780128169544.
4. **Sharma S.** (2019), "*Metallic Glass Based Nanocomposites: Molecular Dynamics Study Of Properties*", Boca Raton, CRC press, ISBN: 9780367076702.

His research interests include molecular dynamics, mechanics of composite materials, metallic glasses, multi-scale modeling, strength of materials, materials science and engineering, fracture mechanics, mechanical vibrations, and finite element modeling.

1

Introduction

An article was published by Professor A. Kelly [1] in Composites Science and Technology titled "Composites in Context." In that article, large-scale social changes which have influenced the development of new materials were reviewed and new materials and processing methods were described and contrasted with some recent advances in composite materials science. Emerging technologies at the time included in situ metal–matrix composites, carbon-fiber reinforced thermoplastic composites, SiC-reinforced aluminum as well as toughening of ceramics through the use of fiber-reinforcement. Tremendous developments have been made in many aspects of composites research and technology during the two decades since the publication of Kelly's paper. Recent advances in producing nanostructured materials with novel material properties have stimulated research to create multifunctional macroscopic engineering materials by designing structures at the nanometer scale. Motivated by the recent enthusiasm in nanotechnology, development of nanocomposites is one of the rapidly evolving areas of composites research.

According to American Ceramic Society, Nanotechnology can be broadly defined as, "The creation, processing, characterization, and utilization of materials, devices, and systems with dimensions of the order of 1–100 nm, exhibiting novel and significantly enhanced physical, chemical, and biological properties, functions, phenomena, and processes due to their nano-scale size." By US National Nanotechnology Initiative (NNI) standards, nanotechnology involves the following according to Hunt [2]:

(a) Research and technology development at the atomic, molecular, or macromolecular levels, approximately 1–100 nm in length (approximately 80 000 smaller than a human hair).
(b) Creation and use of structures, devices, and systems that have novel properties and functions because of their small and/or intermediate size.
(c) Ability to control or manipulate on the atomic scale.

Current interests in nanotechnology encompass nanobiotechnology, nanosystems, nanoelectronics, and nanostructured materials, of which nanocomposites are a significant part. Through nanotechnology, it is envisioned that nanostructured materials will be developed using a bottom-up approach. "More materials and products will be made from the bottom-up, that is, by building them from atoms, molecules, and the nano-scale powders, fibers and other small structural components made from them. This differs from all previous manufacturing, in which raw materials get pressed, cut, molded and otherwise coerced into parts and products." Scientists and engineers working with fiber-reinforced composites have practiced this bottom-up approach in processing and manufacturing for decades. When designing a composite, the material properties are tailored for the desired performance across various

Mechanics of Particle- and Fiber-Reinforced Polymer Nanocomposites: From Nanoscale to Continuum Simulations, First Edition. Sumit Sharma.

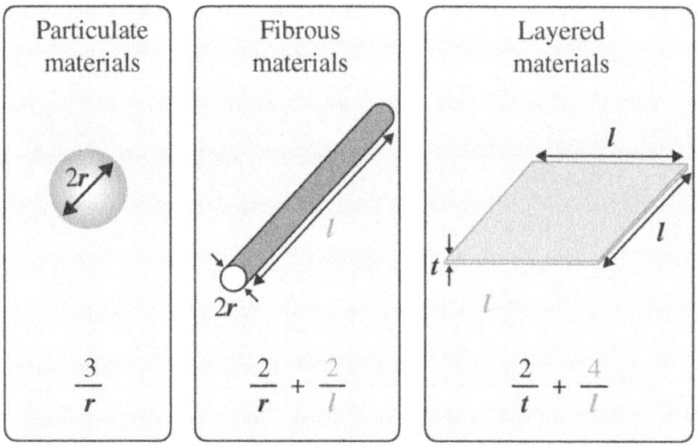

Figure 1.1 Surface area/volume relations for varying reinforcement geometries.

length scales. From selection and processing of matrix and fiber materials, and design and optimization of the fiber/matrix interface/interphase at the submicron scale to the manipulation of yarn bundles in 2-D and 3-D textiles to the lay-up of lamina in laminated composites and finally the net-shape forming of the macroscopic composite part, the integrated approach used in composites processing is a remarkable example in the successful use of the "bottom-up" approach.

Expansion of length scales from meters (finished woven composite parts), micrometers (fiber diameter), submicrometers (fiber/matrix interphase) to nanometers (nanotube diameter) presents tremendous opportunities for innovative approaches in the processing, characterization, and analysis/modeling of the new generation of composite materials. As scientists and engineers seek to make practical materials and devices from nanostructures, understanding material behavior across length scales from the atomistic to macroscopic levels is required. Knowledge of how the nanoscale structure influences the bulk properties will enable design of the nanostructure to create multifunctional composites.

A morphological characteristic that is of fundamental importance in understanding of structure–property relationship of nanocomposites is the surface area/volume ratio of the reinforcement materials. This chapter discusses nanocomposites based upon the three categories of reinforcement materials: particles (silica, metal, and other organic and inorganic particles), layered materials (graphite, layered silicate, and other layered minerals), and fibrous materials (nanofibers and nanotubes). As illustrated in Figure 1.1, the change in particle diameter, layer thickness, or fibrous material diameter from micrometer to nanometer, changes the ratio by three orders in magnitude. At this scale, there is often distinct size dependence of material properties. In addition, with the drastic increase in interfacial area, properties of the composite become dominated more by properties of the interface or interphase.

1.1 Nanoparticle-Reinforced Composites

Particulate composites reinforced with nanosized particles of various materials are perhaps the most widely utilized composites in everyday materials. Particles are typically added to enhance the matrix elastic modulus and yield strength. By scaling the particle size down to the nanometer scale, it has been shown that novel material properties can be obtained.

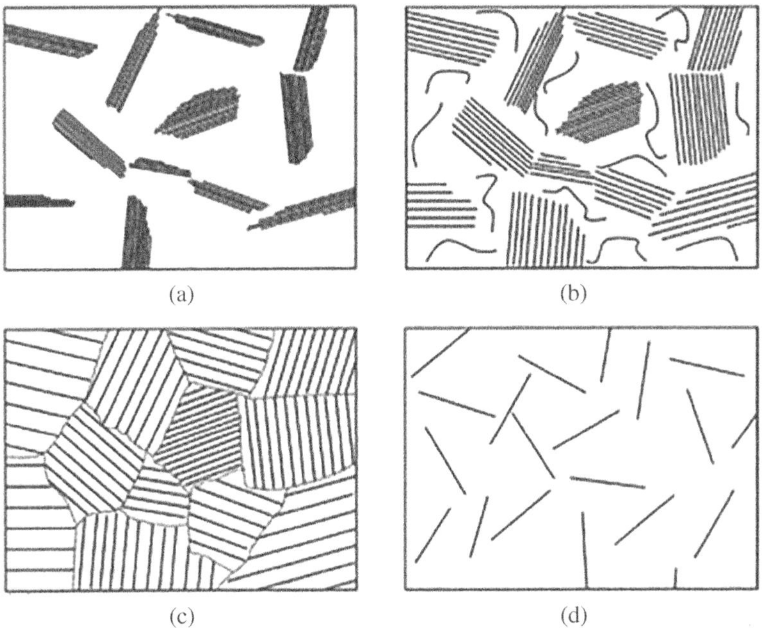

Figure 1.2 Morphologies of polymer/clay nanocomposites: (a) conventional miscible, (b) partially intercalated and exfoliated, (c) fully intercalated and dispersed, and (d) fully exfoliated and dispersed.

1.2 Nanoplatelet-Reinforced Composites

Nanoplatelet-reinforced composites are mainly of two types: clay and graphite. In their bulk state, both clay and graphite exist as layered materials. In order to utilize these material most efficiently, the layers must be separated and dispersed throughout the matrix phase. Morphology of clay/polymer nanocomposites has been shown in Figure 1.2. In the conventional miscible state, interlayer spacing in a clay particle is at its minimum. When polymer resin is inserted into the gallery between the adjacent layers, spacing expands, and it is known as the intercalated state. If the layers are fully separated, clay is considered to be exfoliated. Montmorillonite, saponite, and synthetic mica are commonly used clay materials. Advantages of polymer-based clay nanocomposites include improved stiffness, strength, toughness, and thermal stability as well as reduced gas permeability and coefficient of thermal expansion. In addition to mechanical properties, thermal stability, fire resistance, and gas barrier properties of polymer/clay nanocomposites can be enhanced through the addition of nanometer-scale reinforcement.

1.3 Nanofiber-Reinforced Composites

Vapor grown carbon nanofibers (CNF) have been used to reinforce a variety of polymers, including polypropylene, polycarbonate, nylon, poly(ether sulfone), poly(ethylene terephthalate), poly(phenylene sulfide), acrylonitrile-butadiene-styrene (ABS), and epoxy. CNFs are known to have wide-ranging morphologies from structures with a disordered bamboo-like structure to highly graphitized "cup stacked" structures as shown in Figure 1.3,

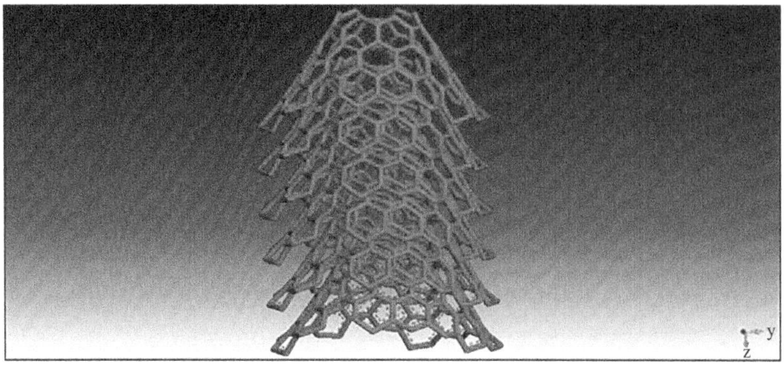

Figure 1.3 Molecular model of stacked cones.

where the conical shells of nanofiber are nested within each other. CNFs typically have diameters of the order of 50–200 nm. CNFs have transport and mechanical properties that approach the theoretical values of single-crystal graphite, similar to the fullerenes, but they can be made in high volumes at low cost, ultimately lower than that of conventional carbon fibers. In equivalent production volumes, CNFs are projected to have a cost comparable to E-glass on a per-pound basis, yet they possess properties that far exceed those of glass and are equal to, or exceed, those of much more costly commercial carbon fiber.

1.4 Carbon Nanotube-Reinforced Composites

Carbon nanotubes (CNTs) are regarded as the ultimate carbon fiber. A single-wall carbon nanotube (SWCNT) consists of a single layer of graphene rolled into a cylinder and capped with fullerenes. These discoveries promoted mass production of chemical vapor grown carbon materials such as CNFs and CNTs. Density of a SWCNT is about 1.33–1.40 g/cm^3, which is just one-half of the density of aluminum. Elastic modulus of SWCNT is comparable to that of diamond (1.2 TPa). Reported tensile strength of SWCNT is much higher than that of high-strength steel (2 GPa). Tremendous resilience of SWCNT in sustaining bending to large angles and restraightening without damage is distinctively different from the plastic deformation of metals and brittle fracture of carbon fibers at much lower strain when subjected to the same type of deformation. Electric current carrying capability is estimated to be 1×10^9 A/cm^2, whereas copper wires burn out at about 1×10^6 A/cm^2. Thermal conductivity of SWCNT is predicted to be 6000 W/m K at room temperature; this is nearly double the thermal conductivity of diamond of 3320 W/m K. SWCNTs are stable up to 2800 °C in vacuum and 750 °C in air, whereas metal wires in microchips melt at 600–1000 °C. SWCNTs have great potential in field emission applications because they can activate phosphors at 1–3 V if electrodes are spaced 1 μm apart. Traditional Mo tips require fields of 10–100 V/μm and have very limited lifetimes.

Outstanding thermal and electric properties combined with their high-specific stiffness and strength, and very large aspect ratios have stimulated the development of nanotube-reinforced composites for both structural and functional applications. Morphology of a CNT is defined by the orientation and magnitude of the chiral vector in a graphene sheet, which is "wrapped up" to form the SWCNT. Two limiting configurations, the armchair and the zigzag nanotubes have been shown in Figure 1.4.

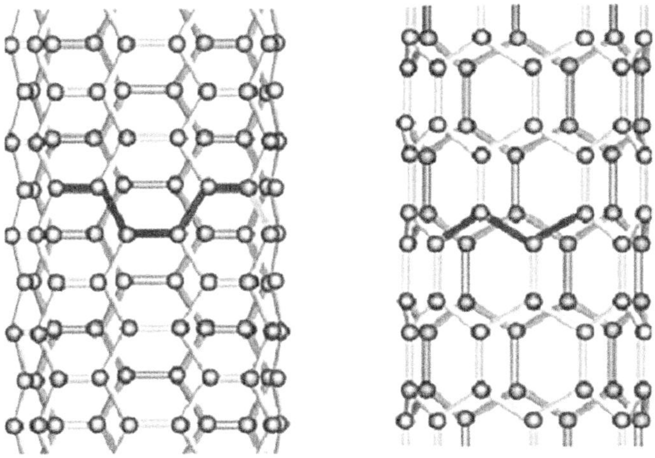
(a) (b)

Figure 1.4 Atomic structures of (a) armchair and (b) zigzag carbon nanotubes.

1.5 Nanomaterials

Nanostructured materials are single-phase or multiphase polycrystalline solids with a typical average grain size of a few nanometers (1 nm = 10^{-9} m), typically less than 100 nm. Such materials exhibit properties that are substantially different from and often superior to those of conventional coarse-grained materials due to their unique microstructure. Since the grain sizes are so small, a significant volume fraction of the atoms resides in grain boundaries. Thus, the material is characterized by a large number of interfaces in which the atomic arrangements are different from those of the crystal lattice. At such small grain sizes, the surfaces start to dominate the bulk behavior of the material and consequently, nanostructured materials exhibit special and completely new and unexpected properties, often not observed in coarse-grained materials. Accordingly, nanostructured materials have been shown to exhibit very high strength, increased diffusivity, reduced sintering temperatures, useful catalytic properties, and attractive physical properties. Due to their novel and improved properties and varied potential applications in different fields of technology, these materials are attracting increasing attention from researchers all over the world.

A theoretical interpretation of a nanomaterial must be based on some primary concept. All materials are composed of granules, which in turn consist of atoms. These granules may be visible or invisible to the naked eye, depending on their size, which can range from hundreds of microns to centimeters. The term "material" usually refers to a substance in a solid-state, which can be in either crystalline or glassy phase. Since nanoparticles can form both nanomaterials and nanopowders, it is worth pointing out that in mechanics, materials are assumed to have a stability of shape, while powders are regarded as granular media. Usually, powders are made into materials by application of special techniques such as compression, sintering, irradiation. As already mentioned, the size of granules in nanopowders and nanomaterials range from 10 to 100 nm in at least one dimension (usually in all three). Many authors believe that nanomaterials, i.e. materials with internal structure of nanoscale dimension, are hardly something new to science, just it has been discovered only recently that some formations of oxides, metals, ceramics, and other substances are nanomaterials. For example, black carbon was discovered at the beginning of the nineteenth century. Fumed silica powder, a component of silicon rubber, came into commercial use in 1940. However, it has only recently become clear that these two substances are nanomaterials. It should be noted that the size of a particle is not the only characteristic of nanoparticle, nanocrystal, or nanomaterial. In the opinion of Guz et al. [3], one quite important and specific property of many nanomaterials is that the majority of their atoms are located on the surface of a particle as opposed to conventional materials, in which atoms are distributed over the volume of a particle.

Carbon particles are a well-studied class of nanoparticles. Science has long been aware of the three forms of carbon: diamond, graphite, and amorphous carbon. Highly symmetric molecule of carbon C_{60} was discovered in 1985. It has a spherical form, resembling a football, with carbon atoms on the surface. It contains 60 atoms in five-atom rings separated by six-atom rings. These molecules were named fullerenes and became the subject of extensive and fruitful studies. Scientists who studied fullerenes were awarded the Nobel Prize in Chemistry in 1996. Since then the number of different varieties of fullerenes has increased considerably, reaching many thousands to date. The molecule C_{60} has got its name from R.B. Fuller (1895–1983), an architect who built a house from pentagons and hexagons. The molecule C_{70} not only resembles a rugby ball and includes only six-atom rings, but also bears the name fullerene. Figure 1.5a,b depicts C_{60} and C_{70} molecules. Fullerenes can form crystals called fullerites. These crystals have a face-centered cubic lattice with cavities of two types: tetrahedral and octahedral. Placing ions of potassium, rubidium, or cesium into these cavities produces new nanoparticles with unusual properties. Fullerenes may be deposited onto a surface, forming a monolayer. There are also rope-like fullerene formations. But the most important thing is that fullerene molecules can form carbon nanotubes, which may be considered as related to graphite.

The molecular structure of graphite looks like a sheet of chicken wire, a tessellation of hexagonal rings of carbon. In graphite, these sheets are stacked one over another and can slide past each other. This explains why graphite is soft and greasy and is often used as a lubricant. When graphite lattices are rolled up into a tube, they form nanotubes, molecules with a very large number of atoms, $C_{10\,000}$–$C_{1\,000\,000}$. Nanotubes differ in diameter, length, and the way they are rolled. The internal cavities may also be different, and tubes may have more than one sheet. Atoms form "hemispherical caps" at the ends of fullerene molecule.

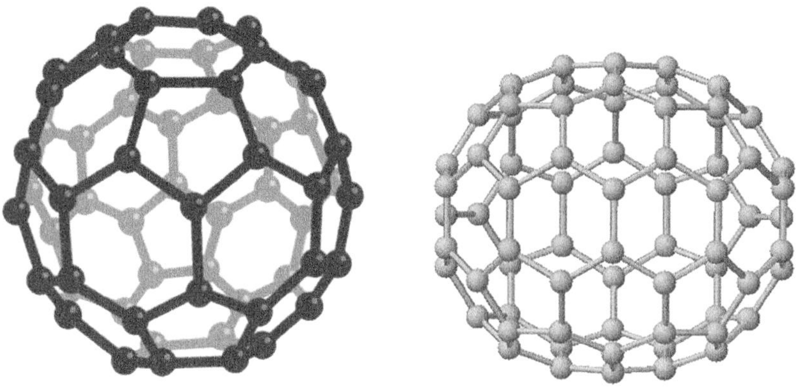

(a) (b)

Figure 1.5 (a) C_{60} fullerene and (b) C_{70} fullerene.

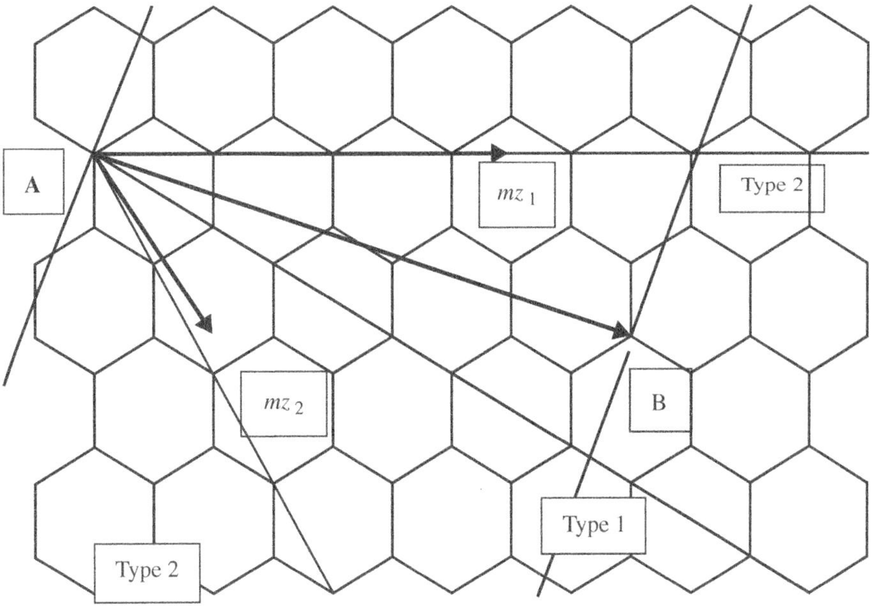

Figure 1.6 A cylindrical nanotube represented as a rolled graphite sheet with a periodic hexagonal structure consisting of regular hexagons with carbon atoms at vertices.

Sheets may be rolled differently forming zigzag, armchair, and chiral structures. Any cylindrical nanotube can be considered as being a rolled graphite sheet. A sheet is schematically depicted as a periodic hexagonal structure or lattice consisting of regular hexagons with carbon atoms at vertices as shown in Figure 1.6. The procedure of rolling a sheet is as follows: choose two atoms A and B that belong to hexagons spaced far enough from each other (it is clear that a tube cannot be made from three hexagons); connect the points with a straight line; cut out a strip from the sheet by two cuts perpendicular to the *AB* line; and roll the strip so as to superpose the points *A* and *B*. Length of *AB* will be the width of the rolled-up strip and simultaneously the length of a parallel circle of the resultant circular cylindrical tube. Such an arbitrary tube is called a chiral tube. Chirality, which can be left or right, is a concept borrowed from stereochemistry. It refers to the property of nonidentity of an object with its mirror image. Three types of chirality are central, axial, and planar, corresponding to three chiral elements, namely center, axis, and plane. Chirality can be measured with the help of a vector;

$$\vec{r} = \vec{AB} = n\vec{e}_1 + m\vec{e}_2 \quad (n, m \in N) \tag{1.1}$$

where $\vec{e1}$ and $\vec{e2}$ are vectors, defining two directions intersecting at *A*. In general, at any vertex of a hexagon there are three directions, important to the lattice. One of them (type 1) is the direction toward the opposite vertex of the hexagon. The straight line crosses the hexagon and then passes along the common side of neighboring hexagons, with this sequence repeating periodically. The side of a hexagon with two adjacent sides resembles an armchair. The other two directions (type 2) pass through *A* and the third (counting from *A*) vertex either on the left or on the right. These two directions define two straight lines that are specific to the lattice; they are differently directed but otherwise identical. The sides connecting all the three vertices (*A*, the second and the third ones) resemble a zigzag, and this sequence repeats periodically. The vector $\vec{e1}$ always lies on a type 2 line (horizontal line in Figure 1.6), while the vector $\vec{e2}$ can be directed arbitrarily.

Chiral structures are distinguished by the numbers n and m and denoted by (n, m). If $\vec{e1}$ and $\vec{e2}$ are codirectional, i.e. B and $\vec{e1}$ lie on the same straight line, then a zigzag chiral structure is obtained. Its typical feature is that m in the notation (n, m) is always equal to zero. If the point B lies on a straight line of type 1, then $\vec{e1}$ lies on the second straight line of type 2 and the type 1 line is the bisector of the angle formed by the type 2 lines. It is an armchair structure and its numbers n and m are always equal. Any other structure is called a chiral structure. Applications of nanotechnology are shown in Figure 1.7. Figure 1.8 shows SWCNTs with different chirality for the particular values of (n, m). The length of nanotubes can be 1000 times (and more) greater than their diameter. They can be single-walled and multiwalled. Multiwalled nanotubes, between 4 and 30 nm in diameter, were discovered by Iijima in 1991. The thinnest single-walled nanotube has wall thickness of 0.4 nm. By comparison, diameter of the carbon atom is equal to 0.15 nm. An assembly of nanotubes may have a diameter of 30–50 nm and a length of more than 50 μm. Single-walled nanotubes may form rough structures or ropes. Ropes consist of nanotubes bundled in an ordered manner to form a lattice. Nanotube ropes may have a diameter of 10–20 nm and a length of 100 μm and more.

Nanotubes are technologically advantageous over conventional carbon fibers, since they are produced from colloidal solutions at room temperatures, whereas carbon fibers require high-temperature environment. Research of nanotubes follows mainly three directions: (i) mechanical and electric properties of nanotubes in polymeric films, (ii) arrangement of nanotubes in polymeric composites, and (iii) properties of the nanotube matrix transition layer.

Major constituents in a fiber-reinforced composite material are the reinforcing fibers and a matrix, which acts as a binder for the fibers. Other constituents that may also be found are coupling agents, coatings, and fillers. Coupling agents and coatings are applied on the fibers to improve their wetting with the matrix as well as to promote bonding across the fiber–matrix interface. Both in turn promote a better load transfer between the fibers and the matrix. Fillers are used with some polymeric matrices primarily to reduce cost and improve their dimensional stability.

Manufacturing of a composite structure starts with the incorporation of a large number of fibers into a thin layer of matrix to form a lamina. The thickness of a lamina is usually in the range of 0.1–1 mm. If continuous (long) fibers are used in making the lamina, they may be arranged either in a unidirectional orientation (i.e. all fibers in one direction, Figure 1.9a), in a bidirectional orientation (i.e. fibers in two directions, usually normal to each other, Figure 1.9b), or in a multidirectional orientation (i.e. fibers in more than two directions, Figure 1.9c). The bi- or multidirectional orientation of fibers is obtained by weaving or other processes used in the textile industry.

1.5.1 Woven Fabric

Basic woven fabrics consist of two sets of yarns interlaced at right angles to create a single layer. Such biaxial or 0/90 fabrics are characterized by the following nomenclature:

(a) *Yarn construction*: May include the strand count as well as the number of strands twisted and plied together to make up the yarn. In case of glass fibers, the strand count is given by the yield expressed in yards per pound or in TEX, which is the mass in grams per 1000 m. For example, if the yarn is designated as 150 2/3, its yield is 150×100 or 15 000 yd/lb.

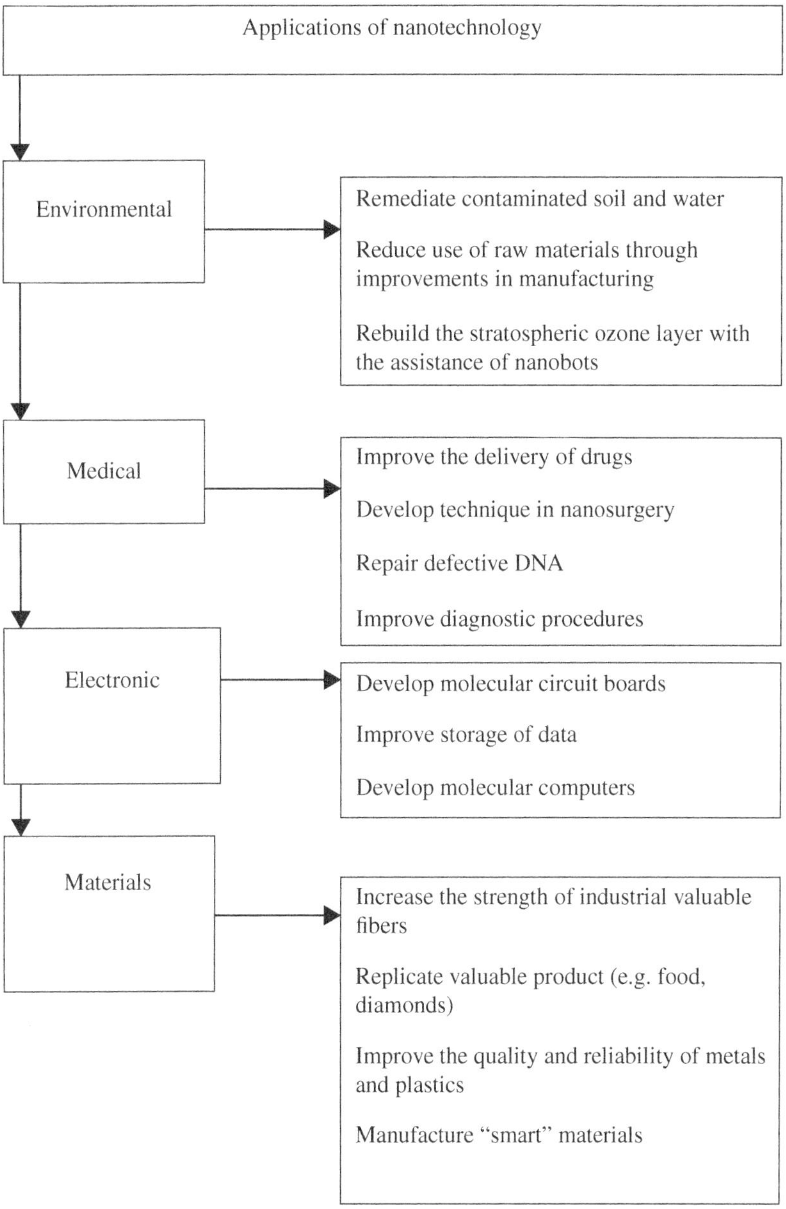

Figure 1.7 Applications of nanotechnology.

The 2/ after 150 indicates that the strands are first twisted in groups of two, and the /3 indicates that three of these groups are plied together to make up the final yarn. The yarns for carbon-fiber fabrics are called tows. They have little or no twist and are designated by the number of filaments in thousands in the tow. Denier (abbreviated as de) is used for designating Kevlar yarns, where 1 denier is equivalent to 1 g/9000 m of yarn.

(b) *Count*: Number of yarns (ends) per unit width in the warp (lengthwise) and fill (crosswise) directions (Figure 1.10). For example, a fabric count of 60×52 means 60 ends per inch in the warp direction and 52 ends per inch in the fill direction.

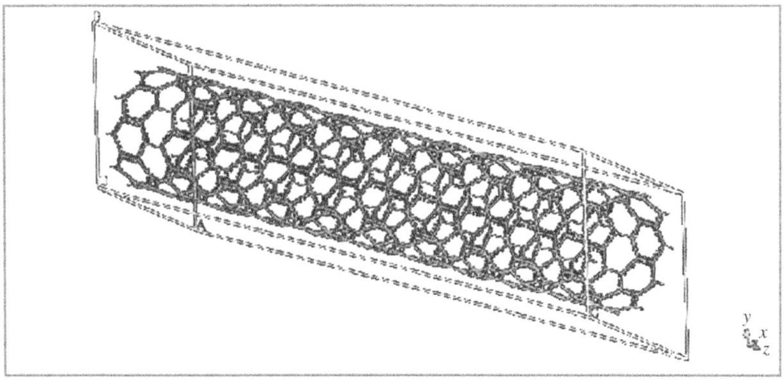

(a)

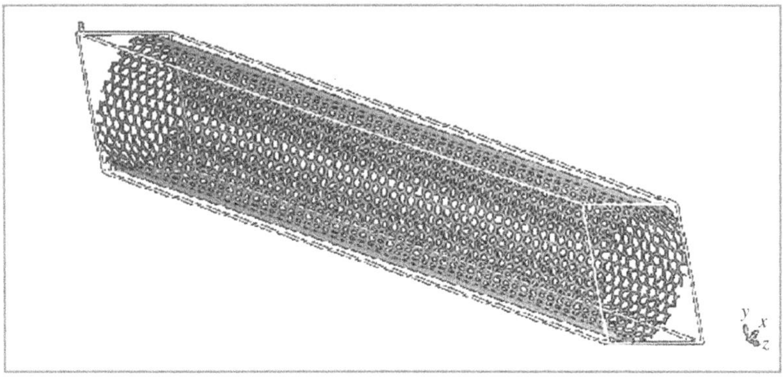

(b)

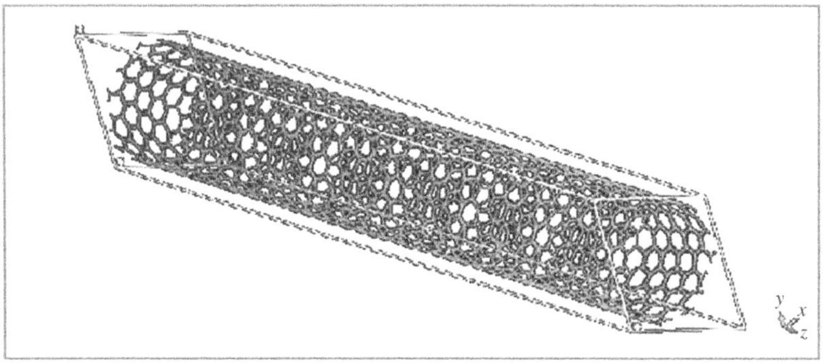

(c)

Figure 1.8 Examples of SWCNTs with different chirality for particular values of (n, m). (a) A zigzag (10,0) SWCNT. (b) An armchair (20,20) SWCNT. (c) A chiral nanotubes (12,6) SWCNT.

(c) *Weight*: Areal weight of the fabric in ounces per square yard or grams per square meter.
(d) *Thickness*: Measured in thousandths of an inch (mil) or in millimeters.
(e) *Weave style*: Specifies the repetitive manner in which the warp and fill yarns are interlaced in the fabric. Common weave styles are shown in Figure 1.11.

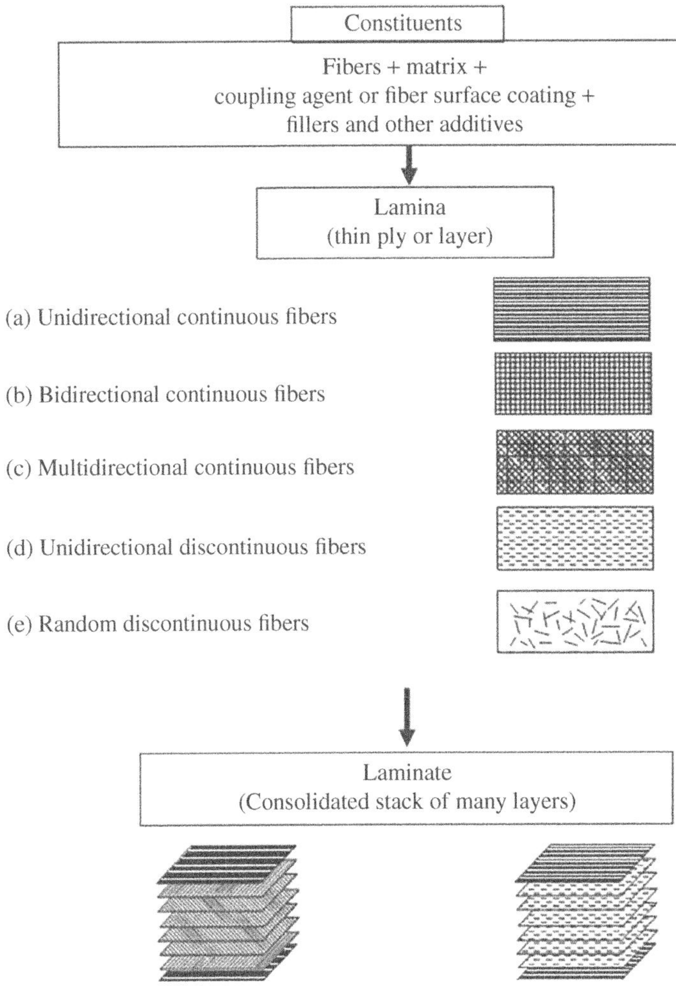

Figure 1.9 Basic building blocks in fiber-reinforced composites.

Figure 1.10 Warp and fill directions of fabrics.

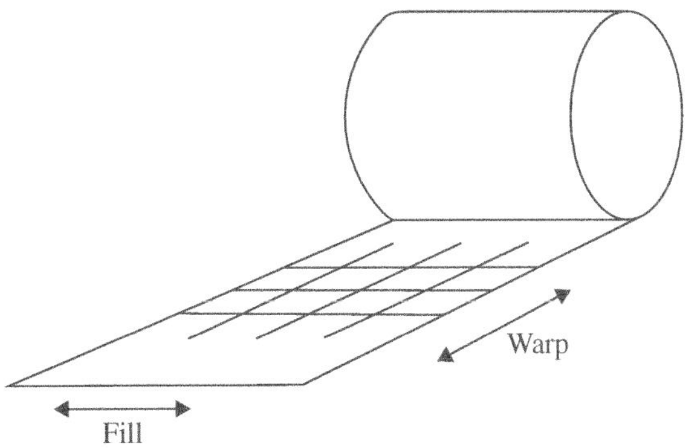

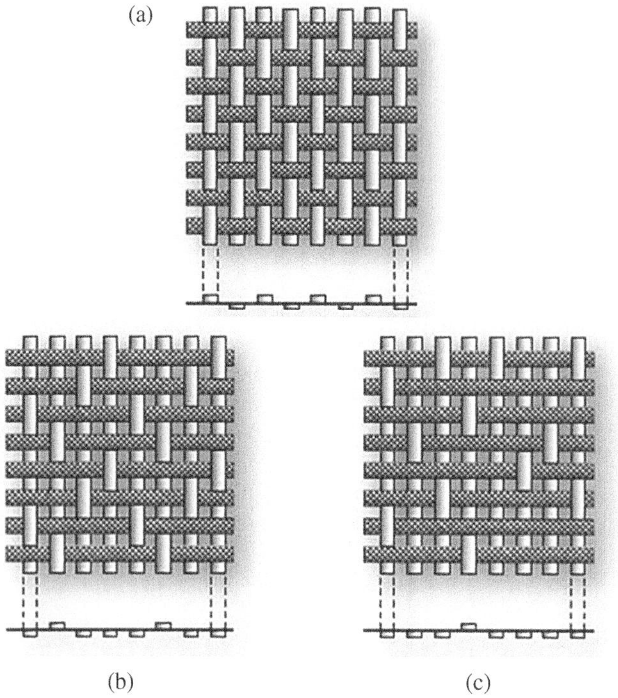

Figure 1.11 Common weave styles: (a) plain weave, (b) crowfoot satin weave and (c) 5 harness satin weave.

(a) Plain weave in which warp and fill yarns are interlaced over and under each other in an alternating fashion.
(b) Basket weave in which a group of two or more warp yarns are interlaced with a group of two or more fill yarns in an alternating fashion.
(c) Satin weave in which each warp yarn weaves over several fill yarns and under one fill yarn.

Common satin weaves are crowfoot satin or four-harness satin in which each warp yarn weaves over three and under one fill yarn, five-harness satin (over four, under one), and eight-harness satin (over seven, under one). For a lamina containing unidirectional fibers, the composite material has the highest strength and modulus in the longitudinal direction of the fibers. However, in the transverse direction, its strength and modulus are very low. For a lamina containing bidirectional fibers, the strength and modulus can be varied using different amounts of fibers in the longitudinal and transverse directions. For a balanced lamina, these properties are the same in both directions.

1.5.2 Fibers

Fibers are the principal constituents in a fiber-reinforced composite material. They occupy the largest volume fraction in a composite laminate and share the major portion of the load acting on a composite structure. Proper selection of the fiber type, fiber volume fraction, fiber length, and fiber orientation is very important, since it influences the following characteristics of a composite laminate:

(i) Density
(ii) Tensile strength and modulus
(iii) Compressive strength and modulus
(iv) Fatigue strength as well as fatigue failure mechanisms
(v) Electrical and thermal conductivities
(vi) Cost

Before discussing the types of fibers, various terms associated with the fibers should be understood first. The main terms associated with the fibers are given below:

1. *Filament*: An individual element.
2. *Strand*: Bundles of 204 filaments or multiple of these.
3. *Roving*: Combination of strands to form thicker parallel bundles.
4. *Yarns*: Strands are twisted to form yarns.
5. *Aspect ratio*: The ratio of length to diameter of a fiber.
6. *Bicomponent fibers*: A fiber made by spinning two compositions concurrently in each capillary of the spinneret.
7. *Blend*: A mix of natural staple fiber such as cotton or wool and synthetic staple fibers such as nylon, polyester. Blends are made to take advantages of the natural and synthetic fibers.
8. *Braiding*: Two or more yarns are intertwined to form an elongated structure. The long direction is called the bias direction or machine direction.
9. *Carding*: Process of making fibers parallel by using rollers covered with needles.
10. *Chopped strands*: Fibers are chopped to various lengths, 3–50 mm, for mixing with resins.
11. *Continuous fibers*: Continuous strands of fibers, generally, available as wound fiber spools.
12. *Cord*: A relatively thick fibrous product made by twisting together two or more plies of yarn.
13. *Covering power*: The ability of fiber to occupy space. Noncircular fibers have greater covering power then circular fibers.
14. *Crimp*: Waviness along the fiber length. Some natural fibers, e.g. wool, have a natural crimp. In synthetic polymeric fibers, crimp can be introduced by passing the filament between rollers having teeth. Crimp can also be introduced by chemical means. This is done by controlling the coagulation of the filament to produce an asymmetrical cross-section.
15. *Denier*: A unit of linear density. It is the weight in grams of 9000 m long yarn. This unit is commonly used in the US textile industry.
16. *Fabric*: A kind of planar fibrous assembly. It allows the high degree of anisotropy characteristic of yarn to be minimized, although not completely eliminated.
17. *Felt*: Homogeneous fibrous structure made by interlocking fibers via application of heat, moisture, and pressure.
18. *Filament*: Continuous fiber, i.e. fiber with aspect ratio approaching infinity.
19. *Fill*: It is similar to weft.
20. *Handle*: Also known as softness of handle. It is a function of denier (or tex), compliance, cross-section, crimp, moisture absorption, and surface roughness of the fiber.
21. *Knitted fabric*: One set of yarn is looped and interlocked to form a planar structure.
22. *Knitting*: This involves drawing loops of yarns over previous loops also called interlooping.

23. *Mat*: Randomly dispersed chopped fibers or continuous fiber strands held together with a binder. The binder can be resin compatible if the mat is to be used to make a polymeric composite.
24. *Microfibers*: Also known as microdenier fibers. These are fibers having less than 1 denier per filament (or less than 0.11 tex per filament). Fabrics made of such microfibers have superior silk-like handle and dense construction. They find applications in stretch fabrics, lingerie, rain wear, and others.
25. *Monofilament*: A large diameter continuous fiber, generally, with a diameter greater than 100 m.
26. *Nonwovens*: Randomly arranged fibers without making fiber yarns. Nonwovens can be formed by spun-bonding, resin-bonding, or needle punching. A planar sheet-like fabric is produced from fibers without going through the yarns spinning step. Chemical bonding and/or mechanical interlocking is achieved. Fibers (continuous or staple) are dispersed in a fluid (i.e. a liquid or air) and laid in a sheet-like planar form on a support and then chemically bonded or mechanically interlocked. Paper is perhaps the best example of a wet laid nonwoven fabric where we generally use wood or cellulosic fibers. In spun-bonded nonwovens, continuous fibers are extruded and collected in random planar network and bonded.
27. *Particle*: Extreme case of a fibrous form: it has a more or less equiaxial form, i.e. the aspect ratio is about 1.
28. *Plaiting*: It is similar to Braiding.
29. *Rayon*: Term used to designate any of the regenerated fibers made by the viscose, cuprammonium, or acetate processes. They are considered to be natural fibers because they are made from regenerated, natural cellulose.
30. *Retting*: A biological process of degrading pectin and lignin associated with vegetable fibers, loosening the stem and fibers, followed by their separation.
31. *Ribbon*: Fiber of rectangular cross-section with width to thickness ratio greater than 4.
32. *Rope*: Linear flexible structure with a minimum diameter of 4 mm. It generally has three strands twisted together in a helix. The rope characteristics are defined by two parameters, unit mass and break length. Unit mass is simply g/m or ktex, while breaking length is the length of rope that will break under the force of its own weight when freely suspended. Thus, break length equals mass at break/unit mass.
33. *Roving*: A bundle of yarns or tows of continuous filaments (twisted or untwisted).
34. *Spinneret*: A vessel with numerous shaped holes at the bottom through which a material in molten state is forced out in the form of fine filaments or threads.
35. *Spunbonding*: Process of producing a bond between nonwoven fibers by heating the fibers to near their melting point.
36. *Staple fiber*: Fibers having short, discrete lengths (10–400 mm long) that can be spun into a yarn are called staple fibers. This spinning quality can be improved if the fiber is imparted a waviness or crimp. Staple fibers are excellent for providing bulkiness for filling, filtration, and so forth. Frequently, staple natural fibers, e.g. cotton or wool, are blended with staple synthetic fibers, e.g. nylon or polyester, to obtain the best of both types.
37. *Tenacity*: A measure of fiber strength that is commonly used in the textile industry. Commonly, the units are gram-force per denier, gram-force per tex, or Newton per tex. It is a specific strength unit, i.e. there is a factor of density involved. Thus, although the tensile strength of glass fiber is more than double that of nylon fiber, both glass and nylon fiber

have a tenacity of about 6 g/den. This is because the density of glass is about twice that of nylon.

38. *Tex*: A unit of linear density. It is the weight in grams of 1000 m of yarn. Tex is commonly used in Europe.
39. *Tow*: Bundle of twisted or untwisted continuous fibers. A tow may contain tens or hundreds of thousands of individual filaments.
40. *Twist*: The angle of twist that individual filaments may have about the yarn axis. Most yarns have filaments twisted because it is easier to handle a twisted yarn than an untwisted one.
41. *Wire*: Metallic filament.
42. *Warp*: Lengthwise yarn in a woven fabric.
43. *Weft*: Transverse yarn in a woven fabric also called fill.
44. *Whisker*: Tiny, whisker-like fiber (a few millimeter long, a few meter in diameter) that is a single crystal and almost free of dislocations. Note that this term involves a material requirement. The small size and crystalline perfection make whiskers extremely strong, approaching the theoretical strength.
45. *Woven fabric*: Flat, drapeable sheet made by interlacing yarns or tows.
46. *Woven roving*: Heavy, drapeable fabric woven from continuous rovings.
47. *Yarn*: A generic term for a bundle of untwisted or twisted fibers (short or continuous). A yarn can be produced from staple fibers by yarn spinning. The yarn spinning process consists of some fiber alignment, followed by locking together by twisting. Continuous synthetic fibers are also used to make yarns. Continuous fibers are easy to align parallel to the yarn axis. Generally, the degree of twist is low, just enough to give some interfilament cohesion.

1.5.3 Types of Fibers

The functions of a reinforcing agent are the following:

1. These are the main load carrying constituents.
2. The reinforcing materials, in general, have significantly higher desired properties. Hence, they contribute the desired properties to the composite.
3. It transfers the strength and stiffness to the matrix material.

The matrix also performs various functions. These functions are listed below:

1. The matrix material holds the fibers together.
2. The matrix plays an important role to keep the fibers at desired positions. The desired distribution of the fibers is very important from micromechanical point of view.
3. The matrix keeps the fibers separate from each other so that the mechanical abrasion between them does not occur.
4. It transfers the load uniformly between fibers. Further, in case a fiber is broken or fiber is discontinuous, then it helps to redistribute the load in the vicinity of the break site.
5. It provides protection to fibers from environmental effects.
6. It provides better finish to the final product.
7. The matrix material enhances some of the properties of the resulting material and structural component (that fiber alone is not able to impart). For example, such properties are transverse strength of a lamina, impact resistance, and so forth.

The fibers that are used in the fabrication of a composite can be divided into two broad categories as follows:

1. Natural fibers and
2. Advanced fibers

1. *Natural fibers*: The natural fibers are divided into the following three subcategories:
   (i) *Animal fibers*: Silk, wool, spider silk, sinew, camel hair, and so forth.
   (ii) *Plant/vegetable fibers*: Cotton (seed), jute (stem), hemp (stem), sisal (leaf), ramie, bamboo, maze, sugarcane, banana, kapok, coir, abaca, kenaf, flax, raffia palm, and so forth.
   (iii) *Mineral fibers*: Asbestos, basalt, mineral wool, glass wool, and so forth.

2. *Advanced fibers*: An advanced fiber is defined as a fiber which has a high-specific stiffness (that is, ratio of Young's modulus to the density of the material, E/ρ) and a high-specific strength (that is, the ratio of ultimate strength to the density of the material, S/ρ). The fibers made from following materials are the advanced fibers:
   (i) Carbon and/or Graphite
   (ii) Glass fibers
   (iii) Alumina
   (iv) Aramid
   (v) Silicon carbide
   (vi) Sapphire

The materials of the advanced fibers are lighter than the conventional metals. These materials occupy higher position as compared to metals in the periodic table. Thus, these materials have higher specific properties (property per unit weight) than that of metals.

1.5.4 Boron Fiber

This fiber was first introduced by Talley in 1959 [4]. In commercial production of boron fibers, the method of chemical vapor deposition (CVD) is used. The CVD is a process in which one material is deposited onto a substrate to produce near theoretical density and small grain size for the deposited material. In CVD, the material is deposited on a thin filament. The material grows on this substrate and produces a thicker filament. The size of the final filament is such that it could not be produced by drawing or other conventional methods of producing fibers. It is the fine and dense structure of the deposited material which determines the strength and modulus of the fiber. In the fabrication of boron fiber by CVD, the boron-trichloride is mixed with hydrogen and boron is deposited according to the reaction:

$$2BCl_3(g) + 3H_2(g) \rightarrow 2B(s) + 6HCl(g) \tag{1.2}$$

In the process, the passage takes place for couple of minutes. During this process, the atoms diffuse into the tungsten core to produce the complete boridization and the production of WB_4 and W_2B_5. In the beginning, the tungsten fiber of 12 μm diameter is used, which increases to 12 μm. This step induces significant residual stresses in the fiber. The core is subjected to compression and the neighboring boron mantle is subjected to tension. The CVD method for boron fiber fabrication has been shown in Figure 1.12.

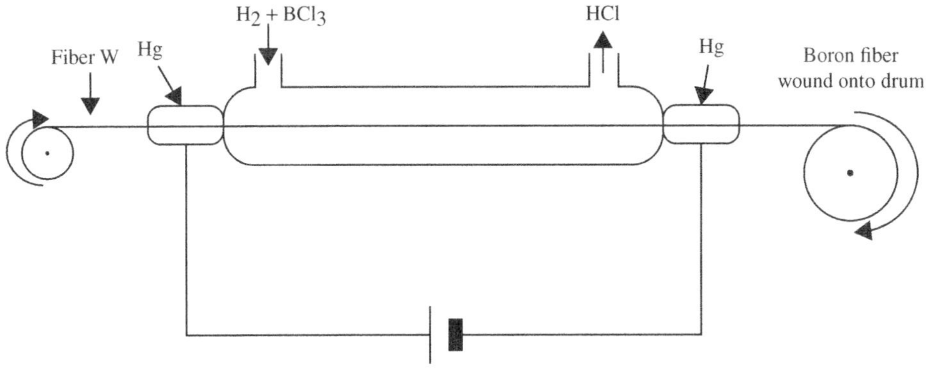

Figure 1.12 Schematic of chemical vapor deposition (CVD) process for fabrication of boron fiber.

1.5.5 Carbon Fiber

The first carbon fiber for commercial use was fabricated by Thomas Edison. There are two types of precursor materials used for C fiber fabrication:

(i) Polyacrylonitrile (PAN)
(ii) Rayon pitch, that is, the residue of petroleum refining

1.5.5.1 Fabrication of C Fiber Using PAN

A common method of manufacture involves heating the spun PAN filaments to approximately 300 °C in air, which breaks many of the hydrogen bonds and oxidizes in the material. The oxidized PAN is then placed into a furnace having an inert atmosphere of a gas such as argon, and heated to approximately 2000 °C, which induces graphitization of the material, changing the molecular bond structure. When heated in the correct conditions, these chains bond side-to-side (ladder polymers), forming narrow graphene sheets which eventually merge to form a single, columnar filament. The result is usually 93–95% carbon. Lower-quality fiber can be manufactured using pitch or rayon as the precursor instead of PAN. The carbon can become further enhanced, as high modulus, or high-strength carbon, through the heat treatment processes. Carbon heated in the range of 1500–2000 °C (carbonization) exhibits the highest tensile strength (820 000 psi, 5650 MPa or N/mm^2), while carbon fiber heated from 2500 to 3000 °C (graphitizing) exhibits a higher modulus of elasticity (77 000 000 psi or 531 GPa or 531 kN/mm^2). There are a number of types of PAN precursor fibers. They are all acrylic-based fibers and contain at least 85% of acrylonitrile. The balance may include secondary polymers or residual spin bath chemicals left over from initial PAN precursor fiber processing. There are three stages of C fiber fabrication. These are

(i) Stabilization
(ii) Carbonization
(iii) Graphitization

PAN precursor, if heated to a high temperature to promote carbonization, will not yield fibers of high strength and stiffness unless a preoxidation, or stabilization, step is used during processing. A thermally stable structure is obtained during this step, so upon further heating the original fiber architecture is retained. Stabilization generally calls for a heat treatment in the range 200–300 °C in an oxygen containing atmosphere. This leads to the polymer backbone of the precursor undergoing a series of chemical reactions that ultimately result in the formation of poly-naphthyridine, a substance with the preferred structural form for the formation of graphite. Cross-linking also occurs, induced either by oxidizing agents or by other catalysts. Upon further heating, this precursor structure gives rise to graphitic nuclei whose basal planes of the carbon atoms are oriented parallel to the direction of the polymer chains. Significant shrinkage (e.g. up to 40%) occurs during stabilization; this can be reduced by stretching the fibers during the heat treatment or by infiltrating PAN fibers with fine silica aggregates. The silica particles are lodged in the interstices between and around the fibers and essentially lock the fiber in place. Using such methods can reduce shrinkage to about 20%. Several chemical treatments can be used to speed up the oxidation process. For instance, treating PAN fibers with a di-ethanol amine or tri-ethanol amine solution prior to heat treatment has been found to substantially reduce the time required for oxidation.

The carbonization step involves heating the stabilized precursor fiber to temperatures up to 1000 °C in an inert or mildly oxidative atmosphere. Carbonization can take anywhere from a few minutes to several hours. In one type of process, fibers possessing high modulus and high strength are obtained by first heating PAN fibers in an oxidizing atmosphere until they are permeated with oxygen. The fibers are then heated further, to initiate carbonization, while being held under tension in a nonoxidizing atmosphere. Finally, to increase the ultimate tensile strength, the fibers are heat-treated in an inert atmosphere between 1300 and 1800 °C. Research has shown that if carbonization is done in an oxidizing atmosphere, then the optimal atmosphere should contain between 50 and 170 ppm oxygen. In this study, argon was used as the carrier gas, with the oxygen content varying between 2.8 and 1500 ppm.

Graphitization occurs by heating the carbonized fiber to high temperature (up to 3000 °C) in an inert atmosphere. The process can last anywhere from 1 to 20 minutes. Tensioning the fibers during graphitization improves ultimate mechanical properties and reduces any residual shrinkage. Precursor fiber begins the process, and tension is applied to the precursor while heating it to an elevated temperature. The tension is reduced before the preoxidation phase and is increased during preoxidation. The fiber is then continuously heated and graphitized under high tension. The graphitized fiber is wound onto a take-up reel and dried. Processing speeds can reach as high as 45 m/h, but are typically between 6 and 12 m/h. All the stages along with temperature and tension profiles have been shown in Figure 1.13.

Figure 1.13 also shows other important steps that can take place after the fibers are dried. First, after graphitization, special coating materials, called sizings, are applied to the fiber surface. Sizings, or fiber surface treatments, are used to provide lubrication and to protect the fibers during subsequent processing and handling. Other chemicals are also applied during the sizing operation to assist in bonding the fibers to the matrix. Second, the fibers must eventually be combined with the matrix material. If the matrix can be made into a liquid or semiliquid form, then after the fibers are dried and surface treatment has been applied, the fibers can be combined with the matrix. In one type of process, the fibers are unwound from the take-up reel, the sizings are applied, the liquid matrix material is applied to a fiber, and the wetted fibers are rewound onto a large drum. The drum is usually covered with a sheet of paper coated to prevent the fibers from sticking to the drum and to act as a backing paper for holding the fiber-matrix system together. Generally, the winding of the fibers as they are impregnated with the matrix in liquid form and the concurrent winding of the backing paper are a continuous process; the result is layer after layer of impregnated fibers are wound on the drum and separated by sheets of backing paper. This matrix-impregnated material is

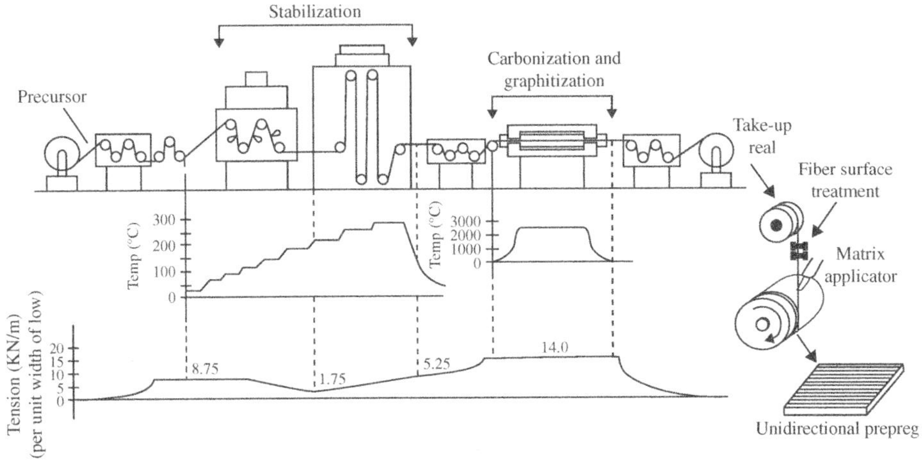

Figure 1.13 Fabrication of C and graphite fibers from PAN precursor. The figure also shows the temperature and tension profiles used during the fabrication process.

popularly known as unidirectional "prepreg," and it is generally the form in which material is received from the supplier. Thus, prepreg is short for *preimpregnated*, meaning impregnated before the user receives the material. The preimpregnated material on the drum is generally cut to some standard width, ranging from, say, 3 mm wide to 300 mm or wider. The narrower forms are often referred to as tapes and are used in machines that automatically fabricate composite structural components.

1.5.5.2 Fabrication of C Fiber Using Pitch

Two particular factors have led to the use of pitch as a precursor: higher yields and faster production rates. However, pitch-derived fibers are more brittle than those derived from PAN, and they have a higher density, leading to lower specific properties. In addition, the steps leading to pitch fibers are slightly different from the steps leading to PAN-derived fibers. The process of producing specific types of pitch from petroleum products is critically important to the successful production of high-modulus and high-strength carbon-based fibers. The basic process is a distillation of residual oils left over from the thermal or catalytic cracking of crude oil. The residuals from asphalt production, natural asphalt, shale oil, or coal tar can also be used. These heavy oil-based products are introduced into a reactor or series of reactors, where they are heated to temperatures between 350 and 500 °C. Thermal cracking, polymerization, and condensation occur, and the gases and light oils that are released are taken out of the reactor through a condenser. The resulting material is even heavier, has higher carbon content, and is the basis for pitch fiber precursor. The preferred raw material for pitch fiber production is liquid crystal, or mesophase, pitch. The mesophase is a highly anisotropic substance in the form of crystals, called spherulites, mixed in an isotropic pitch medium. The spherulites consist of a collection of relatively long molecules with their long axes normal relative to the boundary of the sphere. Under the influence of heat, the spherulites continue to grow and expand at the expense of the isotropic pitch surrounding them. An interesting feature of mesophase pitch is that it softens above 350 °C, and it can be mechanically deformed in this state. When the mesophase content reaches about 75%, the carbon substance can be subjected to fiber-forming techniques such as melt spinning.

Table 1.1 Properties of C-based fibers.

Property	PAN Intermediate modulus	PAN High modulus	PAN Ultra-high modulus	Rayon
Diameter (μm)	8–9	7–10	7–10	6.5
Density (kg/m^3)	1780–1820	1670–1900	1860	1530–1660
Tensile modulus (GPa)	228–276	331–400	517	41–393
Tensile strength (MPa)	2410–2930	2070–2900	1720	620–2200
Elongation (%)	1.0	0.5	0.3–0.4	1.5–2.5
Coefficient of thermal expansion (×10^{-6}/°C)				
Along fiber direction	−0.1 to −0.5	−0.5 to −1.2	−1.0	—
Perpendicular to fiber direction	7–12	7–12	—	—
Thermal conductivity (W/m/°C)	20	70–105	140	38
Specific heat (J/kg/°K)	950	925	—	—

The melt spinning of pitch fibers from the raw material can occur after the carbon content reaches the range 91–96.5% and the mean molecular weight is at least 400. After pitch fibers are spun, they are subjected to an oxidizing gas at a temperature below the spinning temperature, or they are subjected to another chemical treatment that renders them infusible. In one method, for example, pitch fibers are treated for seven hours at a temperature of 100 °C with air containing ozone; thereafter, the temperature is raised at 1 °C/min up to 300 °C. This stage of processing is critically important to guarantee that pitch fibers will retain their shape under heat treatment during carbonization and graphitization. However, if they are exposed to oxidization for too long, the fibers become brittle. The carbonization of pitch fibers occurs at somewhat higher temperatures. The heating rate between 100 and 500 °C is critical to prevent fiber rupture from released volatiles. A typical heating schedule would call for heating from 100 to 500 °C at 5 °C/h and from 500 to 1100 °C at 10 °C/h. The cool down from 1100 °C is usually controlled to be less than 30 °C/h. If desired, the carbonized pitch fibers can be further heated in an inert atmosphere to produce a graphitic microstructure. Graphitization temperatures are typically between 2500 and 3300 °C. Total graphitization times are generally very short, on the order of a few minutes. The properties of the three types of PAN fibers and the rayon fiber have been given in Table 1.1 for comparison.

1.5.6 Glass Fiber

Silica, SiO_2, forms the basis of nearly all commercial glasses. It exists in the form of a polymer $(SiO_2)_n$. It does not melt, but gradually softens until reaching a temperature of 2000 °C, after which it begins to decompose. When silica is heated until fluid like and then cooled, it forms a random glassy structure. Only prolonged heating above 1200 °C will induce crystallization (i.e. a quartz-type structure). Using silica as a glass is perfectly suitable for many industrial applications. However, its drawback is the high-processing temperatures needed to form the glass and work it into useful shapes. Other types of glasses were developed to decrease the complexity of processing and increase the commercialization of glass in fiber form. The main types of glass are as given below:

(i) Type A, a soda-lime glass, was the first used, and it is still retained for a few minor applications.
(ii) Type E, a borosilicate glass, was developed for better resistance to attack by water and mild chemical concentrations.
(iii) Relative to Type E, Type C glass has a much improved durability when exposed to acids and alkalis.
(iv) The increased strength and stiffness of Type S glass makes it a natural choice for use in high-performance applications, where higher specific strength and specific stiffness are important.

Type E: **Borosilicate glass** is a type of glass with silica and boron trioxide as the main glass-forming constituents. Borosilicate glasses are known for having very low coefficients of thermal expansion ($\sim 3 \times 10^{-6}/°C$ at 20 °C), making them resistant to thermal shock, more so than any other common glass. Such glass-made is less subject to thermal stress and is commonly used for the construction of reagent bottles. Borosilicate glass is sold under such trade names as Simax, Borcam, Borosil, Suprax, Kimax, Heatex, Pyrex, Endural, Schott, or Refmex, Kimble.

Type C: **Alkali-lime glass** with high boron oxide content.

Type R-glass: Alumino silicate glass without MgO and CaO with high mechanical requirements as reinforcement.

S-glass: Alumino silicate glass without CaO but with high MgO content with high-tensile strength.

Table 1.2 presents the four predominant glass compositions used to form continuous glass fibers. The basic glass fiber fabrication process consists of five main components:

(i) A fiber-drawing furnace consisting of a heat source, tank, and platinum alloy bushing. Raw material is fed into the furnace tank through an inlet, and fibers are drawn out through tiny nozzles in the bushing. By convention, the number of nozzles is usually 200 or a multiple thereof. Special cooling fins are located immediately under the bushing to stabilize the fiber-drawing process. This drawing process can produce material with aligned properties. Glass fibers tend to be isotropic.
(ii) A light water spray below the bushing to cool the fibers.
(iii) A sizing applicator.
(iv) A gathering shoe to collect the individual fibers and combine them into tows.
(v) A collet or winding mandrel to collect the tows. The winding of the collet produces tensioning of the tows, which draws the fibers through the bushing nozzles.

The wound fiber tows at this point are referred to as glass fiber cakes. The sizings applied to the fibers are almost all aqueous based, and thus the cake has a typical water content of about 10%. Before the fiber is shipped for end use, this water content has to be substantially reduced. This is accomplished by oven drying. There are two main objectives during oven drying: (i) water content must be reduced to less than 0.1% and (ii) the dry fiber must be subjected to heat treatment to allow conglomerates of sizing particles to flow within the tow to impart certain handling characteristics. A nominal drying schedule calls for a temperature between 115 and 125 °C for four to ten hours. If the fibers are to be preimpregnated, the dried fibers are unwound, the matrix material is applied, and the wetted fibers are rewound on a

Table 1.2 Composition of different types of glass.

Constituent	Type A: Soda-lime glass (%)	Type C: Chemical glass (%)	Type E: Electrical glass (%)	Type S: Structural glass (%)
SiO_2	72.0	65.0	55.2	65.0
Al_2O_3	2.5	4.0	14.8	25.0
B_2O_3	0.5	5.0	7.3	—
MgO	0.9	3.0	3.3	10.0
CaO	9.0	14.0	18.7	—
Na_2O	12.5	8.5	0.3	—
K_2O	1.5	—	0.2	—
Fe_2O_3	0.5	0.5	0.3	—
F_2	—	—	0.3	—

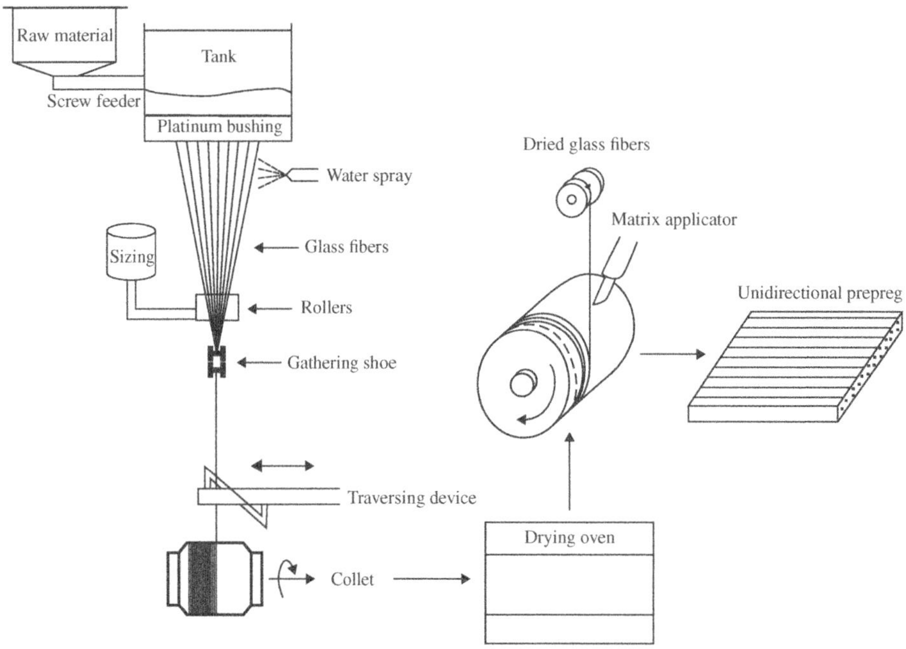

Figure 1.14 Glass fiber fabrication procedure.

drum. Figure 1.14 shows the glass fiber fabrication procedure. Table 1.3 shows the properties of different types of glass fibers.

1.5.7 Aramid (Kevlar) Fiber

Kevlar is perhaps the most common polymer fiber. It was developed by the DuPont Co. in 1968 and is an aromatic polyamide called poly(paraphenylene terephthalamide). The aromatic rings make the fiber fairly rigid. Kevlar fibers are manufactured by the extrusion and spinning processes. A solution of the polymer and a solvent (hexamethylphosphoramide) is

Table 1.3 Properties of glass fibers.

Property	Glass type E	C	S
Diameter (μm)	8–14	—	10
Density (kg/m^3)	2540	2490	2490
Tensile modulus (GPa)	72.4	68.9	85.5
Tensile strength (MPa)	3450	3160	4590
Elongation (%)	1.8–3.2	4.8	5.7
Coefficient of thermal expansion (×10^{-6}/°C)	5.0	7.2	5.6
Thermal conductivity (W/m/°C)	1.3	—	—
Specific heat (J/kg/°K)	840	780	940

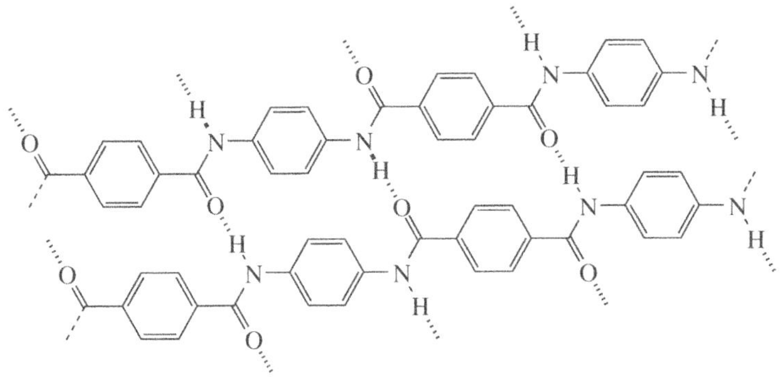

Figure 1.15 Microstructure of Kevlar fiber.

held at a low temperature, between −50 and −80 °C, before being extruded into a hot-walled cylinder at 200 °C. The solvent then evaporates and the fibers are wound onto a drum. The fibers at this stage have low strength and stiffness. The fibers are subsequently subjected to hot stretching to align the polymer chains along the axis of the fiber. Afterward, the aligned fibers show significant increase in strength and stiffness. Figure 1.15 shows a schematic of its structure. The polymer molecules form rigid planar sheets, and the sheets are stacked on top of each other with only weak hydrogen bonding between them. These sheets are folded in the axial direction and are oriented radially, with the fold generally occurring along the hydrogen-bonding line.

Aramid fibers are highly crystalline aromatic polyamide fibers that have the lowest density and the highest tensile strength-to-weight ratio among the current reinforcing fibers. Kevlar 49 is the trade name of one of the aramid fibers available in the market. As a reinforcement, aramid fibers are used in many marine and aerospace applications where lightweight, high-tensile strength, and resistance to impact damage (e.g. caused by accidentally dropping a hand tool) are important. Similar to carbon fibers, they also have a negative coefficient of thermal expansion in the longitudinal direction, which is used in designing low thermal expansion composite panels. The major disadvantages of aramid fiber-reinforced composites are their low compressive strengths and difficulty in cutting or machining.

1.5.8 Matrices

The role of a matrix has been discussed in an earlier section. In this section, different types of matrix materials have been discussed. The matrix materials can be classified as follows:

(A) Polymeric matrices
  (i) Thermoset polymers
    (a) *Epoxies*: principally used in aerospace and aircraft applications
    (b) *Polyesters, vinyl esters*: commonly used in automotive, marine, chemical, and electrical applications
    (c) *Phenolics*: used in bulk molding compounds
    (d) *Polyimides, polybenzimidazoles (PBI), polyphenylquinoxaline (PPQ)*: for high-temperature aerospace applications (temperature range: 250–400 °C)
    (e) Cyanate ester
  (ii) Thermoplastic polymers
    (a) Nylons (such as nylon 6, nylon 6,6), thermoplastic polyesters, such as poly-ethylene-terephthalate (PET).
    (b) Polycarbonate (PC), polyacetals: used with discontinuous fibers in injection-molded articles.
    (c) Polyamide-imide (PAI), polyether ether ketone (PEEK), polysulfone (PSUL), polyphenylene sulfide (PPS), polyetherimide (PEI): suitable for moderately high-temperature applications with continuous fibers.
(B) Metallic matrices
  Aluminum and its alloys, titanium alloys, magnesium alloys, copper-based alloys, nickel-based superalloys, stainless steel: suitable for high-temperature applications (temperature range: 300–500 °C)
(C) Ceramic matrices
  Aluminum oxide (Al_2O_3), carbon, silicon carbide (SiC), silicon nitride (Si_3N_4): suitable for high-temperature applications

1.5.8.1 Polymer Matrix Composite

A polymer is defined as a long-chain molecule containing one or more repeating units of atoms, joined together by strong covalent bonds. A polymeric material (commonly called a plastic) is a collection of a large number of polymer molecules of similar chemical structure (but not of equal length). In the solid state, these molecules are frozen in space, either in a random fashion in amorphous polymers or in a mixture of random fashion and orderly fashion (folded chains) in semicrystalline polymers. In a thermoplastic polymer, individual molecules are not chemically joined together. They are held in place by weak secondary bonds or intermolecular forces, such as van der Waals bonds and hydrogen bonds. With the application of heat, these secondary bonds in a solid thermoplastic polymer can be temporarily broken and the molecules can now be moved relative to each other or flow to a new configuration if pressure is applied on them. On cooling, the molecules can be frozen in their new configuration and the secondary bonds are restored, resulting in a new solid shape. Thus, a thermoplastic polymer can be heat-softened, melted, and reshaped (or postformed) as many times as desired.

In a thermoset polymer, on the other hand, the molecules are chemically joined together by cross-links, forming a rigid, three-dimensional network structure. Once these cross-links are formed during the polymerization reaction (also called the curing reaction), the thermoset polymer cannot be melted by the application of heat. However, if the number of crosslinks is low, it may still be possible to soften them at elevated temperatures. There are two unique characteristics of polymeric solids that are not observed in metals under ordinary conditions, namely, that their mechanical properties depend strongly on both the ambient temperature and the loading rate.

1.5.8.2 Metal Matrix Composites

Metal matrix has the advantage over polymeric matrix in applications requiring a long-term resistance to severe environments, such as high temperature. The yield strength and modulus of most metals are higher than those for polymers, and this is an important consideration for applications requiring high transverse strength and modulus as well as compressive strength for the composite. Another advantage of using metals is that they can be plastically deformed and strengthened by a variety of thermal and mechanical treatments. However, metals have a number of disadvantages, namely, they have high densities, high-melting points (therefore, high process temperatures), and a tendency toward corrosion at the fiber-matrix interface.

The two most commonly used metal matrices are based on aluminum and titanium. Both of these metals have comparatively low densities and are available in a variety of alloy forms. Although magnesium is even lighter, its great affinity toward oxygen promotes atmospheric corrosion and makes it less suitable for many applications. Beryllium is the lightest of all structural metals and has a tensile modulus higher than that of steel. However, it suffers from extreme brittleness, which is the reason for its exclusion as a potential matrix material. Nickel- and cobalt-based super-alloys have also been used as matrix; however, the alloying elements in these materials tend to accentuate the oxidation of fibers at elevated temperatures. Aluminum and its alloys have attracted the most attention as matrix material in metal matrix composites (MMCs). Commercially, pure aluminum has been used for its good corrosion resistance. Aluminum alloys, such as 201, 6061, and 1100, have been used for their higher tensile strength–weight ratios. Carbon fiber is used with aluminum alloys; however, at typical fabrication temperatures of 500 °C or higher, carbon reacts with aluminum to form aluminum carbide (Al_4C_3), which severely degrades the mechanical properties of the composite. Protective coatings of either titanium boride (TiB_2) or sodium has been used on carbon fibers to reduce the problem of fiber degradation as well as to improve their wetting with the aluminum alloy matrix. Carbon fiber-reinforced aluminum composites are inherently prone to galvanic corrosion in which carbon fibers act as a cathode owing to a corrosion potential of 1 V higher than that of aluminum. A more common reinforcement for aluminum alloys is SiC.

Titanium alloys that are most useful in MMCs are α, β alloys (e.g. Ti–6Al–9V) and metastable β-alloys (e.g. Ti–10V–2Fe–3Al). These titanium alloys have higher tensile strength–weight ratios as well as better strength retentions at 400–500 °C in comparison to those of aluminum alloys. The thermal expansion coefficient of titanium alloys is closer to that of reinforcing fibers, which reduces the thermal mismatch between them. One of the problems with titanium alloys is their high reactivity with boron and Al_2O_3 fibers at normal fabrication temperatures. Borsic (boron fibers coated with silicon carbide) and silicon carbide (SiC) fibers show less reactivity with titanium. Improved tensile strength retention is obtained by coating boron and SiC fibers with carbon-rich layers.

1.5.8.3 Ceramic Matrix Composites

Ceramics are known for their high-temperature stability, high-thermal shock resistance, high modulus, high hardness, high-corrosion resistance, and low density. However, they are brittle materials and possess low resistance to crack propagation, which is manifested in their low fracture toughness. The primary reason for reinforcing a ceramic matrix is to increase its fracture toughness.

Structural ceramics used as matrix materials can be categorized as either oxides or nonoxides. Alumina (Al_2O_3) and mullite (Al_2O_3–SiO_2) are the two most commonly used oxide ceramics. They are known for their thermal and chemical stability. The common nonoxide ceramics are silicon carbide (SiC), silicon nitride (Si_3N_4), boron carbide (B_4C), and aluminum nitride (AlN). Of these SiC has found wider applications, particularly where high modulus is desired. It also has an excellent high-temperature resistance. Si_3N_4 is considered for applications requiring high strength and AlN is of interest because of its high thermal conductivity. The reinforcements used in ceramic matrix composites (CMCs) are SiC, Si_3N_4, AlN, and other ceramic fibers. Of these SiC has been the most commonly used reinforcement because of its thermal stability and compatibility with a broad range of both oxide and nonoxide ceramic matrices. The forms in which the reinforcement is used in CMCs include whiskers (with length to diameter ratio as high as 500), platelets, particulates, and both monofilament and multifilament continuous fibers.

1.6 Manufacturing Methods

1.6.1 Polymer Matrix Composites

Many techniques, originally developed for making glass fiber-reinforced polymer matrix composites (PMCs) can also be used with other fibers. Glass fiber-reinforced polymer composites represent the largest class of PMCs. Polymeric matrix materials can be conveniently classified as thermosets and thermoplastics. Recall that thermosets harden on curing. Curing or cross-linking occurs in thermosets by appropriate chemical agents and/or application of heat and pressure. Conventionally, thermal energy (heating to 200 °C or above) is provided for this purpose. This process, however, brings in the problems of thermal gradients, residual stresses, and long curing times. Residual stresses can cause serious problems in nonsymmetric or very thick PMC laminates, where they may be relieved by warping of the laminate, fiber waviness, matrix microcracking, and ply delamination. Electron beam curing offers an alternative that avoids these problems. It is a nonthermal curing process that requires much shorter cure time cycles. Curing by electron beam occurs by electron-initiated reactions at a selectable cure temperature. Different methods of fabrication of PMCs-first thermoset-based composites and then thermoplastic-based composites, have been discussed below.

1.6.1.1 Thermoset Matrix Composites

There are many processing methods for composites with thermoset matrix materials including epoxy, unsaturated polyester, and vinyl ester.

Hand Layup and Spray Techniques

Hand layup and spray techniques are perhaps the simplest polymer processing techniques. Fibers can be laid onto a mold by hand and the resin (unsaturated polyester is one of the most common) is sprayed or brushed on. Frequently, resin and fibers (chopped) are sprayed together onto the mold surface. In both the cases, the deposited layers are densified with rollers. Figure 1.16 shows schematics of these processes. Accelerators and catalysts are frequently used. Curing may be done at room temperature or at a moderately high temperature

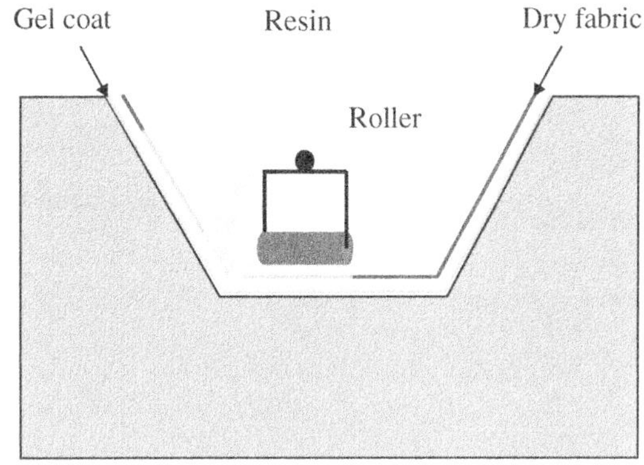

Figure 1.16 The hand layup process.

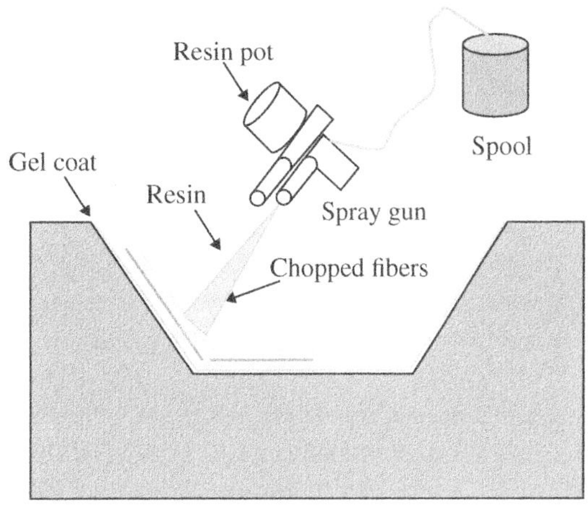

Figure 1.17 The spray-up process.

in an oven. Figures 1.16 and 1.17 show the hand layup and the spray techniques, respectively. The advantages of the hand layup process are as follows:

(a) It results in low-cost tooling with the use of room-temperature cure resins.
(b) The process is simple to use.
(c) Any combination of fibers and matrix materials can be used.
(d) Higher fiber contents and longer fibers can be used as compared to other processes. The disadvantages of the hand layup process are as follows.
(a) Since the process is worked by hands, there are safety and hazard considerations.
(b) The resin needs to be less viscous so that it can be easily worked by hands.
(c) The quality of the final product is highly skill-dependent of the laborers.
(d) Uniform distribution of resin inside the fabric is not possible. It leads to voids in the laminates.
(e) Possibility of diluting the contents.

The hand layup process is suitable for the fabrication of wind-turbine blades, boats, and architectural moldings.

The **spray-up process** offers the following advantages:

(a) It is suitable for small to medium-volume parts.
(b) It is a very economical process for making small to large parts.
(c) It utilizes low-cost tooling as well as low-cost material systems.

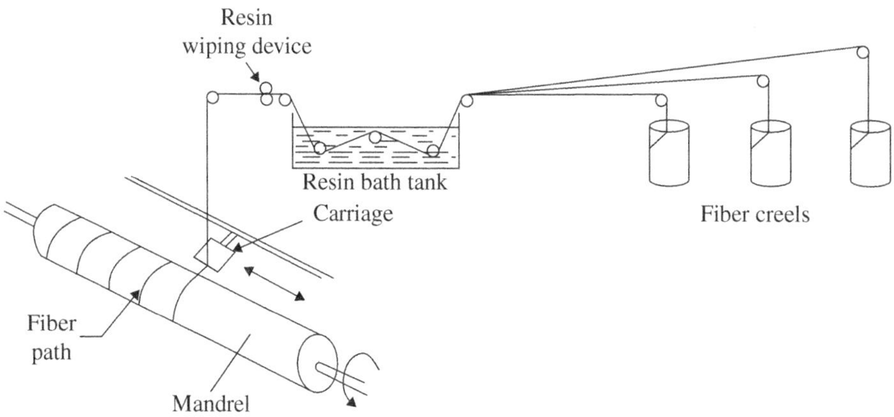

Figure 1.18 Filament winding process.

The following are some of the limitations of the spray-up process:

(a) It is not suitable for making parts that have high structural requirements.
(b) It is difficult to control the fiber volume fraction as well as the thickness. These parameters highly depend on operator skill.
(c) Because of its open mold nature, styrene emission is a concern.
(d) The process offers a good surface finish on one side and a rough surface finish on the other side.
(e) The process is not suitable for parts where dimensional accuracy and process repeatability are prime concerns.
(f) Cores, when needed, have to be inserted manually.
(g) Only short fibers can be used in this process.
(h) Similar to wet/hand layup process, the resins need to be of low viscosity so that it can be sprayed.

The spray-up process has the following applications: Making simple enclosures, lightly loaded structural panels, e.g. caravan bodies, truck fairings, bathtubs, shower trays, and some small dinghies.

Filament Winding

This process is an automated process. This process is used in the fabrication of components or structures made with flexible fibers. This process is primarily used for hollow, generally circular or oval-sectioned components. Fiber tows are passed through a resin bath before being wound onto a mandrel in a variety of orientations, controlled by the fiber feeding mechanism, and rate of rotation of the mandrel. The wound component is then cured in an oven or autoclave. One can use resins such as epoxy, polyester, vinylester, and phenolic along with any fiber. The fiber can be directly drawn from creel, nonwoven, or stitched into a fabric form. The filament winding process is shown in Figure 1.18.

The advantages of the filament winding process are as follows:

(a) Resin content is controlled by nips or dies.
(b) The process can be very fast. The process is economic.
(c) Complex fiber patterns can be attained for better load bearing of the structure.

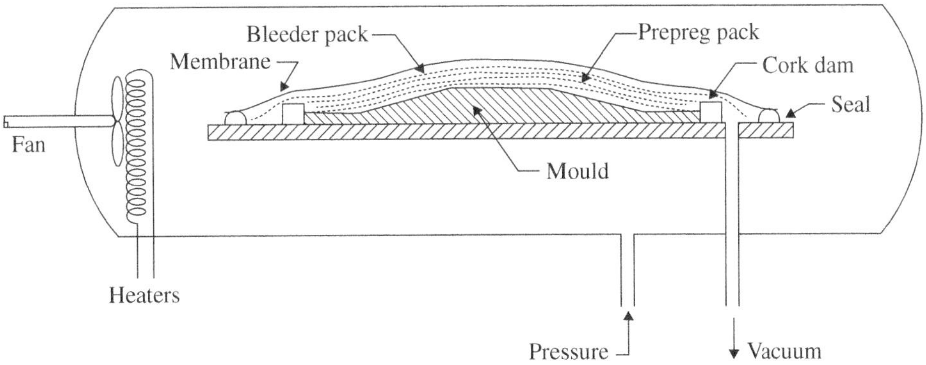

Figure 1.19 The setup inside an autoclave molding machine.

The disadvantages of this process are the following:

(a) Resins with low viscosity are needed.
(b) The process is limited to convex-shaped components.
(c) Fiber cannot easily be laid exactly along the length of a component.
(d) Mandrel costs for large components can be high.
(e) The external surface of the component is not smoothly finished.

The filament winding process is used for making pressure bottles, rocket motor casing, chemical storage tanks, pipelines, gas cylinders, firefighters, breathing tanks, and so forth.

Autoclave Curing

An autoclave is a closed vessel for controlling temperature and pressure is used for curing polymeric matrix composites. Composites to be cured are prepared either through hand layup or machine placement of individual laminae in the form of fibers tape which has been impregnated with resin. Components are then placed in an autoclave and subjected to a controlled cycle of temperature and pressure. After curing, the composite is "solidified." One can use the fibers such as carbon, glass, aramid, and so forth, along with any resin.

Autoclave-based processing of PMCs results in a very high-quality product. That is the reason it is used to make components in the aerospace field. An autoclave is a closed vessel (round or cylindrical) in which processes (physical and/or chemical) occur under simultaneous application of high temperature and pressure. Heat and pressure are applied to appropriately stacked prepregs. The combined action of heat and pressure consolidates the laminae, removes the entrapped air, and helps cure the polymeric matrix. Autoclave processing of composites thus involves a number of phenomena: chemical reaction (curing of the thermoset matrix), flow of the resin, and heat transfer.

Figure 1.19 shows schematically the setup in an autoclave to make a laminated composite. Figure 1.20 shows the autoclave molding machine employing the vacuum bagging process discussed in the next section.

The advantages of autoclave curing process are highlighted below:

(a) Large components can be fabricated.
(b) Since the curing of matrix material is carried out under controlled environment, the resin distribution is better as compared to hand or spray layup processes.
(c) Less possibility of dilution with foreign particles.

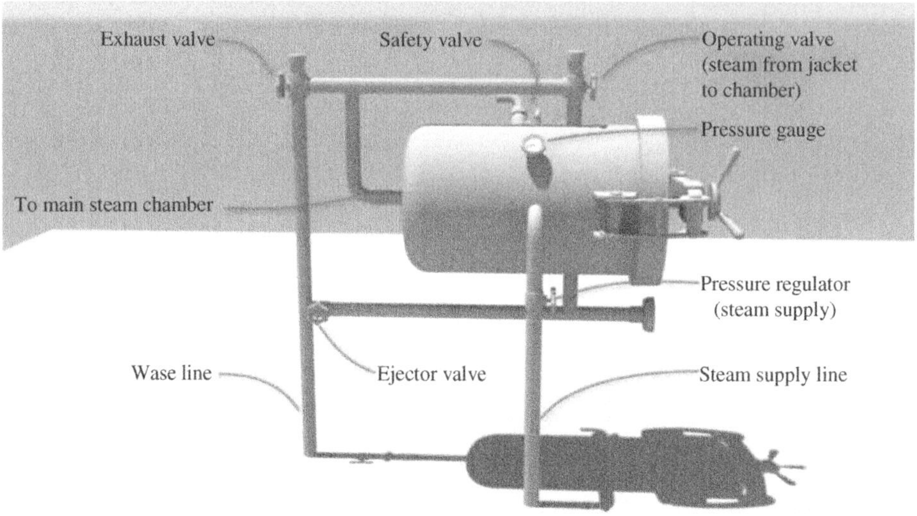

Figure 1.20 An autoclave molding machine.

(d) Better surface finish.

Disadvantages of this process include the following:

(a) Initial cost of tooling is high.
(b) Running and maintenance cost is high.
(c) Not suitable for small products.

The process is suitable for aerospace, automobile parts such as wing box, chassis, bumpers.

Vacuum Bagging Process

This is basically an extension of the wet layup process described above where pressure is applied to the laminate once laid-up in order to improve its consolidation. This is achieved by sealing a plastic film over the wet laid-up laminate and onto the tool. The air under the bag is extracted by a vacuum pump and thus up to one atmosphere of pressure can be applied to the laminate to consolidate it. The process has the following advantages:

(a) Higher-fiber content laminates can usually be achieved than with standard wet layup techniques.
(b) Lower void contents are achieved than with wet layup.
(c) Better fiber wet-out due to pressure and resin flow throughout structural fibers, with excess into bagging materials.
(d) *Health and safety*: The vacuum bag reduces the amount of volatiles emitted during cure.

The process has some limitations which are highlighted below.

(a) The extra process adds cost both in labor and in disposable bagging materials.
(b) A higher level of skill is required by the operators.
(c) Mixing and control of resin content are still largely determined by the operator skill.

The application of a vacuum to assist in compressing the plies together (debulking) has proven to be valuable in wet layups and necessary in prepreg layups. The vacuum provides the dual advantage of pressing the layers together and simultaneously withdrawing the excess volatiles. These volatiles could be residual solvent, low-molecular weight resin components, absorbed moisture, or trapped air. The method of applying the vacuum that has been developed for composites allows the volatiles to escape freely and also permits good debulking. The vacuum bagging process is used for making large one-off cruising boats, race car components, core-bonding in production boats, etc.

The procedure for building up such an assembly is as follows (Figure 1.21):

(a) Prepare the mold by coating it with an appropriate mold release.
(b) Remove prepreg materials from the freezer and bring them to room temperature before opening the protective bag to prevent contamination from water condensation.
(c) Build up the part by placing layers of prepreg on top of each other in the prescribed pattern and to the prescribed thickness. This buildup can either be done manually or be automated. The buildup may also involve inclusion of any inserts, ribs, or other structural members and the additional prepreg layers necessary to serve as anchoring and support for these inclusions.
(d) Place the release film or peel ply material over the part. The release material is generally porous to permit excess resin to flow through, while leaving an impression on the part suitable for secondary bonding without further surface preparation.
(e) The bleeder material is a mat that absorbs the excess resin. Common bleeder materials are polyester felt or mat, fiberglass (which may be coated with Teflon or mold release), and cotton. The important characteristic is that the bleeder material should have good absorption qualities and not compact under the pressure of an autoclave. The resin content of the final part is dependent on the ability of the bleeder material to absorb a measured amount of resin, as well as on temperature, pressure, viscosity, and amount of bleeder. The bleeder should be conformable so that it does not cause wrinkles in the assembly when it is placed under vacuum.
(f) The barrier is a layer that limits the upward movement of the resin and prevents it from reaching or clogging the breather and vacuum lines. The barrier material must not, therefore, allow the passage of resin, but must allow air to pass. In many bagging assemblies, this material is omitted. In autoclave processing, such an omission may lead to resin-plugged vacuum lines and pumps. When the barrier is omitted in wet layup applications, resin traps are often used.
(g) The breather material acts as a distributor for air, escaping volatiles, and gases, as well as a buffer between bag wrinkles and the part surfaces. Breather layers should be highly porous and must not collapse under vacuum, temperature, or pressure. Typical materials are fiberglass, polyester felt, and cotton, although the bleeder materials are sometimes used. For near-net or phenolic resin systems, the use of breather is sometimes omitted when the thickness of the bleeder is great and experience has shown that a separate breather system is not required.
(h) Vacuum bag sealing tape or sealant is applied around the mold. This sticky material makes an airtight seal with the bagging film.
(i) Thermocouples or other monitoring devices are inserted into the assembly. To ensure that the places where the thermocouple wires enter the assembly will be airtight, the outer insulation is often scraped off the thermocouple wire (especially if it has cloth insulation). The wires are then embedded in the sealant material. Thermocouples are usually placed in the outer area of a part that will be trimmed off.

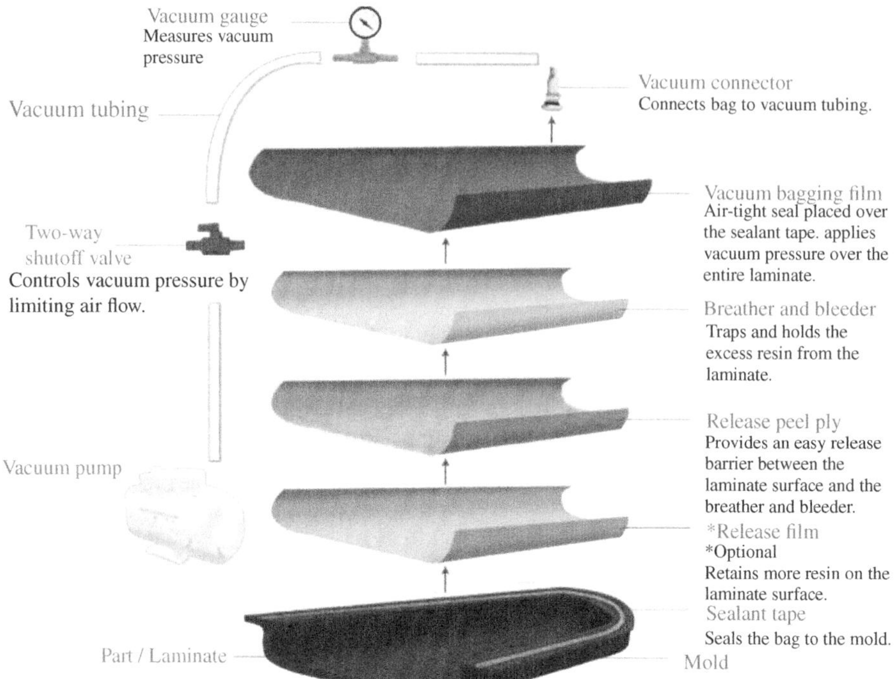

Figure 1.21 Components of vacuum bagging.

(j) The vacuum bag is then laid over the assembly and a vacuum port is attached through the bag, providing a fitting for attaching the vacuum hose. The bag is then pressed against the sealant to give an airtight seal all around. Vacuum bags can be made of any plastic film material that is strong enough to hold a vacuum, fit and conform to the assembly, and withstand the cure temperature without degrading. In practice, the most common material is nylon or a coextruded nylon that has been heat-stabilized.

The most common problems associated with vacuum bag processing are material quality and bag leaks. Vacuum bagging materials are often procured under vague material, construction, and performance requirements. Poor industry standards often lead to products with varying performance. For instance, uncontrolled or unregulated vacuum bag quality is the leading cause of bag failure, typified by film degradation or decomposition above 82 °C. Bag leaks most often occur at the sealant vacuum bag interface. The second most common cause of failure is handling damage to the nylon film before cure. Nylon film is hygroscopic and subject to moisture change in relation to the relative humidity of its environment. Film becomes dry and "brittle" when its moisture content falls below 2%. Dry film is susceptible to damage/cracking when excessively handled or abused (wadded).

Bridging is also a common problem in vacuum bag molding. Bridging occurs when, because of the shape of the part, the vacuum bagging materials are not pressed against all of the part surface. Some areas, therefore, are not properly pressed. (For instance, for a part with a narrow, deep channel, bagging materials may form a material bridge across the top of the channel, preventing the transfer of pressure to the bottom of the channel.) If the nylon bag itself bridges over a narrow gap, the film may be stretched beyond its limits during cure

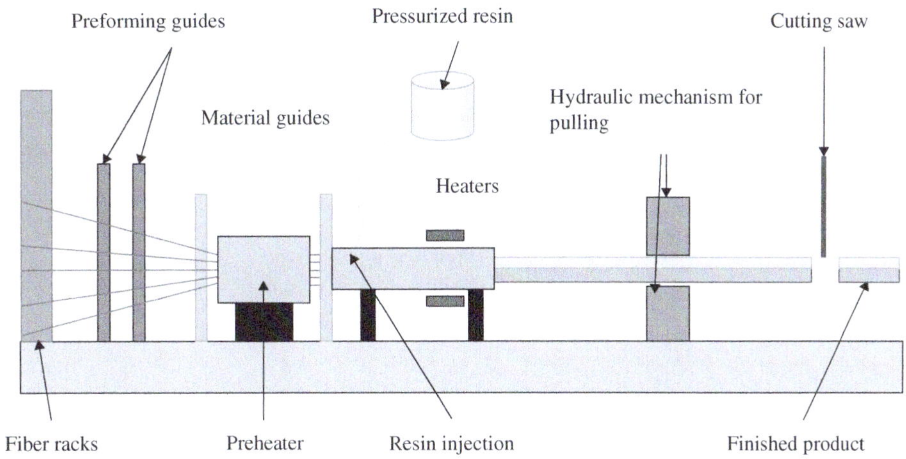

Figure 1.22 Pultrusion process.

and may burst. Good bagging techniques, such as intentionally putting pleats into the bag during assembly, allow enough excess film to be present to provide bag conformity against all surfaces. Another technique is the placement of preformed rubber pads under the bag in corners and channels to fill areas where bridging is possible. This technique is particularly useful with vacuum bags made of rubber.

Pultrusion

It is a continuous process in which composites in the form of fibers and fabrics are pulled through a bath of liquid resin. Then the fibers wetted with resin are pulled through a heated die. The die plays important roles like completing the impregnation and controlling the resin. Further, the material is cured to its final shape. The die shape used in this process is nothing the replica of the final product. Finally, the finished product is cut to length. In this process, the fabrics may also be introduced into the die. The fabrics provide a fiber direction other 0°. Further, a variant of this method to produce a profile with some variation in the cross-section is available. This is known as pulforming. The pultrusion process is shown in Figure 1.22. The process has the following advantages:

(a) The process is suitable for mass production.
(b) The process is fast and economic.
(c) Resin content can be accurately controlled.
(d) Fiber cost is minimized as it can be taken directly from a creel.
(e) The surface finish of the product is good.
(f) Structural properties of product can be very good as the profiles have very straight fibers.

The process has the following disadvantages:

(a) Limited to constant or near constant cross-section components.
(b) Heated die costs can be high.
(c) Products with small cross-sections alone can be fabricated.

The pultrusion process is used for making beams and girders used in roof structures, bridges, ladders, frameworks, and so forth.

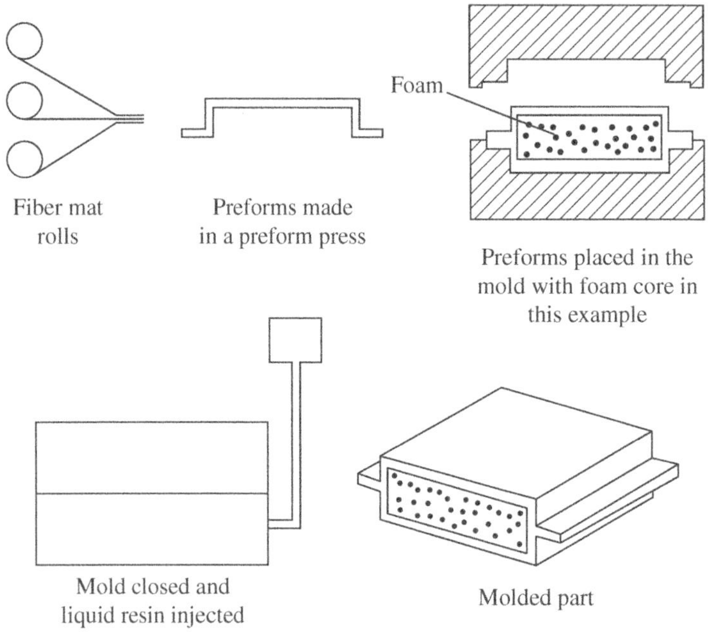

Figure 1.23 Resin transfer molding (RTM) process.

Resin Transfer Molding (RTM)

In RTM, several layers of dry continuous strand mat, woven roving, or cloth are placed in the bottom half of a two-part mold, the mold is closed, and a catalyzed liquid resin is injected into the mold via a centrally located sprue. The resin injection point is usually at the lowest point of the mold cavity. The injection pressure is in the range of 69–690 kPa (10–100 psi). As the resin flows and spreads throughout the mold, it fills the space between the fiber yarns in the dry fiber preform, displaces the entrapped air through the air vents in the mold, and coats the fibers. Depending on the type of the resin–catalyst system used, curing is performed either at room temperature or at an elevated temperature in an air-circulating oven. After the cured part is pulled out of the mold, it is often necessary to trim the part at the outer edges to conform to the exact dimensions. Instead of using flat-reinforcing layers, such as a continuous strand mat, the starting material in an RTM process can be a preform that already has the shape of the desired product (Figure 1.23). The advantages of using a preform are good moldability with complicated shapes (particularly with deep draws) and the elimination of the trimming operation, which is often the most labor-intensive step in an RTM process. In this process, the resins such as epoxy, polyester, vinylester, and phenolic can be used. Further, one use the high-temperature resins such as bismaleimides can be used at elevated process temperatures. The fibers of any type can be used. The stitched materials work well in this process since the gaps allow rapid resin transport. Some specially developed fabrics can assist with resin flow.

The RTM process has been successfully used in molding such parts as cabinet walls, chair or bench seats, hoppers, water tanks, bathtubs, and boat hulls. It also offers a cost-saving alternative to the labor-intensive bag-molding process or the capital-intensive compression-molding process. It is particularly suitable for producing low- to mid-volume parts, say 5000–50 000 parts a year. There are several variations of the basic RTM process. In

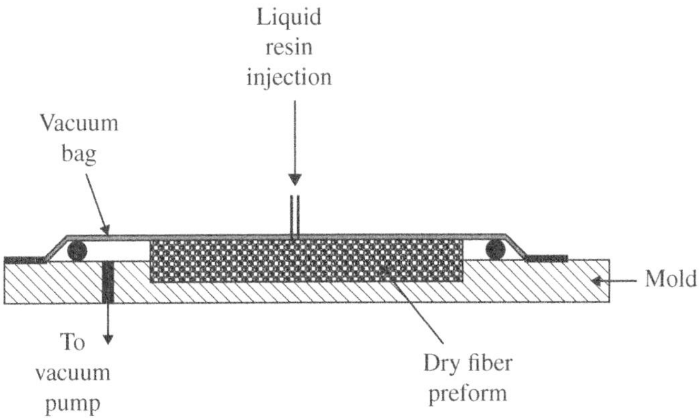

Figure 1.24 Vacuum-assisted resin transfer molding process (VARTM).

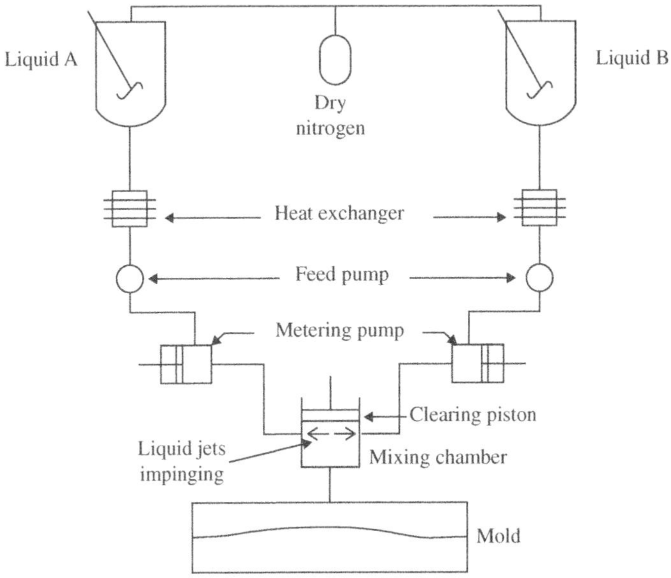

Figure 1.25 Structural reaction injection molding process.

one of these variations, known as vacuum-assisted RTM (VARTM), shown in Figure 1.24, vacuum is used in addition to the resin injection system to pull the liquid resin into the preform. Another manufacturing process very similar to RTM is called structural reaction injection molding (SRIM). It also uses dry fiber preform that is placed in the mold before resin injection. The difference in RTM and SRIM is mainly in the resin reactivity, which is much higher for SRIM resins than for RTM resins. SRIM is based on the reaction injection molding (RIM) technology in which two highly reactive, low-viscosity liquid streams are impinged on each other at high speeds in a mixing chamber immediately before injecting the liquid mix into a closed mold cavity (Figure 1.25).

The RTM process has the following advantages:

(a) The process is very efficient.
(b) Suitable for complex shapes.

(c) High-fiber volume laminates can be obtained with very low void contents.
(d) Good health and safety, and environmental control due to enclosure of resin.
(e) Possible labor reductions.
(f) Both sides of the component have a molded surface. Hence, the final product gets a superior surface finish.
(g) Better reproducibility.
(h) Relatively low clamping pressure and ability to induce inserts.

The RTM process has the following disadvantages:

(a) Matched tooling is expensive and heavy in order to withstand pressures.
(b) Generally limited to smaller components.
(c) Unimpregnated areas can occur resulting in very expensive scrap parts.

1.6.1.2 Thermoplastic Matrix Composites

Thermoplastic matrix composites have several advantages and disadvantages over thermoset matrix composites. We first list these and then describe some of the important processes used to form thermoplastic matrix composites. The advantages of thermoplastic matrix composites include the following:

(a) Refrigeration is not necessary with a thermoplastic matrix.
(b) Parts can be made and joined by heating.
(c) Parts can be remolded, and any scrap can be recycled.
(d) Thermoplastics have better toughness and impact resistance than thermosets. This can generally also be translated into thermoplastic matrix composites.

The disadvantages include the following:

(a) The processing temperatures are generally higher than those with thermosets.
(b) Thermoplastics are stiff and boardy, i.e. they lack the tackiness of the partially cured epoxies.

The processing techniques for thermoplastic matrix composites have been discussed below.

Film Stacking

Laminae of thermoplastic matrix containing fibers with very low resin content (~15 w/o) are used in this process. Low resin content is used because these are very boardy materials. The laminae are stacked alternately with thin films of pure polymer matrix material. This stack of laminae consists of fibers impregnated with insufficient matrix and polymer films of complementary weight to give the desired fiber volume fraction in the end product. These are then consolidated by simultaneous application of heat and pressure. The impregnation of thermoplastic matrix takes place under the simultaneous application of heat and pressure; the magnitude of pressure and temperature must be sufficient to force the polymeric melt to flow into and through the reinforcement preform. The rate of penetration of a fluid into the fibrous preform structure is described by Darcy's law. Darcy's law says that flow rate is directly proportional to the applied pressure and inversely proportional to the viscosity. Thus, increasing the applied pressure and decreasing the viscosity of the molten polymer (i.e. increasing the temperature) helps in the processing. The process is shown in Figure 1.26.

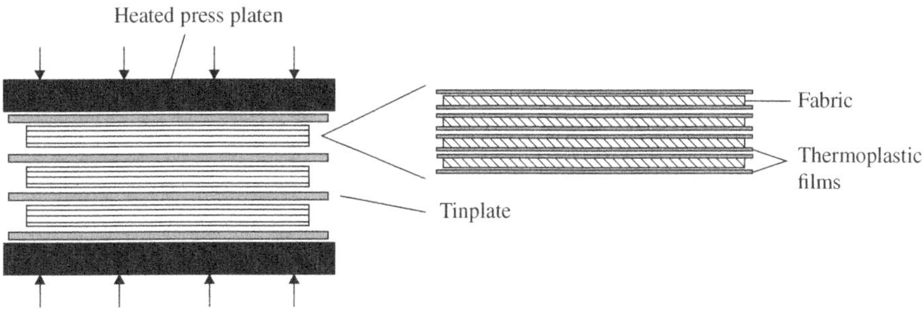

Figure 1.26 Film stacking process.

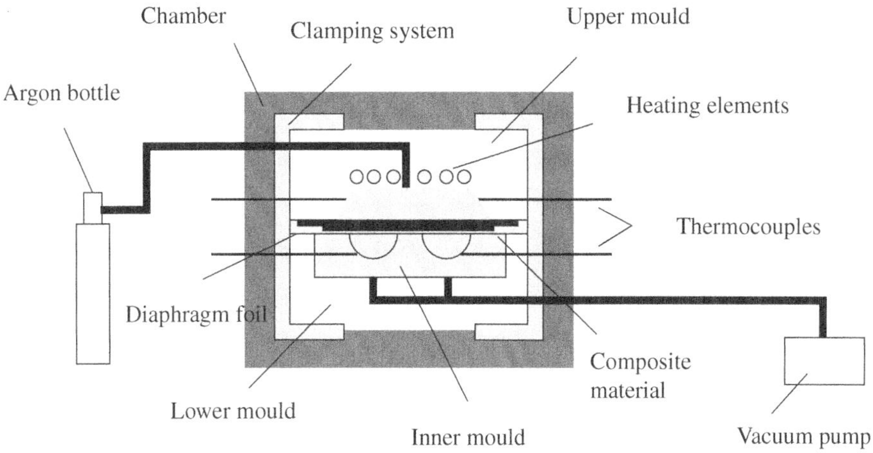

Figure 1.27 Diaphragm forming process.

Diaphragm Forming

This process involves the sandwiching of freely floating thermoplastic prepreg layers between two diaphragms. The air between the diaphragms is evacuated and thermoplastic laminate is heated above the melting point of the matrix. Pressure is applied to one side, which deforms the diaphragms and makes them take the shape of the mold. The laminate layers are freely floating and very flexible above the melting point of the matrix, thus they readily conform to the mold shape. After the completion of the forming process, the mold is cooled, the diaphragms are stripped off, and the composite is obtained. One of the advantages of this technique is that components with double curvatures can be formed. The diaphragms are the key to the forming process, and their stiff ness is a very critical parameter. Compliant diaphragms do the job for simple components. For very complex shapes requiring high-molding pressures, stiff diaphragms are needed. At high pressures, a significant transverse squeezing flow can result, and this can produce undesirable thickness variations in the final composite. The process is shown in Figure 1.27.

Thermoplastic Tape Laying

Thermoplastic tape laying machines are also available, although they are not as common as the thermoset ting tape laying machines. Figure 1.28 shows the schematic of one such machine. A controllable tape head has the tape dispensing and shim dispensing/take-up reels

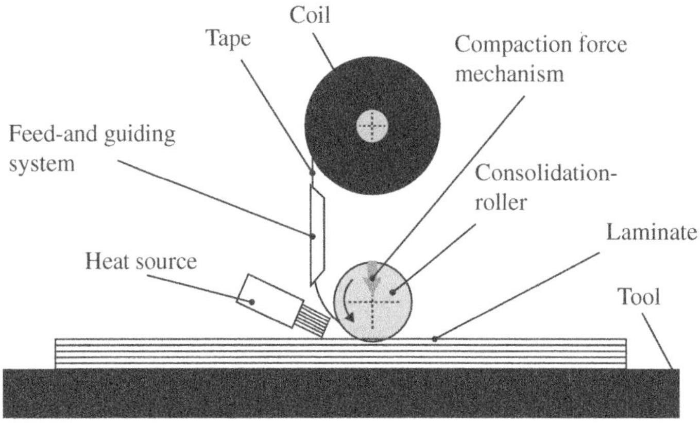

Figure 1.28 Tape laying process for thermoplastic matrix composites.

and heating shoes. The hot head dispenses thermoplastic tape from a supply reel. There are three heating and two cooling/compaction shoes. The hot shoes heat the tape to molten state. The cold shoes cool the tape instantly to a solid state.

Sheet Molding Compound

There are some common PMCs that do not contain long, continuous fibers; hence, we describe them separately in this section. Sheet molding compound (SMC) is the name given to a composite that consists of a polyester resin containing short glass fibers plus some additives called fillers. The additives generally consist of fine calcium carbonate particles and mica flakes. Sometimes calcium carbonate powder is substituted by hollow glass microspheres, which results in a lower density, but makes it more expensive. Figure 1.29 shows a schematic of the SMC processing. Polyester resin can be replaced by vinyl ester to further reduce the weight, but again with a cost penalty. SMC is used in making some auto body parts, such as bumper beams, radiator support panels, and many others. It has been used in the Corvette sports car for many decades. Polypropylene resin can be reinforced with calcium carbonate particles, mica flakes, or glass fibers. Such composites, though structurally not as important as, say, carb on fiber/epoxy composites, do show improved mechanical properties vis-à-vis unreinforced resin. Characteristics such as strength, stiffness, and service temperature are improved. These materials are used in automotive parts, appliances, electrical components, and so forth.

1.6.2 Metal-Matrix Composites

Many processes for fabricating MMCs are available. For the most part, these processes involve processing in the liquid and solid state. Some processes may involve a variety of deposition techniques or an in situ process of incorporating a reinforcement phase.

1.6.2.1 Liquid-State Processes

Metals with melting temperatures that are not too high, such as aluminum, can be incorporated easily as a matrix by liquid route. The liquid-state processes have been discussed below.

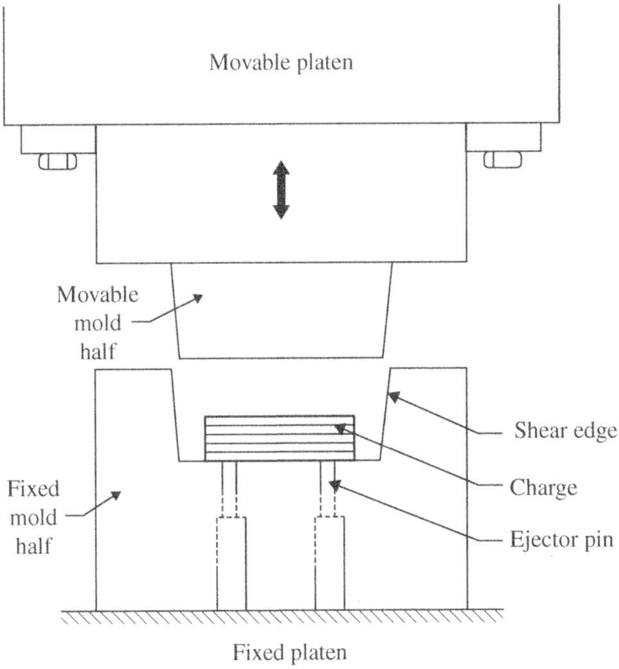

Figure 1.29 Sheet molding compound.

Casting or Liquid Infiltration

Casting, or liquid infiltration, involves infiltration of a fiber bundle by liquid metal. It is not easy to make MMCs by simple liquid-phase infiltration, mainly because of difficulties with wetting of ceramic reinforcement by the molten metal. When the infiltration of a fiber preform occurs readily, reactions between the fiber and the molten metal can significantly degrade fiber properties. Fiber coatings applied prior to infiltration, which improve wetting and control reactions, have been developed and can result in some improvements. In this case, however, the disadvantage is that the fiber coatings must not be exposed to air prior to infiltration because surface oxidation will alter the positive effects of coating. One commercially successful liquid infiltration process involving particulate reinforcement is the Duralcan process. Figure 1.30 shows a schematic of this process. Ceramic particles and ingot-grade aluminum are mixed and melted. The ceramic particles are given a proprietary treatment. The melt is stirred just above the liquidus temperature – generally between 600 and 700 °C. The melt is then converted into one of the following four forms: extrusion blank, foundry ingot, rolling bloom, or rolling ingot. The Duralcan process of making particulate composites by liquid metal casting involves the use of 8–12 μm particles. Too small particles, e.g. 2–3 μm, will result in a very large interface region and thus a very viscous melt. In foundry-grade MMCs, high Si aluminum alloys (e.g. A356) are used, while in wrought MMC, Al–Mg type alloys (e.g. 6061) are used. Alumina particles are typically used in foundry alloys, while silicon carbide particles are used in the wrought aluminum alloys. For making continuous fiber-reinforced MMCs, tows of fibers are passed through a liquid metal bath, where the individual fibers are wet by the molten metal, excess metal is wiped off, and a composite wire is produced. A bundle of such wires can be consolidated by extrusion to make a composite.

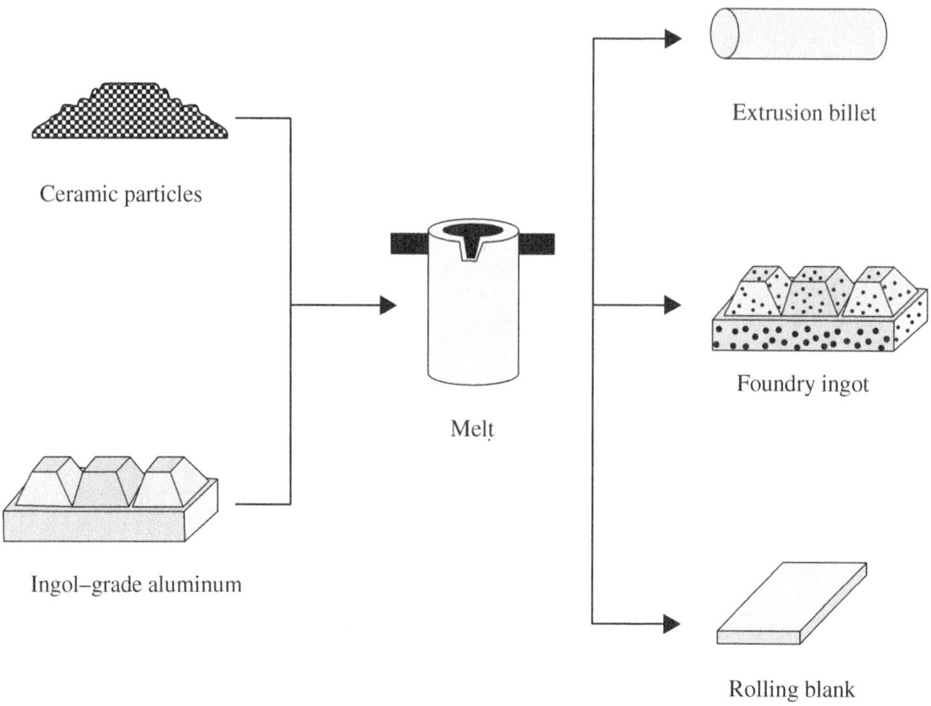

Figure 1.30 Duralcan process for metal-matrix composites.

Squeeze Casting

Squeeze casting, or pressure infiltration, involves forcing the liquid metal into a fibrous preform. Figure 1.31 shows two processes of making a fibrous preform. In the press forming process, an aqueous slurry of fibers is well agitated and poured into a mold, pressure is applied to squeeze the water out, and the preform is dried (Figure 1.31a). In other process, suction is applied to a well-agitated mixture of whisker, binder, and water. This is followed by demolding and drying of the fiber preform (Figure 1.31b). A schematic of the squeeze casting process is shown in Figure 1.32. Pressure is applied until the solidification is complete. By forcing the molten metal through small pores of fibrous preform, this method obviates the requirement of good wettability of the reinforcement by the molten metal.

Composites fabricated with this method have minimal reaction between the reinforcement and molten metal because of short dwell time at high temperature and are free from common casting defects such as porosity and shrinkage cavities. Squeeze casting is really an old process, also called liquid metal forging in earlier versions. It was developed to obtain pore-free, fine-grained aluminum alloy components with superior properties than conventional permanent mold casting. In particular, the process has been used in the case of aluminum alloys that are difficult to cast by conventional methods, for example, silicon-free alloys used in diesel engine pistons where high-temperature strength is required. Inserts of nickel-containing cast iron, called Ni-resist, in the upper groove area of pistons have also been produced by the squeeze casting technique to provide wear resistance. Use of ceramic fiber-reinforced MMCs at locations of high wear and high thermal stress has resulted in a product much superior to the Ni-resist cast iron inserts.

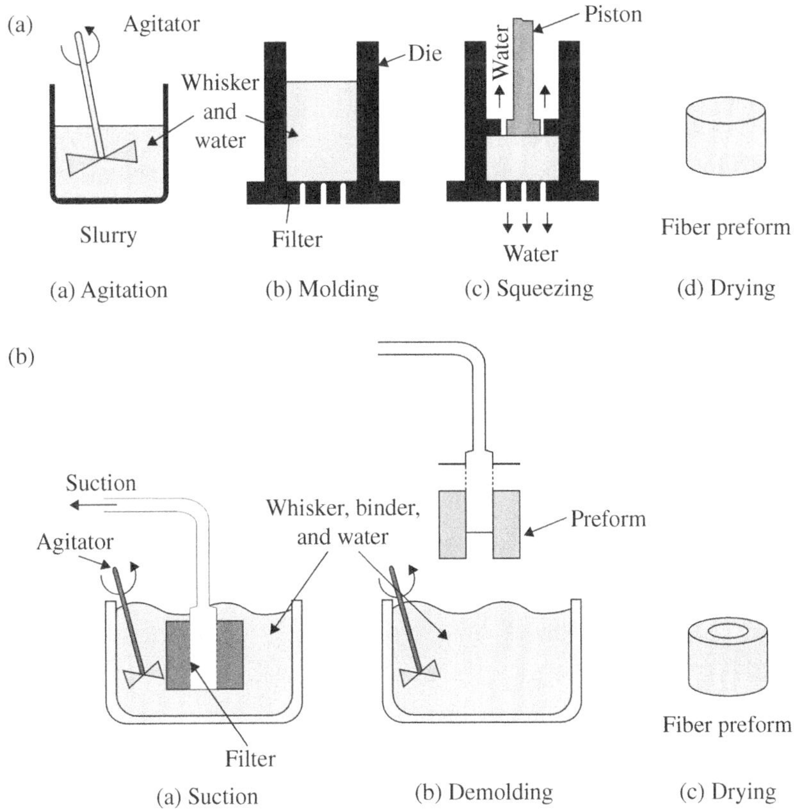

Figure 1.31 (a) Press forming and (b) suction forming of a preform.

The squeeze casting technique, shown in Figure 1.32, has been quite popular in making composites with selective reinforcement. A porous fiber preform (generally of discontinuous Saffil-type Al_2O_3 fibers) is inserted into the die. Molten metal (aluminum) is poured into the preheated die located on the bed of a hydraulic press. The applied pressure (70–100 MPa) makes the molten aluminum penetrate the fiber preform and bond the fibers. Infiltration of a fibrous preform by means of a pressurized inert gas is another variant of liquid metal infiltration technique. The process is conducted in the controlled environment of a pressure vessel and rather high-fiber volume fractions; complex-shaped structures are obtainable. Although commonly, aluminum matrix composites are made by this technique, alumina fiber-reinforced intermetallic matrix composites (e.g. TiAl, Ni_3Al, and Fe_3Al matrix materials) have been prepared by pressure casting. The technique involves melting the matrix alloy in a crucible in vacuum, while the fibrous preform is heated separately. The molten matrix material (at about 100 °C above the T_m) is poured onto the fibers, and argon gas is introduced simultaneously. Argon gas pressure forces the melt to infiltrate the preform. The melt generally contains additives to aid in wetting the fibers.

Centrifugal Casting

One of the disadvantages of MMCs with ceramic reinforcement is that they are typically more difficult to machine than the unreinforced alloy. In centrifugal casting, optimal placement of

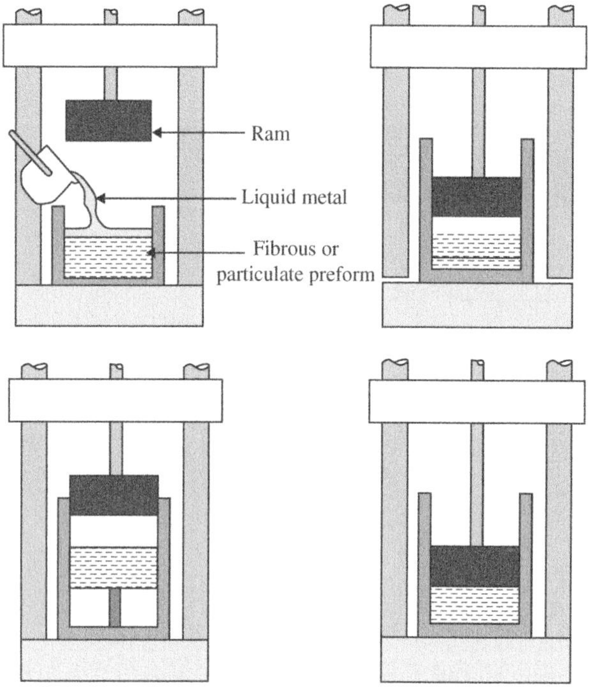

Figure 1.32 Squeeze casting process for fabrication of MMCs.

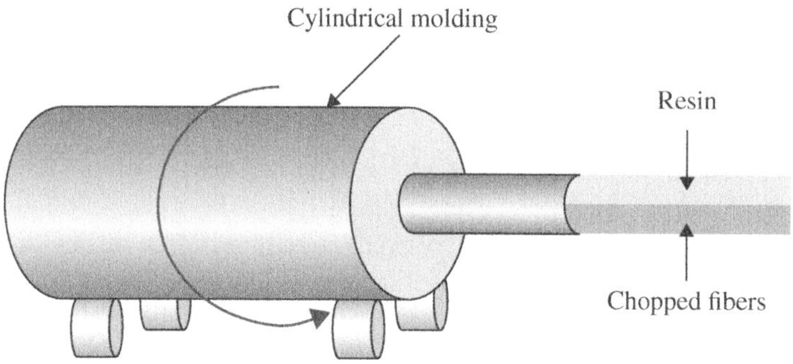

Figure 1.33 Centrifugal casting process for MMCs.

the reinforcement can be achieved by inducing a centrifugal force immediately during casting which allows one to intentionally obtain a gradient in reinforcement volume fraction. Figure 1.33 shows the centrifugal casting process. In brake rotors, for example, wear resistance is needed on the rotor face, but not in the hub area. Thus, in areas where reinforcement is not as crucial, such as in the hub area, easier machining may be obtained.

Spray Forming

Spray-forming of particulate MMCs involves the use of spray techniques that have been used for some time to produce monolithic alloys. A spray gun is used to atomize a molten aluminum alloy matrix. Ceramic particles, such as silicon carbide, are injected into this stream. Usually, the ceramic particles are preheated to dry them. Figure 1.34 shows a

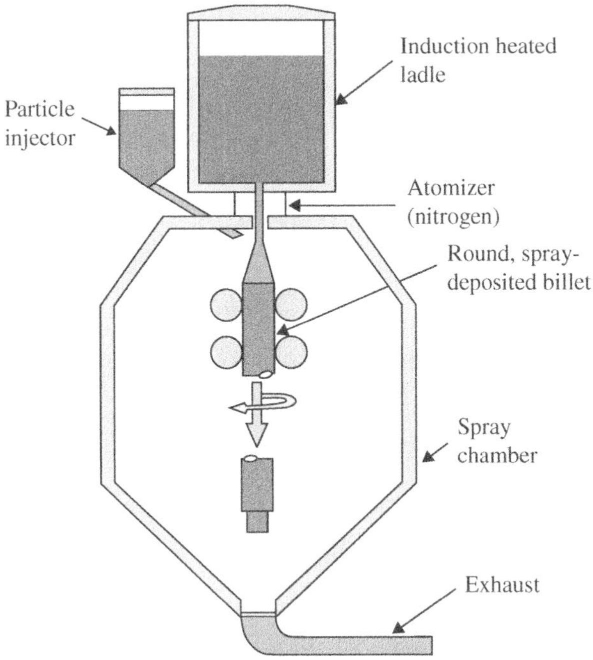

Figure 1.34 Spray forming process for MMCs.

schematic of this process. An optimum particle size is required for an efficient transfer. Whiskers, for example, are too fine to be transferred. The preform produced in this way is generally quite porous. The cosprayed MMC is subjected to scalping, consolidation, and secondary finishing processes, thus making it a wrought material. The process is totally computer-controlled and quite fast. It also should be noted that the process is essentially a liquid metallurgy process. One avoids the formation of deleterious reaction products because the time of flight is extremely short. Silicon carbide particles of an aspect ratio (length/diameter) between 3 and 4 and volume fractions up to 20% have been incorporated into aluminum alloys. A great advantage of the process is the flexibility that it affords in making different types of composites. For example, one can make in situ laminates using two sprayers or one can have selective reinforcement. This process, however, is quite expensive, mainly because of the high cost of the capital equipment.

1.6.2.2 Solid-State Processes

Many solid-state techniques are available. Some of the important ones have been described below.

Diffusion Bonding

Diffusion bonding is a common solid-state welding technique used to join similar or dissimilar metals. Inter-diffusion of atoms from clean metal surfaces in contact at an elevated temperature leads to welding. There are many variants of the basic diffusion bonding process; however, all of them involve a step of simultaneous application of pressure and high temperature. Matrix alloy foil and fiber arrays, composite wire, or monolayer laminae are stacked in a predetermined order. Figure 1.35 shows a schematic of one such diffusion bonding process, also called the foil-fiber-foil process. Vacuum hot pressing is a most important step in

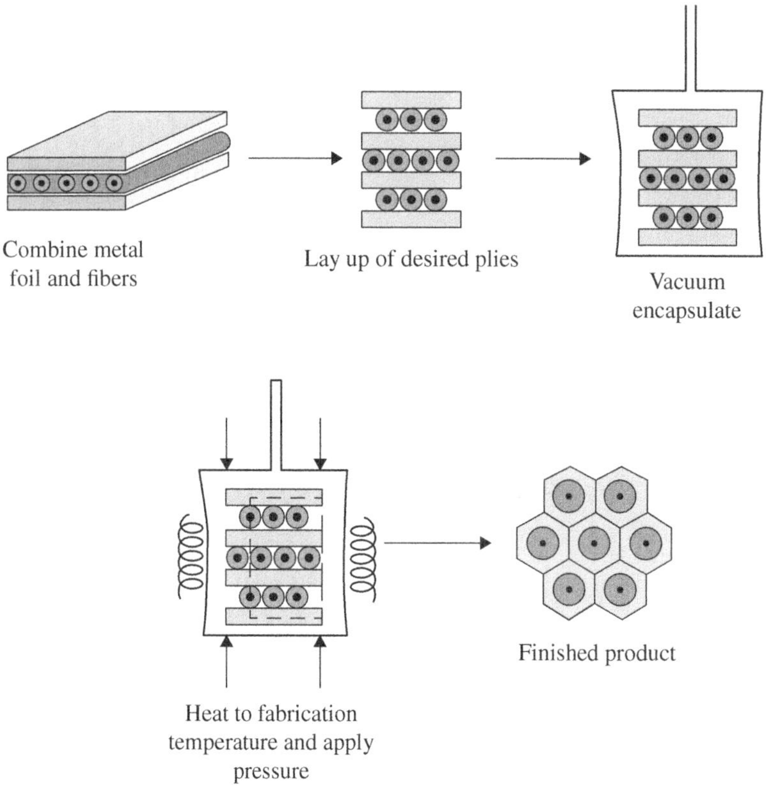

Figure 1.35 Diffusion bonding process: (a) apply metal foil and cut to shape, (b) lay-up desired plies, (c) vacuum encapsulate and heat to fabrication temperature, (d) apply pressure and hold for consolidation cycle, and (e) cool, remove, and clean part.

the diffusion bonding processes for MMCs. The major advantages of this technique are the ability to process a wide variety of matrix metals and control of fiber orientation and volume fraction. Among the disadvantages are processing times of several hours, high-processing temperatures and pressures, all of which make the process quite expensive; besides the fact only objects of limited size can be produced. Hot isostatic pressing (HIP), instead of uniaxial pressing, can also be used. In HIP, gas pressure against a can consolidates the composite piece contained inside the can. With HIP, it is relatively easy to apply high pressures at elevated temperatures over variable geometries.

Deformation Processing

Deformation processing of MMCs involves mechanical processing (swaging, extrusion, drawing, or rolling) of a ductile two-phase material. The two phases codeform, causing the minor phase to elongate and become fibrous in nature within the matrix. These materials are sometimes referred to as in situ composites. The properties of a deformation-processed composite depend largely on the characteristics of the starting material, which is usually a billet of two-phase alloy that has been prepared by casting or powder metallurgy methods. Roll bonding is a common technique used to produce a laminated composite consisting of different

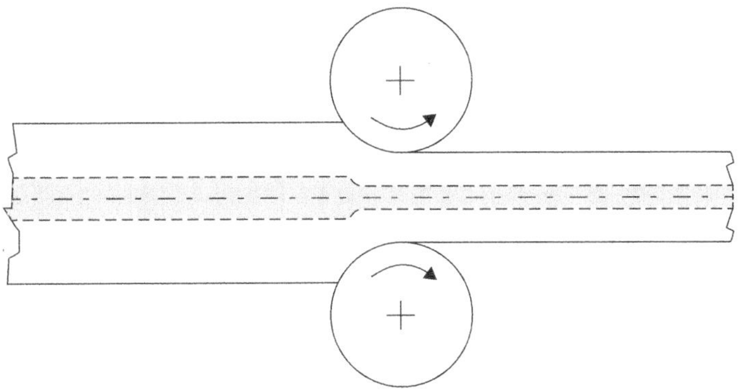

Figure 1.36 Roll-bonding process for fabrication of MMCs.

metals in the sheet form. Such composites are called sheet-laminated MMCs. Other examples of deformation-processed MMCs are the niobium based conventional filamentary superconductors and the high-T_c superconductors. Figure 1.36 shows a roll bonding technique for making a laminated MMC, which produces a metallurgical bond.

Powder Processing

Powder processing methods in conjunction with deformation processing are used to fabricate particulate or short fiber-reinforced composites. This typically involves cold pressing and sintering, or hot pressing to fabricate primarily particle-or whisker-reinforced MMCs. The matrix and the reinforcement powders are blended to produce a homogeneous distribution. The blending stage is followed by cold pressing to produce what is called a green body, which is about 80% dense and can be easily handled. The cold-pressed green body is canned in a sealed container and degassed to remove any absorbed moisture from the particle surfaces. The material is hot pressed, uniaxially or isostatically, to produce a fully dense composite and extruded. The rigid particles or fibers cause the matrix to be deformed significantly. The process has been shown in Figure 1.37. During hot extrusion, dynamic recrystallization takes place at the particle/matrix interface, yielding randomly oriented grains near the interface, and relatively textured grains far from the interface.

Sinter-Forging

Sinter-forging is a novel and low-cost deformation processing technique. In sinter forging, a powder mixture of reinforcement and matrix is cold compacted, sintered, and forged to nearly full density. The main advantage of this technique is that forging is conducted to produce a near-net shape material, and machining operations and material waste are minimized. The low-cost, sinter-forged composites have tensile and fatigue properties that are comparable to those of materials produced by extrusion. Figure 1.38 shows a schematic of the sinter-forging process.

Deposition Techniques

Deposition techniques for MMC fabrication involve coating individual fibers in a tow with the matrix material needed to form the composite followed by diffusion bonding to form a

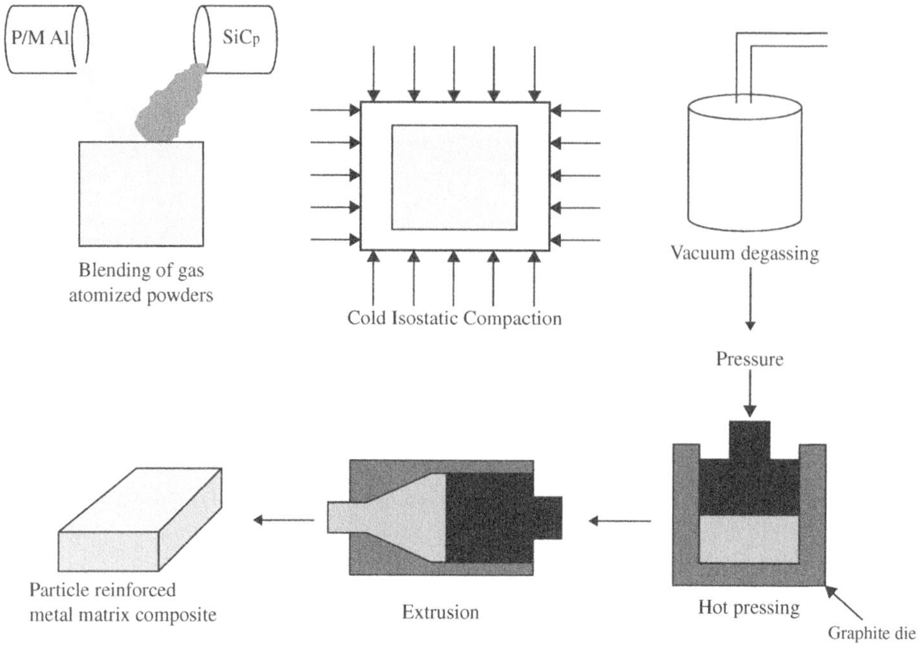

Figure 1.37 Powder processing, hot pressing, and extrusion process for fabricating particulate or short fiber-reinforced MMCs.

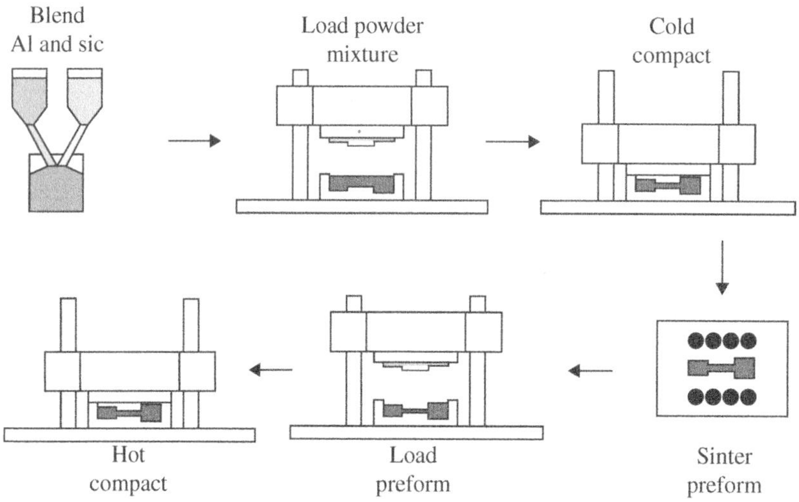

Figure 1.38 Sinter-forging process.

consolidated composite plate or structural shape. The main disadvantage of using deposition techniques is that they are very time-consuming. However, there are several advantages:

(a) The degree of interfacial bonding is easily controllable; interfacial diffusion barriers and compliant coatings can be formed on the fiber prior to matrix deposition or graded interfaces can be formed.

(b) Thin, monolayer tapes can be produced by filament winding; these are easier to handle and mold into structural shapes than other precursor forms-unidirectional or angle-plied composites can be easily fabricated in this way.

Several deposition techniques are available: immersion plating, electroplating, spray deposition, CVD, and physical vapor deposition (PVD). Dipping or immersion plating is similar to infiltration casting except that fiber tows are continuously passed through baths of molten metal, slurry, sol, or organometallic precursors. Electroplating produces a coating from a solution containing the ion of the desired material in the presence of an electric current. Fibers are wound on a mandrel, which serves as the cathode, and placed into the plating bath with an anode of the desired matrix material. The advantage of this method is that the temperatures involved are moderate and no damage is done to the fibers. Problems with electroplating involve void formation between fibers and between fiber layers, adhesion of the deposit to the fibers may be poor, and there are limited numbers of alloy matrices available for this processing.

A spray deposition operation, typically, consists of winding fibers onto a foil-coated drum and spraying molten metal onto them to form a monotape. The source of molten metal may be powder or wire feedstock, which is melted in a flame, arc, or plasma torch. The advantages of spray deposition are easy control of fiber alignment and rapid solidification of the molten matrix. In a CVD process, a vaporized component decomposes or reacts with another vaporized chemical on the substrate to form a coating on that substrate. The processing is generally carried out at elevated temperatures.

1.6.2.3 In Situ Processes

In in situ techniques, one forms the reinforcement phase in situ. The composite material is produced in one step from an appropriate starting alloy, thus avoiding the difficulties inherent in combining the separate components as in a typical composite processing. Controlled unidirectional solidification of a eutectic alloy is a classic example of in situ processing. Unidirectional solidification of a eutectic alloy can result in one phase being distributed in the form of fibers or ribbons in the other. One can control the fineness of distribution of the reinforcement phase by simply controlling the solidification rate. The solidification rate in practice, however, is limited to a range of 1–5 cm/h because of the need to maintain a stable growth front. The stable growth front requires a high-temperature gradient. Figure 1.39 shows the in situ processing by controlled unidirectional solidification of a eutectic alloy.

1.6.3 Ceramic Matrix Composites

CMCs can be processed either by conventional powder processing techniques used for making polycrystalline ceramics or by some new techniques developed specifically for making CMCs. Some of the important processing techniques for CMCs have been discussed below.

1.6.3.1 Cold Pressing and Sintering

Cold pressing of the matrix powder and fiber followed by sintering is a carryover from conventional processing of ceramics. Generally, in the sintering step, the matrix shrinks considerably and the resulting composite has many cracks. In addition to this general problem of shrinkage associated with sintering of any ceramic, certain other problems arise when we put high-aspect ratio (length/diameter) reinforcements in a glass or ceramic matrix material

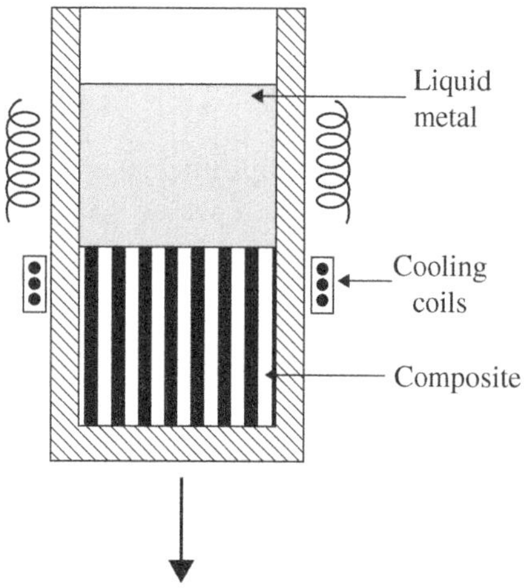

Figure 1.39 In situ processing by controlled unidirectional solidification of a eutectic alloy.

and try to sinter. Fibers and whiskers can form a network that may inhibit the sintering process. Depending on the difference in thermal expansion coefficients of the reinforcement and matrix, a hydrostatic tensile stress may develop in the matrix on cooling, which will counter the driving force (surface energy minimization) for sintering. Thus, the densification rate of the matrix will, in general, be retarded in the presence of reinforcements. Whiskers or fibers may also give rise to the phenomenon of bridging, which is a function of the orientation and aspect ratio of the reinforcement.

1.6.3.2 Hot Pressing

Some form of hot pressing is frequently resorted to in the consolidation stage of CMCs. This is because a simultaneous application of pressure and high temperature can accelerate the rate of densification and a pore-free and fine-grained compact can be obtained. A common variant, called the slurry infiltration process, is one of the most important technique used to produce continuous fiber-reinforced glass and glass–ceramic composites. The slurry infiltration process involves two stages:

(i) Incorporation of a reinforcing phase into an unconsolidated matrix.
(ii) Consolidation of matrix by hot pressing.

Figure 1.40 shows a schematic of this process. In addition to incorporation of the reinforcing phase, the first stage involves some kind of fiber alignment. A fiber tow or a fiber preform is impregnated with matrix-containing slurry by passing it through a slurry tank. The impregnated fiber tow or preform sheets are similar to the prepregs used in PMCs. The slurry consists of the matrix powder, a carrier liquid (water or alcohol), and an organic binder. The organic binder is burned out prior to consolidation. Wetting agents may be added to ease the infiltration of the fiber tow or preform. The impregnated tow or prepreg is wound on a drum and dried. This is followed by cutting and stacking of the prepregs and consolidation in a hot press. The process has the advantage that, as in PMCs, the prepregs can be arranged in a variety of stacking sequences. The slurry infiltration process is well suited for glass or glass–CMCs, mainly because the processing temperatures for these materials are lower than those used for crystalline matrix materials and glassy phase has good flow properties. Any hot pressing process has certain limitations in producing complex shapes. The fibers should suffer little or no damage during handling. Application of a very high pressure can easily

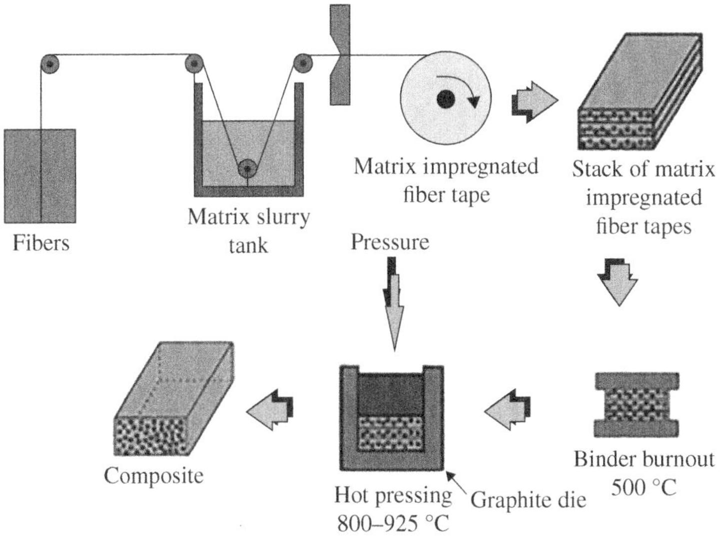

Figure 1.40 Slurry-impregnation method.

damage fibers. Refractory particles of a crystalline ceramic can damage fibers by mechanical contact. The reinforcement can also suffer damage from reaction with the matrix at very high-processing temperatures. The matrix should have as little porosity as possible in the final product as porosity in a structural ceramic material is highly undesirable. To this end, it is important to completely remove the fugitive binder and use a matrix powder particle smaller than the fiber diameter. The hot pressing operational parameters are also important. Precise control within a narrow working temperature range, minimization of the processing time, and utilization of a pressure low enough to avoid fiber damage are important factors in this final consolidation part of the process. Fiber damage and any fiber/matrix interfacial reaction, along with its detrimental effect on the bond strength, are unavoidable attributes of the hot pressing operation.

1.6.3.3 Reaction Bonding

Reaction bonding processes similar to the ones used for monolithic ceramics can be used to make CMCs. Reaction bonding process has the following advantages:

(a) Problems with matrix shrinkage during densification can be avoided.
(b) Rather large volume fractions of whiskers or fiber can be used.
(c) Multidirectional, continuous fiber preforms can be used.
(d) The reaction bonding temperatures for most systems are generally lower than the sintering temperatures so that fiber degradation can be avoided.

One great disadvantage of this process is that high porosity is difficult to avoid. A hybrid process involving a combination of hot pressing and reaction bonding technique can also be used. A silicon cloth is prepared by attrition milling of a mixture of silicon powder, a polymer binder, and an organic solvent to obtain a dough of proper consistency. This dough is then rolled to make a silicon cloth of desired thickness. Fiber mats are made by filament winding of silicon carbide with a fugitive binder. The fiber mats and silicon cloth are stacked in an alternate sequence, debinderized (the step of binder removal), and hot pressed in a molybdenum die in a nitrogen or vacuum environment. The temperature and pressure are adjusted to produce a preform. At this stage, the silicon matrix is converted to silicon nitride by transferring the composite to a nitriding furnace between 1100 and 1400 °C. Typically, the silicon nitride matrix has about 30% porosity, which is not unexpected in reaction bonded silicon nitride. Reaction bonding process can also be used to make oxide fiber/oxide matrix composites, which are very attractive for high-temperature applications in air. Specifically, the technique has been applied to alumina and mullite.

1.6.3.4 Infiltration

Infiltration of a preform made of reinforcement can be done with a matrix material in solid, liquid, or gaseous form.

Liquid Infiltration

This technique is very similar to liquid polymer or liquid metal infiltration (Figure 1.41). Proper control of the fluidity of the liquid matrix is, of course, the key to this technique. It yields a high-density matrix, i.e. no pores in the matrix. Almost any reinforcement geometry can be used to produce a virtually flaw-free composite. The temperatures involved, however, are much higher than those encountered in polymer or metal processing. Processing at such high temperatures can lead to deleterious chemical reactions between the reinforcement and the matrix. Thermal expansion mismatch between the reinforcement and the matrix, the rather large temperature interval between the processing temperature and room temperature, and the low strain to failure of ceramics can add up to a formidable set of problems in producing a crack-free CMC. Viscosities of ceramic melts are generally quite high, which makes the infiltration of preforms rather difficult. Wettability of the reinforcement by the molten ceramic is another item to be considered. A preform made of reinforcement in any form (for example, fiber, whisker, or particle) having a network of pores can be infiltrated by a ceramic melt by using capillary pressure. Application of pressure or processing in vacuum can aid in the infiltration process. Penetration will also be easier if the contact angle is low (i.e. better wettability), and the surface energy and the pore radius are large. If the radius of the channel is made too large, the capillarity effect will be lost. The advantages are as follows:

(a) The matrix is formed in a single processing step.
(b) A homogeneous matrix can be obtained.

The disadvantages of infiltration techniques are as follows:

(a) High-melting points of ceramics mean a greater likelihood of reaction between the melt and the reinforcement.
(b) Ceramics have higher melt viscosities than metals; therefore, infiltration of preforms is relatively difficult.
(c) The matrix is likely to crack because of the differential shrinkage between the matrix and the reinforcement on solidification. This can be minimized by choosing components with nearly equal coefficients of thermal expansion.

Gaseous Infiltration

Using a reactive gas mixture which deposits a ceramic material within a preform of the reinforcement, commonly known as chemical vapor deposition (CVD) or chemical vapor

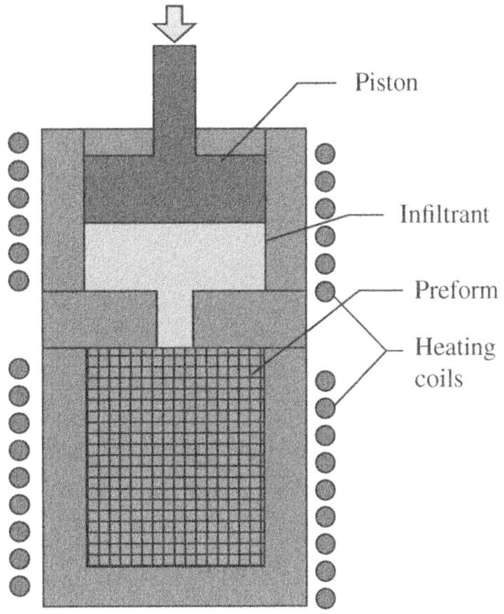

Figure 1.41 Melt-infiltration process.

infiltration (CVI), and typically performed at high temperature; all of which have particular advantages or disadvantages for different types of matrix material. For those matrix materials which can be produced by reaction between gases, such as carbon and silicon carbide, a gas phase route can substitute for the liquid impregnation route. This tends to be a slow process, because if deposition is allowed to occur too quickly, it mostly occurs at the external surface, blocking penetration. Developments in the technique have concentrated on ensuring that deposition occurs internally to the component, by forcing reactive gas flow through the walls of the component (CVI), using temperature gradients, and on speeding the process up, e.g. by using microwave heating which also has the advantage of improving deposition internally to the preform. However, as with liquid infiltration, full densification of the matrix never occurs as access to remaining pores becomes blocked, but the method can be used for complex shapes.

In gas phase routes, i.e. the so-called CVI processes, the reinforcement (usually as a multidirectional preform) is densified by the matrix deposited from a gaseous precursor, e.g. an hydrocarbon for carbon or a mixture of methyltrichlorosilane and hydrogen for silicon carbide. There are several versions of the CVI-process. It is now well established that a fiber coating, referred to as the interphase, has to be deposited on the fiber prior to the infiltration of the matrix in order to control the fiber-matrix bonding and the mechanical behavior of the composite. Chemical vapor impregnation is a method of infiltrating fiber architectures with matrix particles via the vapor phase. Although this is a similar process to CVD in terms of gas reaction, decomposition conditions in CVI are chosen for in-depth decomposition rather than coating the surface of the substrate. Several techniques have been developed to introduce the reaction gases into the fiber architecture, such as temperature gradient, pressure gradient and pulse CVI. Figure 1.42 shows the schematic of chemical vapor impregnation technique.

1.6.3.5 Polymer Infiltration and Pyrolysis

Just as polymeric precursors can be used to make ceramic fibers, we can use polymeric precursors to form a ceramic matrix in a composite. This technique is called polymer infiltration

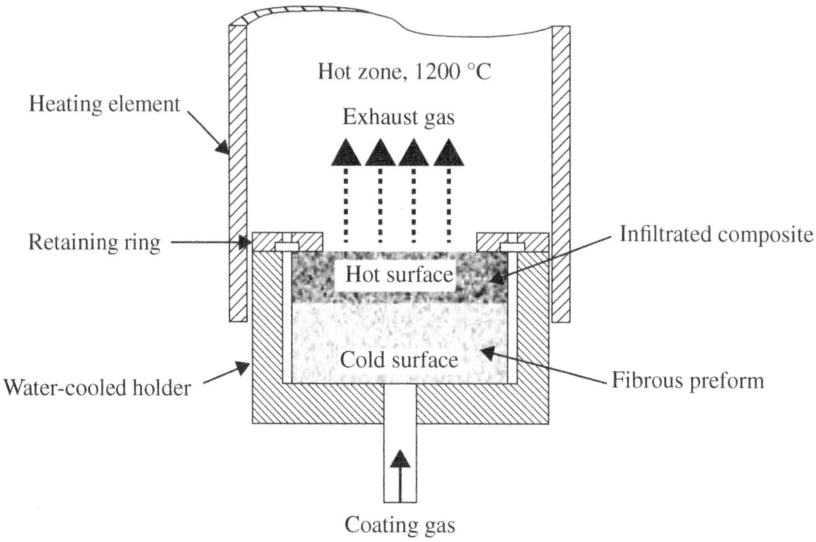

Figure 1.42 Chemical vapor impregnation technique.

and pyrolysis (PIP). It is an attractive processing route because of its relatively low cost, small amounts of residual porosity, and minimal degradation of the fibers. Moreover, this approach allows near net-shape molding and fabrication technology that is able to produce nearly fully dense composites. In PIP, the fibers are infiltrated with an organic polymer, which is heated to fairly high temperatures and pyrolyzed to form a ceramic matrix. Due to the relatively low yield of polymer to ceramic, multiple infiltrations are used to densify the composite. Polymeric precursors for ceramic matrices allow one to use conventional polymer composite fabrication technology that is readily available, and take advantage of processes used to make PMCs. These include complex shape forming and fabrication. Furthermore, by processing and pyrolyzing at lower temperatures (compared to sintering and hot-pressing, for example) one can avoid fiber degradation and the formation of unwanted reaction products at the fiber/matrix interface. Among the desirable characteristics of a pre-ceramic polymer are the following:

(a) High-ceramic yield from polymer precursor.
(b) Precursor that yields a ceramic with low free carbon content (which will oxidize at high temperatures).
(c) Controllable molecular weight, which allows for solvent solubility and control over viscosity for fabrication purposes.
(d) Low temperature crosslinking of the polymer, which allows resin to harden and maintain its dimensions during the pyrolysis process.
(e) Low cost and toxicity.

Most preceramic polymer precursors are formed from chloroorganosilicon compounds to form poly(silanes), poly(carbosilanes), poly(silazanes), poly(borosilanes), poly(silsesquioxanes), and poly(carbosiloxanes). The synthesis reaction involves the dechlorination of the chlorinated silane monomers. Since a lot of the chlorosilane monomers are formed as by-products in the silicone industry, they are inexpensive and readily available. The monomers can be further controlled by an appropriate amount of branching which

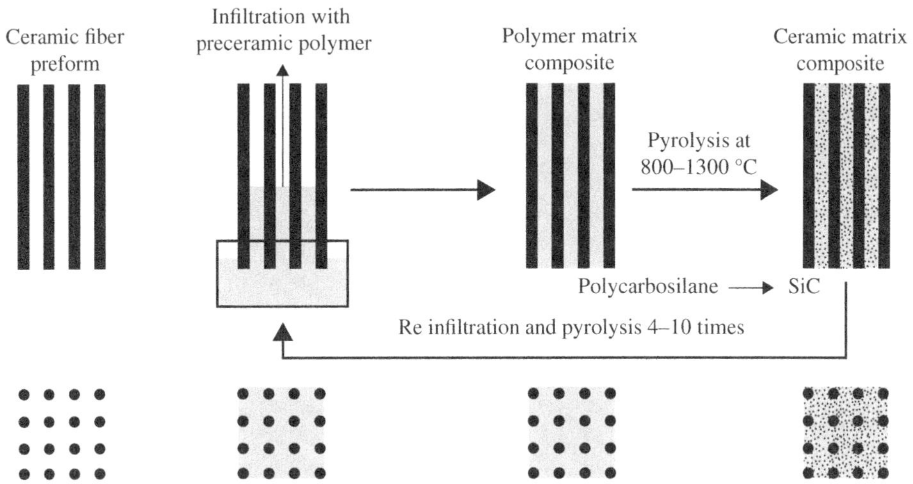

Figure 1.43 Polymer infiltration and pyrolysis technique.

controls important properties such as the viscosity of the precursor as well as the amount of ceramic yield. All silicon-based polymer precursors lead to an amorphous ceramic matrix, where silicon atoms are tetrahedrally arranged with nonsilicon atoms. This arrangement is similar to that found in amorphous silica. High-temperature treatments typically lead to crystallization and slight densification of the matrix, which results in shrinkage. At high temperatures, the amorphous ceramic begins to form small domains of crystalline phase, which are more thermodynamically stable. Typically, the range of the molecular weight of the polymer is tailored, followed by shaping of the product. The polymer is then crosslinked and finally pyrolyzed in an inert or reactive atmosphere (e.g. NH_3) at temperatures between 1000 and 1400 °C. The pyrolysis step can be further subdivided into three steps:

(a) In the first step, between 550 and 880 °C, an amorphous hydrogenated compound of the type $Si(C_aO_bN_cB_d)$ is formed.
(b) The second step involves nucleation of crystalline precipitates such as SiC, Si_3N_4, and SiO_2 at temperatures between 1200 and 1600 °C. Grain coarsening may also result from consumption of any residual amorphous phase and reduction of the amount of oxygen due to vaporization of SiO and CO. Porosity is typically on the order of 5–20 vol% with pore sizes of the order of 1–50 nm. It should be noted that the average pore size and volume fraction of pores decrease with increasing pyrolysis temperature, since the amount of densification (and shrinkage) becomes irreversible at temperatures above the maximum pyrolysis temperature. Figure 1.43 shows a schematic of the PIP process.

Among the disadvantages of PIP, we can list:

(a) Low yield accompanies the polymer-to-ceramic transformation.
(b) Large shrinkage, which typically causes cracking in the matrix during fabrication. Due to shrinkage and weight loss during pyrolysis, residual porosity after a single impregnation is on the order of 20–30%.
(c) To reduce the amount of residual porosity, multiple impregnations are needed. Reimpregnation is typically conducted with a very low viscosity prepolymer so that the slurry may wet and infiltrate the small micropores that exist in the preform. Usually, reimpregnation is done by immersing the part in the liquid polymer in a vacuum bag, while higher

viscosity polymers require pressure impregnation. Typically, the amount of porosity will reduce from 35% to less than 10% after about five impregnations. Significant gas evolution also occurs during pyrolysis. Thus, it is advisable to allow these volatile gases to slowly diffuse out of the matrix, especially for thicker parts. Typically, pyrolysis cycles ramp to somewhere between 800 and 1400 °C over periods of one to two days, to avoid delamination. Recall that pyrolysis must be done at a temperature below the crystallization temperature of the matrix (or large volume changes will occur) and below the degradation temperature of the reinforcing fibers. The pyrolysis atmosphere is most commonly argon or nitrogen, although in ammonia, we obtain a pure amorphous silicon nitride with low amounts of free carbon.

References

1 Kelly, A. (1985). Composites in context. *Composites Science and Technology* 23 (3): 171–199.
2 Hunt, W.H. (2004). Nano-materials: nomenclature, novelty and necessity. *Journal of Materials Science* 56: 13–18.
3 Guz, I.A., Rodger, A.A., Guz, A.N., and Rushchitsky, J.J. (2007). Developing the mechanical models for nano-materials. *Composites Part A: Applied Science and Manufacturing* 38: 1234–1250.
4 Talley, C.P. (1959). Mechanical properties of glassy boron. *Journal of Applied Physics* 30: 1114.

2

Literature Review of Different Modeling Methods

2.1 Material Development

Each distinct age in the development of humankind has been associated with advances in materials technology. Some historians have linked key technological and societal events with the materials technology that was prevalent during the "stone age," "bronze age," and so forth, but the last 350 years have seen many advances in materials technology that have helped shape our world today. Much of this groundbreaking work (Table 2.1) was because of perseverant research scientists and engineers finding solutions after long periods of experimentation and development. Within the last 20 years, many research institutions have recognized the need for a more systematic approach to new materials development that employs a multiscale modeling approach. This approach was one that would combine interdisciplinary research, new advances in computational modeling and simulation, and critical laboratory experiments to rapidly reduce the time from concept to end product. The general consensus is that this new paradigm by which all future materials research would be conducted and has come to be known simply as "Computational Materials." Traditionally, research institutions have relied on a discipline-oriented approach to material development and design with new materials. It is recognized, however, that within the scope of materials and structures research, the breadth of length and time scales may range more than 12 orders of magnitude and different scientific and engineering disciplines are involved at each level.

To help address this wide-ranging interdisciplinary research, Computational Materials programs have been formulated with the specific goal of exploiting the tremendous physical and mechanical properties of new nanomaterials by understanding materials at atomic, molecular, and supramolecular levels. Computational Materials draws from physics and chemistry, but focuses on constitutive descriptions of materials that are useful in formulating macroscopic models of material performance. In particular, how the Computational Materials approach utilizes multiscale analysis methods to establish the technology for the scale-up of nanostructured materials into engineering level and multifunctional materials for advanced applications such as next generation aircraft and spacecraft, is shown in Figure 2.1. Benefits of Computational Materials approach are threefold. First, it encourages a reduced reliance on costly trial and error, or serendipity, of "Edisonian" approach to materials research. Second, it increases the confidence that new materials will possess the desired properties when scaled up from the laboratory level, so that lead-time for the introduction of new technologies is reduced. Third, the Computational Materials approach lowers the likelihood of conservative or compromised designs that might have resulted from reliance on less-than-perfect materials.

Mechanics of Particle- and Fiber-Reinforced Polymer Nanocomposites: From Nanoscale to Continuum Simulations, First Edition. Sumit Sharma.

Table 2.1 Significant events in material development over the last 350 years.

Year	Event
1665	Robert Hooke – material microstructure
1808	John Dalton – atomic theory
1824	Portland cement
1839	Vulcanization
1856	Large-scale steel production
1869	Mendeleev and Meyer – periodic table of the chemical elements
1886	Aluminum
1900	Max Planck – quantum mechanics
1909	Bakelite
1921	A.A. Griffith – fracture strength
1928	Staudinger – polymers (small molecules that link to form chains)
1955	Synthetic diamond
1970	Optical fibers
1985	First university initiatives attempt computational materials design
1985	Bucky balls (C_{60}) discovered at Rice University
1991	Carbon nanotubes discovered by Sumio Iijima

2.2 Nanostructured Materials

Nanotechnology is still at its early stages of development, since even till date, we have not been able to find the most efficient manufacturing processes that can make nanomaterials. The prefix "nano" means one billionth part of something. In Greek, nano means "dwarf," in nanotechnology and nanomechanics, "nano" refers to the length scale of 1 nm or 10^{-9} m. Composite structures whose characteristic dimensions are found at the nanoscale are called nanocomposites. A morphological characteristic that is of fundamental importance in the understanding of structure–property relationship of nanocomposites is the surface area/volume ratio of the reinforcement materials according to Chou et al. [1]. Nanocomposites can be described based upon the three categories of reinforcement materials: particles (silica, metal, and other organic and inorganic particles), layered materials (graphite, layered silicate, and other layered minerals), and fibrous materials (nanofibers and nanotubes). From a "bottom–up" perspective, multiscale approach should consider the intrinsic attributes of the constituent materials for the system of study. Relationship between time and length scales for multiscale simulation methodology is shown in Figure 2.2.

Much of the current work focuses on the use of nanostructured materials. The origins of focused research into nanostructured materials can be traced back to a seminal lecture given by Richard Feynman in 1959. In this lecture, he proposed an approach to "the problem of manipulating and controlling things on a small scale." The scale he referred to was not the microscopic scale that was familiar to scientists of the day but the unexplored atomistic scale. Over the subsequent years, this idea was refined and eventually resulted in the announcement of the National Nanotechnology Initiative in 2000. Nanostructured materials

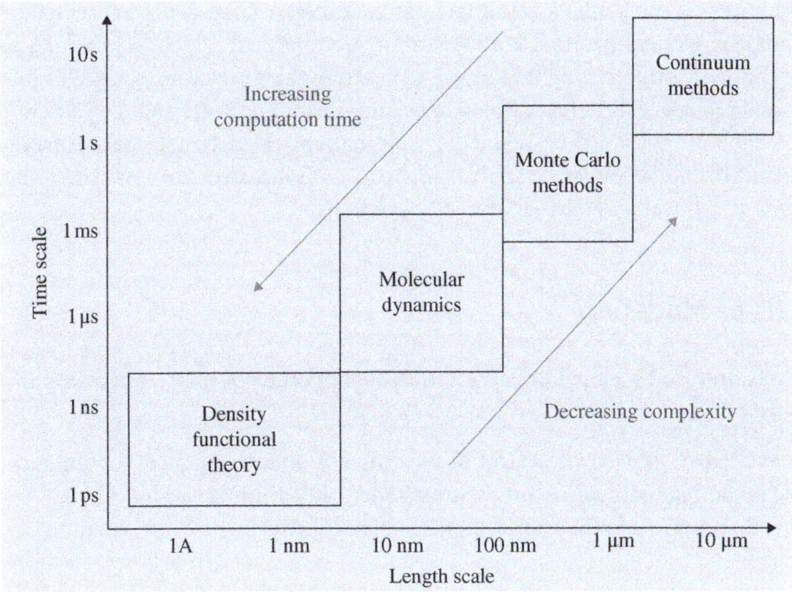

Figure 2.1 Range of length and time scales of key simulation methods.

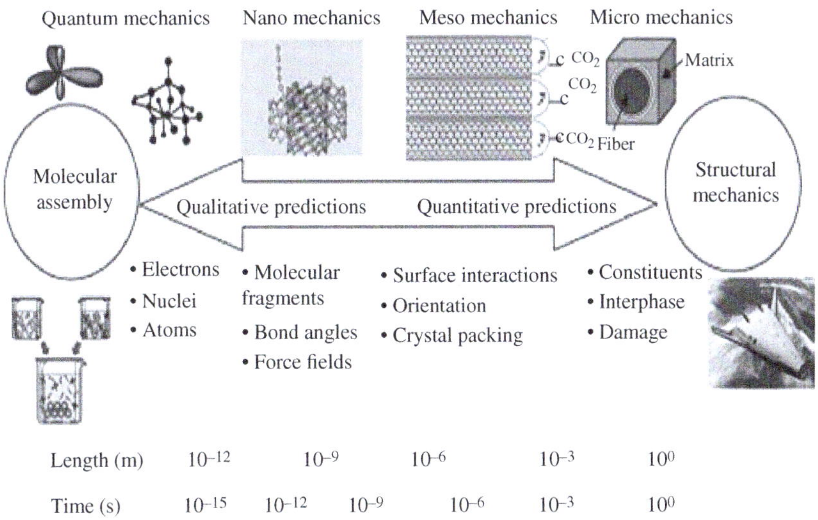

Figure 2.2 Schematic illustration of relationships between time and length scales for multiscale simulation methodology.

based on carbon nanotubes (CNTs) and related carbon structures are of current interest for much of the materials community. Although at the time of their discoveries, other materials with well-defined nanoscopic structure were known, investigators were intrigued to find that these new forms of carbon could be viewed as either individual molecules or as potential structural materials. This realization in turn energized a whole new culture of nanotechnology research accompanied by worldwide efforts to synthesize nanomaterials and to use them to create multifunctional composite materials. More broadly then, nanotechnology presents the vision of working at the molecular level, atom by atom, to create large structures with fundamentally new molecular organization. With regards to aerospace, the objectives within the National Nanotechnology Initiative include advances in ultralight, ultrastrong, space durable materials for very large space structures (telescopes, antennae, and solar sails), spacecraft electronics for greater autonomy and on-board decision-making, microsystems based on biological principles, utilization of in situ resources to create complex structures in space, and biologically inspired architectures for long-duration missions.

2.3 Methods of Modeling

Following methods are used for modeling the different properties of nanocomposites and these are described below:

(i) *Continuum methods*: These methods can be classified as either analytical or computational. These modeling methods assume the existence of continuum for all calculations and generally do not include the chemical interactions between the constituent phases of the composite.
(ii) *Molecular modeling*: Molecular modeling assumes a noncontinuous composition of the material, which makes it a powerful tool for studying atomic interactions at the nanometer length scale. In recent years, molecular modeling has emerged as an important tool in the prediction of physical material properties such as elastic response, atomic structure, vibrational frequencies, heat of reaction, electric permittivity, and binding energies.

Mechanical properties of nanostructured materials can be determined by a select set of computational methods. These modeling methods span a wide range of length and time scales. For the smallest length and time scales, Computational Chemistry techniques are primarily used to predict atomic structure using first-principles theory. For largest length and time scales, Computational Mechanics is used to predict the mechanical behavior of materials and engineering structures. Computational Chemistry and Computational Mechanics modeling methods are based on thoroughly established principles that have been developed in science and engineering. However, intermediate length and time scales do not have general modeling methods that are as well developed as those on the smallest and largest time and length scales. Therefore, multiscale modeling techniques are employed, which take advantage of Computational Chemistry and Computational Mechanics methods simultaneously for prediction of structure and properties of materials. Each modeling method has broad classes of relevant modeling tools. The Quantum mechanical and Nanomechanical modeling tools assume the presence of a discrete molecular structure of matter. Micromechanical and Structural Mechanics assume the presence of a continuous material structure. The continuum-based methods primarily include techniques such as the finite element method (FEM) and the micromechanics approach developed for composite materials. Specific Micromechanical techniques include Eshelby approach, Mori–Tanaka method, and Halpin–Tsai method. The molecular modeling tools include Molecular Dynamics, Monte Carlo (MC), and Ab initio techniques. Each of these molecular based and continuum modeling methods are described below.

2.3.1 Atomistic, Molecular Methods

At the atomistic or molecular level, reliance is on molecular mechanics, molecular dynamics (MD), and coarse-grained simulation. Molecular models encompassing thousands and perhaps millions of atoms can be solved by these methods and used to predict fundamental, molecular-level material behavior. The methods are both static and dynamic. For example, molecular mechanics can establish the minimum-energy structure statically and MD can resolve the nanosecond-scale evolution of a molecule or molecular assembly. These approaches can model both bonded and nonbonded forces (e.g. van der Waals and electrostatic), but are not parameterized for bond cleavage.

MD method was first introduced by Alder and Wainwright [2] to study the interactions of hard spheres. Many important insights concerning the behavior of simple liquids emerged from their studies. The next major advance was when Rahman [3] carried out the first simulation by using a realistic potential for liquid argon. These simulations and their results provide a typical example of MD simulation results: structural information, transport phenomena, and time dependence of physical properties. Simulation results could be average quantities at a thermodynamic state point or the development of a structure-based property in time. In a nonequilibrium simulation, where the system is subjected to a temporary perturbation, the response of the material can be analyzed. Connections to measurable quantities are made through thermodynamics and statistical mechanics. Number of simulation techniques has greatly expanded; there exist now many specialized techniques for particular problems, including mixed quantum mechanical–classical simulations given by Car and Parrinello [4].

In classical MD, particle movement in simulation is driven by the forces on each particle which is described by a set of functions, the force field, used to describe how the particles interact. In quantum MD, the particles move according to the ab initio-derived forces. A variation of this approach is to use the ab initio-derived forces in small areas of the system to provide a very detailed description of a specific area the physical system. MD simulation techniques are widely used to help interpret experimental results from X-ray crystallography and nuclear magnetic resonance spectroscopy. MD simulations generate information at the nanolevel, including atomic positions and velocities. The conversion of this information to macroscopic observables such as pressure, energy, heat capacities, and so forth, requires statistical mechanics. In a MD simulation, the time-dependent behavior of the molecular system is obtained by integrating Newton's equations of motion. The result of the simulation is a time series of conformations or the path followed by each atom. Most MD simulations are performed under conditions of constant number of atoms, volume, and energy (N,V,E) or constant number of atoms, temperature, and pressure (N,T,P) to better simulate experimental conditions. Basic steps in the MD simulation are given as follows:

(i) Establish initial coordinates.
(ii) Minimize the structure.
(iii) Assign initial velocities.
(iv) Establish dynamics of the thermal conditions.
(v) Perform equilibration dynamics.
(vi) Rescale the velocities and check if the temperature is correct.
(vii) Perform dynamic analysis of trajectories.

Current generation force fields (or potential energy functions) provide a reasonably good compromise between accuracy and computational efficiency. They are often found empirically and calibrated to experimental results (e.g. X-ray crystallography) and quantum mechanical calculations of small model compounds. The development of parameter sets, that define these force fields, may require extensive optimization and is an area of continuing research. One of the most important limitations imposed on a force field is that no drastic changes in electronic structure are allowed, i.e. no events like bond making or breaking can be modeled. The most time consuming part of a MD simulation is the calculation of the nonbonded terms in the potential energy function, e.g. the electrostatic and van der Waals forces. In principle, the nonbonded energy terms between every pair of atoms should be evaluated. This requirement would imply that the number of computations increases as the square of the number of atoms for a pair-wise model. To speed up the computation, the interactions between two atoms separated by a distance greater than a predefined distance, the cut-off distance, are ignored.

MD simulations are by their nature mechanical because the particles are driven to move by the forces acting on them. In an MD simulation, the force, velocity, and position of each atom are known for each configuration as the simulation trajectory evolves in time. From this information, elastic mechanical constitutive properties can be calculated by using MD. One basic use of MD in the determination of mechanical properties is to use it for generation of representative volume elements (RVEs) of the material. The RVE should contain any structural information not readily available from computational mechanics. MD simulation includes all the atomistic degrees of freedom and can be parameterized for a description of surface-to-surface interactions within the material system. This application of MD has recently been developed to study polymer nanocomposites. In these material systems, there are at least two components the nanostructured inclusion and the polymer. Because there are two components, the details of how the components interact affect the mechanical behavior. Assumptions such as perfect bonding between the components or arbitrary placement of atoms are avoided. Instead, a representative structure based on a well-parameterized molecular force field is generated. Information from the force fields can then be used at other levels to describe the atomistic interactions. Once an RVE of the material is obtained from MD, it may then be used for the computation of mechanical properties.

2.3.2 Coarse Grain Methods

Although MD methods provide the kind of details necessary to resolve molecular structure and localized interactions, this fidelity comes with a price. Namely, both the size and time scales of the model are limited by numerical and computational boundaries. To help overcome these limitations, coarse-grained methods are available that represent molecular chains as simpler models. A comparison to MD has shown up to 4 orders of magnitude decrease in CPU time through the use of the simpler models given by Lopez [5]. Although the coarse-grain models lack the atomistic detail of MD, they do preserve many of the important aspects of the chemical structure and allow for simulation of material behavior above the nanoscale. The connection to the more-detailed atomistic model can be made directly through an atomistic-to-coarse-grain mapping procedure that when reversed allows one to model well-equilibrated atomistic structures by performing this equilibration by using the coarse-grain model. This mapping and reverse mapping helps to overcome the time-scale upper limits of MD simulations. Several approaches to coarse-graining have been proposed and include both continuous and lattice models. The continuous models seem to be preferable for dynamic problems such as might occur when considering dynamic changes in volume. As outlined by Kremer and Muller-Plathe [6], the systematic development of the coarse-grain model requires three principal steps:

(i) Determine the degree of coarse-graining and the geometry of the model.
(ii) Choose the form of the intra- and interchain potentials.
(iii) Optimize the free parameters, especially for the nonbonded interactions.

Coarse-grain models are often implemented by MC simulations to provide a timely solution. The MC method is used to simulate stochastic events and provide statistical approaches to numerical integration as shown by Binder [7]. As given by Raabe [8], there are three characteristic steps in the MC simulation that are given as follows:

(i) Translate the physical problem into an analogous probabilistic or statistical model.
(ii) Solve the probabilistic model by a numerical sampling experiment.
(iii) Analyze the resultant data by using statistical methods.

MC simulation methods are roughly grouped into four categories: weighted and non-weighted sampling methods, lattice type, spin model, and energy operator. As a specific example, consider coarse-grain modeling of polymers. Polymers are long-chain molecules possessing structural detail across a wide range of length scales (10^{-10}–10^{-6} m). At the smaller length scale, are the details associated with the chemical structure of the monomer. These include number and arrangement of atoms, bond lengths, and angles. At the larger length scale is the conformation of the polymer due to the characteristic of a long-chain molecule. Associated with this range of length scales is a corresponding range of time scales. The relevant time scales range from bond length vibrations (10^{-13} seconds) to conformational rearrangements ($>10^{-4}$ seconds). Due to this range of length and time scales, a variety of approaches have been used to model polymers.

In order to study the universal conformational features of long-chains molecules, simplified coarse-grained models have been developed. In recent years, efforts have been made to bridge the gap between coarse grain and fully atomistic simulation types. These bridging methods attempt to address the problem of simulation across a range of scales through various mapping and reverse-mapping techniques. Extensive reviews of these various techniques are presented by Baschnagel [9] and Muller-Plathe [10]. Primary application for coarse grain polymer modeling involves studying processes which occur on longer time scales than is possible to study with atomistic simulations. MD simulation is typically used to equilibrate atomistic models of small molecules as shown by Allen and Tildesley [11]. Due to the long relaxation times of polymer conformations, however, the time scale for equilibration can be prohibitive with atomistic models of polymers. Multiscale modeling techniques can be used to address this issue. A schematic for this process is shown in Figure 2.3 with a representative polymer chain shown with a periodic boundary cell. Atomistic model at position A is mapped to a coarse-grained model at position B. This is equilibrated with computationally faster MC simulation to position C. Reverse-mapping to position D recovers the atomistic model. The slow MD simulation (A to D) is circumvented with multiscale mapping/reverse mapping procedure.

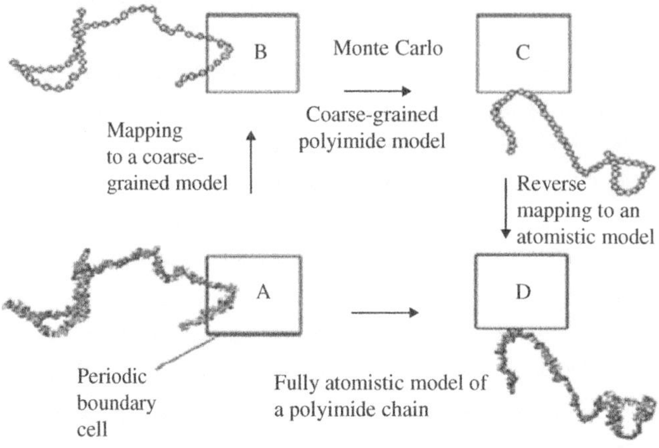

Figure 2.3 Schematic illustration of the procedure for coarse graining and reverse mapping.

2.3.3 Continuum Methods

With proper understanding of the molecular structure and nature of materials, the behavior of collections of molecules and atoms can be homogenized. At the continuum level, the observed macroscopic behavior is explained by disregarding the discrete atomistic and molecular structure and assuming that the material is continuously distributed throughout its volume. The continuum material is assumed to have an average density and can be subjected to body forces such as gravity and surface forces such as the contact between two bodies. The continuum can be assumed to obey several fundamental laws. The first, continuity, is derived from the conservation of mass. The second, equilibrium, is derived from momentum considerations and Newton's second law. The third, the moment of momentum principle, is based on the model that the time rate of change of angular momentum with respect to an arbitrary point is equal to the resultant moment. The next two laws, conservation of energy and entropy are based on the first and second laws of thermodynamics, respectively. These laws provide the basis for the continuum model and must be coupled with the appropriate constitutive equations and equations of state to provide all the equations necessary for solving a continuum problem. State of the continuum system is described by several thermodynamic and kinematic state variables. Equations of state provide the relationships between the nonindependent state variables.

Continuum method relates the deformation of a continuous medium to the external forces acting on the medium and the resulting internal stress and strain. Computational approaches range from simple closed-form analytical expressions to micromechanics to complex structural mechanics calculations based on beam and shell theory. The continuum-mechanics methods rely on describing the geometry (i.e. a physical model), and must have a constitutive relationship to achieve a solution.

(a) *Micromechanics*: One approach to the homogenization of a multiconstituent material is through the combination of the continuum method and a micromechanics model to provide a transition from the microscale to the macroscale. Micromechanics assumes small-deformation continuum mechanics as outlined in Section 2.3.3. Continuum mechanics, in general, assumes uniform material properties within the boundaries of the problem. At the microscale, this assumption of uniformity may not hold and hence the micromechanics method is used to express the continuum quantities associated with an infinitesimal material element in terms of the parameters that characterize the structure and properties of the microconstituents of the element as shown by Nemat-Nasser and Hori [12]. A central theme of micromechanics models is the development of a RVE that is a statistical representation of the local continuum properties. In this sense, the RVE may include material boundaries, voids, and defects apparent at the microscale. The RVE is constructed to ensure that the length scale is consistent with the smallest constituent that has a first-order effect on the macroscopic behavior. The RVE is then used in a repeating or periodic nature in the full-scale model. The approach to the micromechanics solution therefore requires a RVE and a suitable averaging technique. Micromechanics method can also account for interfaces between constituents, discontinuities, and coupled mechanical and nonmechanical properties.

(b) *Finite element method (FEM)*: FEM has a long history of development for a wide variety of applications including problems in mechanical, biological, and geological systems. FEM provides a numerical, approximate solution to initial-value and boundary-value problems including time-dependent processes. The method uses a variational technique for solving the differential equations wherein the continuous problem described by the differential equation is cast into the equivalent variation form and the solution is found to be a linear combination of approximation functions. In FEM, the physical shape of the domain of interest is broken into simple subdomains (elements) that are interconnected and fill the entire domain without overlaps. Following steps outline the FEM approach:
   (i) Replace the continuum domain with an assemblage of subdomains.
   (ii) Select the appropriate constitutive laws.
   (iii) Select the interpolation functions necessary to map the element topology.
   (iv) Describe the problem by using the variational principle and divide the system-level integral into subintegrals over the elements.
   (v) Replace continuum state variables by interpolation functions.
   (vi) Assemble element equations.
   (vii) Assemble global system equations.
   (viii) Solve global system of equations, taking into account the prescribed boundary conditions.
   (ix) Calculate the state equation values from state variables.

2.3.4 Effective Continuum Approach

Effective continuum approach for connecting atomistic models to continuum models uses relevant input from the atomistic simulations and carries forward the critical information to represent the continuum with the intrinsic nanoscale features incorporated into the model. The design of large-scale engineering structures requires a complete knowledge of the bulk-level behavior and properties of a material. For structural analysis, the bulk-level material behavior is described or predicted using continuum-based approaches, such as the micromechanical and FEMs described above. Continuum mechanical parameters, such as Young's modulus or stress, are classically defined with the assumption given by Fung [13] that the material is a mathematical continuum. However, a set of atoms in a molecular modeling simulation, which possess a structure that is in thermodynamic equilibrium, clearly does not resemble a mathematical continuum, but a discrete lattice structure. Therefore, the direct application of continuum-mechanics analyses for molecular models is problematic unless steps are taken to secure their equivalency. Establishing an effective-continuum model for a discrete structure is the only way to reliably describe the behavior of the discrete structure in terms of continuum mechanics-based parameters. Ideally, the behavior of the effective continuum closely resembles that of the atomistic structure under any set of boundary conditions. Early attempts at establishing effective continuum models include those developed for simple crystalline materials by Born and Huang [14] and Kittel [15] and for aerospace lattice structures by Sun et al. [16], Noor [17], and Lee [18]. Many of these studies such as the one given by Eringen [19] incorporate a generalized theory of elasticity which allows displacement and rotational degrees of freedom of infinitesimal material points and/or the concept of energy equivalence of the lattice and effective continuum models.

Prediction of mechanical properties of crystalline materials, such as metals with dislocation defects and grain boundaries, using a combination of atomistic and finite element (FE) models, has been performed with the Quasi-continuum approach given by Tadmor et al. [20] and Shenoy [21]. With this method, the deformation of a RVE of individual atoms is mapped into a finite element model such that the nodes of the finite element model deform in an identical manner as the corresponding points in the molecular model. In addition, the energies of deformation for the atomic and finite element models are the same for identical loading conditions. This process is captured with the Cauchy–Born rule, which hypothesizes that an atomic crystal, which is represented by a continuum, will deform according to the overall continuum deformation gradient. Therefore, the strain energy density at a continuum point can be determined from the energy of the atomistic model.

Similar approaches have also been employed which either map the deformation of atoms from atomistic simulations onto a finite element model (or nonclassical elastic model) or incorporate a handshake region between the atomistic and finite element models. Nakano [22] simulated projectile impact onto a silicon crystal. The Macroscopic Atomistic Ab initio Dynamics (MAAD) approach given by Abraham [23] and Broughton [24] was used to study brittle crack propagation in silicon. The Bridging Scale method developed by Wagner and Liu [25] was used to overcome issues of time scale with simultaneous simulations of molecular and continuum models. This approach has been used to model various atomic lattice structures and CNTs. A bridging domain method developed by Xiao and Belytschko [26] overlaps continuum and molecular domains where the Hamiltonian function is a linear combination of the two models. The atomic-scale finite element method (AFEM) was developed by Liu and Chen [27] and applied to CNTs.

2.4 Literature Review of Different Methods of Modeling

Various properties of nanocomposites have been studied by number of researchers by using different modeling methods. Some of these according to the method used are discussed below:

2.4.1 Micromechanics/FEM

Adams et al. [28] performed a series of experiments to determine the specific damping capacities of composites. They pursued the idea of utilizing the measured quantities subsequently in analytical considerations. However, many of the known formulations to predict damping based on generally valid theories are developed only for certain deformation modes (e.g. torsion of a beam or plate bending) and cannot be employed in a fully three-dimensional analysis. During the last few decades a number of studies have been carried out focusing on experimental and analytical work characterizing the absorbed damping energy of a composite. A frequently applied method based on the complex modulus approach was developed by Hashin [29]. The introduced analytical expressions account for on-axis ply damping.

Damping properties of continuous fiber composites have been studied by a number of researchers. There are relatively few publications on damping of discontinuous fiber composites. However, studies reported by, for instance, McLean and Read [30] indicate that vibration damping of polymeric fiber-reinforced composites may be significantly improved, and possibly can be readily optimized by using as a reinforcement discontinuous fibers rather than continuous ones. McLean and Read [30] and Gibson and Yau [31] indicated that the damping in a given composite can be improved by using discontinuous fibers rather than continuous fibers.

Research work of Gibson and Yau [31] indicates that by varying the fiber aspect ratio and fiber orientation, superior damping and stiffness could be achieved separately. This observation implies that the optimum conditions, in terms of microstructural parameters, for damping may not be necessarily the same for stiffness. Consequently, it is important to study the influence of the various governing microstructural parameters as pertaining to both damping and stiffness. Studies carried out by Gibson and Yau [31], Sun et al. [32], Suarez et al. [33], and Gibson et al. [34] have confirmed that the damping factor of polymeric discontinuous fiber-composites is frequency-dependent, and it relates strongly with the microstructure's particulars, e.g. fiber aspect ratio, fiber volume fraction, and fiber off-axis angle. Also based on the complex modulus approach, Sun et al. [32] and Gibson et al. [34] accomplished work leading to a hysteretic damping model.

Damping properties of continuous fiber composites have been studied by a number of investigators. Sun et al. [32] developed an analytical model based on the force-balance approach to predict damping in randomly oriented short-fiber polymer matrix composites. They showed analytically that there exists an optimum aspect ratio and an optimum fiber orientation for maximum damping. The predicted optimum aspect ratio lies in the range of actual aspect ratios for whiskers and microfibers.

Sun et al. [32] distinguished several mechanisms of dissipative behavior in a fiber reinforced composite:

(i) Viscoelastic behavior of matrix and/or fibers.
(ii) Thermoelastic damping due to cyclic heat flow.
(iii) Coulomb friction due to slip in unbounded regions of the fiber-matrix interface.
(iv) Dissipation caused by microscopic or macroscopic damage in the composite.

Suarez et al. [33] described theoretically and experimentally how very low aspect ratios ($l/d < 20$) are required to produce significant improvements in damping. In addition, a finite element model developed by Hwang and Gibson [35] showed that improved damping was obtained at low-aspect ratios ($l/d < 20$). A possible explanation for improved damping at low-aspect ratios is the presence of shear stress concentrations at the numerous fiber ends in a discontinuous fiber composite and the resulting shear stress transfer to the viscoelastic matrix. However, the optimum aspect ratio for damping is not necessarily the best for stiffness.

Zabaras and Pervez [36] used experimental data on specific damping capacities of unidirectional composite beams to define a modal loss factor of laminated plates. They applied an equivalent viscous damping approximation to introduce the damping mechanism into a finite element analysis. Based on the simplified micromechanical equations given by Saravanos and Chamis [37], the model for homogenizing data of unidirectional fiber reinforced materials is extended by Saravanos and Chamis [38] to damping capacities identified with the six engineering stresses.

Ruoff and Lorents [39] used known elastic properties of graphite sheet to derive the mechanical properties, i.e. tensile and bending stiffness constants for ideal single-walled carbon nanotubes (SWCNTs) and multiwalled carbon nanotubes (MWCNTs). Authors suggested that the thermal expansion of CNTs are essentially isotropic, which can be contrasted with strongly anisotropic expansion in conventional fibers (large diameter fibers) and in graphite. On the other hand, the thermal conductivity may be highly anisotropic and along the long axis higher than any other material. The authors also presented a short discussion of topological constraints to surface chemistry in idealized multiwalled nanotubes (MWNTs) and highlighted the importance of strong interface between nanotube and matrix for the formation of high-strength nanotube-reinforced composites.

Ajayan et al. [40] studied the mechanical behavior of MWCNTs/epoxy composites in both tension and compression and observed the load transfer to the nanotubes in the composite is higher in compression. The value of the compression modulus is higher as compared to tensile modulus because during the load transfer in MWNTs only the outer layer is seen to be in tension, whereas other layers in compression.

Chandra et al. [41] have provided a comprehensive review of damping involving macromechanical, micromechanical, and viscoelastic approaches. According to them, the various composite damping mechanisms are viscoelastic nature of matrix or fiber materials, damping due to interphase, viscoplastic damping, and thermoelastic damping. Chandra et al. [42] have predicted damping coefficients of two-phase continuous fiber-reinforced composites. They studied the effect of shape of fiber cross-section and fiber volume fraction on various damping coefficients using viscoelastic correspondence principle. They also determined loss factors for composites reinforced with cylindrical continuous fiber using 2D micromechanical FEM/strain energy approach. They concluded that longitudinal loss factor is independent of the shape of fiber and transverse loss factor is slightly affected by aspect ratio of fiber from low-to-moderate fiber volume fraction.

Zengin et al. [43] have modeled damping characteristics of MWNTs in epoxy using stick-slip concept. Experiments conducted by Rajoria and Jalili [44] showed an increase of about 140% in damping ratio of single-walled nanotube (SWNT) epoxy composite as compared to plain epoxy composites. Koratkar et al. [45] observed approximately 200% increase in damping ratio by addition of MWNTs in epoxy. One reason behind such damping increase, according to stick-slip mechanism, is the poor adhesion between CNT and polymer matrix. Such good enhancement in damping ratio with low proportion and poor dispersion and alignment of CNTs in epoxy could be further improved by increasing proportion and improving dispersion and alignment of CNTs in epoxy.

Finegan et al. [46] developed an analytical strain energy method and a FEM for micromechanical modeling of composite including the fiber-matrix interphase region. Liu and Chen [47] evaluated the effective mechanical properties of CNT-based composites using a 3D nanoscale RVE based on 3D elasticity theory and solved by the FEM. Formulas to extract the material constants from solutions for the RVE under three loading cases are established using the elasticity. An extended rule of mixtures (ERMs), which can be used to estimate the Young's modulus in the axial direction of the RVE and to validate the numerical solutions for short CNTs, is also derived using the strength of materials theory. Numerical results demonstrate that the reinforcing capabilities of the CNTs in a matrix are significant. It is also observed that with only about 2% and 5% volume fractions of CNTs in a matrix, the stiffness of the composite in the CNTs axial direction can increase as many as 0.7 and 9.7 times for the case of short and long CNT fibers, respectively.

Finegan et al. [46] summarized recent analytical and experimental investigations of the dynamic mechanical properties of carbon nanofiber (CNF)/polypropylene composites. The CNFs have been vapor-grown carbon fibers (VGCF) which are grown catalytically from the gaseous hydrocarbons using metallic catalyst particles. Dynamic mechanical thermal analyzer is used for measuring the storage modulus and damping of nanocomposites. For the analytical model, elastic–viscoelastic correspondence principle is used to transform the elastic Halpin–Tsai equations to complex form. The fiber injection direction is the preferred direction corresponding to which predicted and measured results are validated by the authors. Authors found that although predicted and measured damping values are in the same range, the predicted values do not show the measured decrease in damping with increasing fiber volume fraction. Predictions show that composites having very low-fiber aspect ratios should have higher damping than those having high-fiber aspect ratios, and that the CNF aspect ratios of approximately $l/d = 19$ are in the range which yields the highest predicted damping.

Modified Halpin–Tsai equation was used to predict the macroscopic elastic constants of CNT polymer composites by Thostenson and Chou [48] and Mori–Tanaka method was also applied to CNT polymer composites with long wavy nanotubes by Fisher et al. [49]. However, these two methods have a common shortcoming that they cannot accurately reveal the interactions between neighboring nanotubes and cannot evaluate the microstresses involving with individual nanotubes because of the limitation of their analytical models. Parallel to the above theoretical researches, numerical approaches have also been developed based on the analysis of a basic unit cell by FEM in the past years. In general, numerical approaches can be roughly classified into average-field method and homogenization method. These two methods are available for the evaluation of macrostiffness and microstresses of nanocomposites at the same time.

Liu and Chen [27] evaluated the effective properties of CNT-based composites using a square RVE based on the continuum mechanics and solved by the FEM. Formulas to extract the effective material constants from solutions for the square RVEs under different loading conditions are established based on elasticity. Square RVEs with multiple CNTs are also investigated in evaluating the Young's modulus and Poisson's ratio in the transverse plane. The results obtained from FEM showed that the load-carrying capacity of CNTs in matrix is sufficient. The results showed that with the addition of only about 3.6% volume fraction of the CNTs in a matrix, the stiffness of the composite in the CNT axial direction can increase as much as 33% for the case of long CNT fibers. Authors further used the conventional rule of mixture for long CNTs to find out the effective elastic properties of nanocomposites, whereas ERM for short-fibers-reinforced composites.

Papanikos and Tserpes [50] proposed a three-dimensional finite element (FE) model for armchair, zigzag, and chiral SWCNTs. The model developed is based on assumption that CNTs, when subjected to loading, behave like space-frame structures. The bonds between carbons atoms are considered as connecting load-carrying members, while the carbon atoms as joints of the members. To create the FE models, nodes are placed at the locations of carbon atoms and the bonds between them are modeled using three-dimensional elastic beam elements. The elastic moduli of beam elements are determined by using a linkage between molecular and continuum mechanics. The various factors such as tube wall thickness, tube diameter, and chirality has an impact on elastic properties of CNTs. Experimental and theoretical results obtained from FEM considering the effect of wall thickness have been compared which indicated inverse relation between wall thickness and elastic modulus. Also as tube diameter increases, there is an increase in the storage modulus. Results indicated that the elastic modulus found from chiral structure is higher than zigzag and arm chair structure.

Gao and Li [51] developed a shear lag model for predicting the interfacial stress transfer in CNT-reinforced polymer composites using multiscale approach. The morphological features of the nanocomposites have been predicted by using a square RVE in which CNT capped at both ends are embedded in a concentric composite cylinder. The atomistic structure of the capped nanotube is incorporated in the model by using the molecular structure mechanics, which leads to the determination of the effective Young's modulus of the nanotube. The capped nanotube is equivalently replaced by an effective fiber having the same diameter and length, whose elastic modulus is determined from that of the nanotube under an iso-strain condition. For axis symmetric problems, by using elasticity theory, the continuum-based shear lag formulation has been carried out which provides results in closed-form formulas for calculating the interfacial shear stress. Authors obtained a good agreement between the predicted numerical results with the past computational and experimental data. The results conclude that the aspect ratio of CNTs plays a critical role in designing the nanotube reinforcing polymer composites.

Schulte et al. [52] studied different types of CNTs, such as SWCNTs, MWCNTs, and double-walled CNTs, their influence on the mechanical properties of nanocomposites and surface functionalization of fillers. The authors studied the effect of filler contents, the varying dispersibility, aspect ratio, the specific surface area, and an amino-functionalization on the properties of nanocomposites. It has been found that the reagglomeration of CNTs was reduced by functionalization of CNTs with amino group which improved the dispersibility and stabilized the CNT suspension. The dispersion of the CNTs and the interfacial adhesion to the epoxy matrix is the key issues in development of nanocomposites. Small amount of CNTs in epoxies leads to increased mechanical properties. The most significant improvement of strength (10%), stiffness (15%), and especially fracture toughness (43%) has been attained with amino-functionalized double-walled CNTs at 0.5% filler content.

Srikanth et al. [53] determined energy dissipation in the form of damping capacity using free–free type suspended beam arrangement coupled with circle-fit approach. This technique is based on the classical vibration theory, by which the geometry and material properties of the metallic specimen are related to the measured resonant frequency. Using the fact that the ratio of vibration response to the applied force fits to a circle in the Argand plane for each resonant frequency of the test specimen, the damping factor and natural frequency are predicted accurately for the test specimen. The results revealed that an increase in the alumina content up to 0.4% volume percentage lead to an increase in the damping capacity up to 34%. The addition of alumina particles also improves the other properties of magnesium such as hardness and elastic modulus.

Odegard et al. [54] developed a continuum-based elastic micromechanics model for silica nanoparticle/polyamide composites with various nanoparticle/polyamide interfacial treatments. The model incorporates the molecular structures of the nanoparticle, polyamide, and interfacial regions, which are determined using a molecular modeling method that involves coarse-grained and reverse-mapping techniques. The micromechanics model includes an effective interface between the polyamide and nanoparticle with properties and dimensions that are determined using the results of MD simulations. Model predicted the elastic properties of silica nanoparticle/polyamide composite for large range of nanoparticle radius 10–10 000 Å. For silica nanoparticle of radius equal to 1000 Å, the predicted results showed good agreement with Mori–Tanaka micromechanical approach. Predictions also indicated, silica nanoparticle/polyamide interface addition has a significant effect on the composite mechanical properties for radius below 1000 Å. Elastic and shear moduli predicted from effective interface model showed increasing trend with increase in effective particle size.

Liu et al. [55] developed a new continuum model for the CNT-based composites for large-scale analysis at the microscale in order to characterize such composites. In this new approach, CNTs are treated as rigid fibers in the elastic matrix due to its high stiffness that are at least two orders higher than those of most polymer matrices. A recently developed fast multipole boundary element method (BEM) is employed to solve the boundary integral equations governing this rigid-inclusion problem. The largest model studied contains 16 000 CNT fibers and has the total degree of freedom above 28.8 million. These results clearly showed the effectiveness, efficiency, and promise of the development of fast multiple BEM as a fast and first order numerical tool for large-scale characterizations of CNT composites. Results predicted from model have been validated with MD simulations and multiple approach method.

Haque and Ramasetty [56] utilized the shear lag model to investigate the effects of CNTs aspect ratio, CNTs volume fraction, and matrix modulus on stress transfer in SWCNTs nanocomposites, indicating that no significant increase in CNTs axial stress would occur when the CNT volume fraction increases. In addition, the effect of polymer matrix modulus on axial stress distribution of CNTs is negligible.

Youn and Song [57] predicted the effective elastic properties of CNT-based nanocomposites with asymptotic expansion homogenization (AEH) method. A RVE constructed for AEH method by considering an assumption that nanocomposites showed geometrical periodicity with respect to microscopic scale. For successfully implementing the homogenization method, control volume finite element method (CVFEM) has been adopted. For better interfacial bonding between CNTs and matrix, the CNTs are treated by plasma so that the better load is transferred and the mechanical properties are improved. The homogenization method predicted the effective elasticity tensor of nanocomposite that is validated with analytical and experimental results. Results showed very good agreement with conventional Halpin–Tsai equation and varied linearly as the volume fraction of CNT increased. However, for the increase in the aspect ratio of CNT the homogenization method showed different results from analytical models. The AEH method possessed a large potential for the prediction of elastic and thermal properties of homogeneous composite material because this method handled the geometrical complexity and material anisotropy properly.

Shen et al. [58] studied the load transfer between the walls of MWCNTs using MD simulation and found that loading the outer wall of MWCNTs does not effectively transfer the load to the inner walls. However, when the chemical bonding between the walls is established, it can significantly improve the load transfer to the inner nanotubes.

Lagoudas and Seidel [59] predicted the effective elastic properties of CNT/polymer nanocomposites using different micromechanical techniques. Using the in-plane elastic properties of graphene, the effective properties of CNTs are calculated utilizing a composite cylinders micromechanics technique. These effective properties are then used in the self-consistent and Mori–Tanaka methods to obtain effective elastic properties of composites consisting of aligned SWCNTs or MWCNTs embedded in a polymer matrix. The results revealed that in the CNT embedded direction, the axial Young's modulus is dominated, while all the properties associated with transverse direction are matrix dominated. Multilayer composite cylinder approach has been adopted to find the behavior of nanocomposite with interphase layer. Effect of interphase on the transverse properties is visible only at low-volume fractions. Further, the Tessellation method in conjunction with Mori–Tanaka technique has been used to study the effect of CNTs clustering. CNTs clustering (bundles of CNTs) are formed due to the interatomic forces. Observation showed that clustering enhances the effective properties associated with transverse to CNT alignment direction. Effective properties predicted by Mori–Tanaka method applied to randomly dispersed CNT in matrix are compared with experimental results.

Hubert and Ashrafi [60] investigated the elastic properties of SWCNTs arrays and their composites. The properties of twisted SWCNT nanoarrays or ropes of circular cross-section are predicted through a finite element analysis by applying proper boundary conditions to the model and using strain energy method. The nanoarray properties are then used to describe the properties of twisted SWCNT nanoarray/polymer composites. Effect of volume fraction and aspect ratio of nanoarray as reinforcement for dilute polymer composite systems are examined for aligned and random distribution using conventional micromechanics. Observation showed that for small twist angle (<20°), the axial modulus drops dramatically. Also the transverse properties of the array remain constant for twist angle less than 15°. The authors found that composites reinforced by the aligned SWCNT arrays had higher modulus compared to composites made of SWCNT arrays with a twist angle of 10°. An important conclusion pointed to the fact that there is a possibility of making infinite SWCNT arrays due to the twist angle that provides additional load transfer.

Koratkar et al. [45] studied the behavior of improved damping in epoxy films reinforced with nanoscale fillers. The mechanism for damping improvement is related to frictional energy dissipation arising from interfacial sliding at the nanofiller matrix interface. The combination of extremely large interfacial contact area, high-aspect ratio, many interfaces, and low-density implies that nanoscale fillers could be far more efficient with regard to damping augmentation than traditional fillers. The author investigated three categories of nanofillers such as silica nanoparticles, silicon nanorods, and silicon nanosprings. Results revealed that silicon nanosprings generate greater energy dissipation than nanoparticles or nanorods. The material loss factor of epoxy film (with silicon nanoscale fillers) increased by 150% as compared to pure epoxy. Nanocomposite damping materials developed show that the potential to overcome the limitations of traditional viscoelastic polymers and could help to efficiently deliver significant levels of damping to structural components.

Zalamea et al. [61] employed the shear transfer model as well as the shear lag model to understand the stress transfer from the outermost layer to the interior layers in the MWCNTs. It is pointed out that as the layer number increases, the corresponding efficiency decreases.

Guzman de Villoria and Miravete [62] compared experimental results with different theoretical models such as Voigt–Reuss model, Halpin–Tsai model, Modified Mixture law, and Cox model to estimate the Young's modulus of nanocomposites. The theoretical models assumed the homogeneous dispersion of CNTs in the matrix, but experimentation behavior showed heterogeneous dispersion, which is indicative of gap between theoretical and experimentation results. Due to the heterogeneous dispersion of the nanoreinforcement of the composite and the clusters formation, a new micromechanical model has been developed, dominated dilute suspension of clusters. Main issue of this model is that it explains in an easy way the effect of the dispersion of nanoreinforcement in mechanical properties, taking into account the in homogeneity of the sample. Comparing the theoretical–experimental correlation of this new model with the other models presented, the improvement is significant.

Basu et al. [63] devised a finite element (FE)-based unit-cell model with periodic boundary conditions to calculate the effective $\tan\delta$ of a composite as a function of frequency. Using this method, it is demonstrated that the $\tan\delta$ of the nanocomposite manifests important information about the extent and properties of the interphase region that surrounds each nanoparticle. The interphase is assumed to have higher value of E compared to matrix but may have higher or lower damping factor. Authors found that the presence of agglomeration in nanocomposites introduces error in the interpretation of damping. Predicted results showed that the $\tan\delta$ of the interphase lower than that of the matrix, at some volume fraction, indicative of the size of the interphase, the damping capacity of the nanocomposites is very small.

CNTs have been used with polymers from the date of their inception to make composites having remarkable properties. Kanagaraj et al. [64] tried to enhance the mechanical and tribological properties of the composite materials. The latter have been achieved through the injection molding of high-density polyethylene (HDPE) reinforced with specific volume fraction of CNTs. A considerable improvement on mechanical properties of the material can be observed when the volume fraction of the CNT is increased. The volumetric wear rate is calculated from the Wang's model, Ratner's correlation, and reciprocal of toughness. The results obtained clearly show the linear relationship with the CNT loading which support the microscopic wear model. The author concludes that both Halpin–Tsai and modified series model can be used to predict Young's modulus of CNT–HDPE composites.

Based on above review for FEM/Micromechanics method, following conclusions can be drawn:

(i) Vibration damping of polymeric fiber-reinforced composites may be significantly improved and can be readily optimized by using as reinforcement, discontinuous fibers rather than continuous ones.
(ii) Number of studies have been carried out focusing on experimental and analytical work characterizing the absorbed damping energy of a composite.
(iii) Elastic–viscoelastic correspondence principle and strain energy approach have been widely used for prediction of elastic moduli and loss factors for linear viscoelastic-aligned short-fiber–reinforced nanocomposites.
(iv) Although some work has been carried out on load transfer mechanism in SWCNT and MWCNTs in both tension and compression, very less work has been done for load transfer in shear.
(v) Some work is available on evaluation of elastic moduli and damping along fiber direction, whereas studies considering transverse and shear damping effects are limited.
(vi) Most of the models such as Halpin–Tsai and Cox model cannot accurately reveal the interactions between neighboring nanofibers/nanotubes and also cannot evaluate the microstresses involving with individual nanofibers because of the limitation of their analytical models. In order to evaluate macrostiffness and microstresses of nanocomposites at the same time, either average field method or homogenization method can be used.

(vii) FEM modeling has been used extensively for evaluating elastic moduli of short nanofiber or short CNT-reinforced composites. Also, effect of various parameters such as l/d, V_f, and effect of interphase has been studied using FEM.

(viii) Shear lag model has been used widely to investigate the effects of CNT's aspect ratio, V_f, and matrix modulus on stress transfer in SWCNT's nanocomposites. Less work has been carried out for MWCNTs.

(ix) Since the problem of stress transfer in single short-fiber composites is relatively simple, most existing theories apply to these composites. Because the stress transfer takes place between the only fiber and the surrounding matrix, these models do not include the effect of neighboring fibers on the stress transfer. In real multishort-fiber composites, the neighboring fibers definitely affect the stress transfer from the surrounding matrix to the fiber under consideration. The effect could be significant. Thus, the stress transfer behavior in multishort fiber composites must be different from that in single short-fiber composites. Very few models have been developed to analyze stress transfer in multishort fiber composites.

2.4.2 Effective Continuum

Odegard et al. [65] developed a method for modeling structure–property relationships of nanostructured materials. Basic approach here relates to replacement of the discrete molecular structures with equivalent continuum model in computational chemistry and solid mechanics which has been accomplished by equating the vibration potential energy of a nanostructured material with the strain energy of representative truss and continuum models. This continuum model is also applied to determine the effective continuum geometry of graphene sheet. Further effective thickness is used for developing the RVE of equivalent continuum model.

Odegard et al. [66] presented two approaches, equivalent-continuum modeling and the self-similar approach for predicting the elastic properties of SWCNT/polymer composites. In both cases, the model incorporates information about molecular interactions at the nanometer length scale into a continuum-mechanics based model. The two methods have been compared in terms of assumptions and ranges of validity. It is found that the two approaches can predict elastic properties of SWCNT/polymer composites in a combined range spanning dilute to hyperconcentrated SWCNT volume fractions. Comparison of results using both approaches showed reduction in predicted Young's modulus compared to classical rule of mixture of about 15%.

Buryachenko and Roy [67] modeled nanocomposite as a linearly elastic composite medium, which consists of a homogeneous matrix containing a statistically homogeneous random field of homogeneous prolate spheroidal nanofibers with prescribed random orientation. Estimation of effective elastic moduli of nanocomposites is performed by the version of effective field method developed in the framework of quasicrystalline approximation when the spatial correlations of inclusion location take particular ellipsoidal form. The independent justified choice of shapes of inclusions and correlation holes provides the formulae of effective moduli which are symmetric, completely explicit, and easy to use. The authors noted that the nanofibers are not straight; they are straight only in the form of cylinders with smooth ends rather than prolate spheroids. The assumption that the shape of nanofibers is prolate spheroid, only used for applying the analytical Eshelby solution for modeling the single homogeneous ellipsoidal inclusion inside homogeneous matrix subjected to homogeneous remote loading.

Odegard et al. [54] proposed a technique for developing constitutive models for polymer composite systems reinforced with SWCNTs. It is seen that the size scale of polymer matrix molecules and the CNTs are same due to this reason that the interaction at the interface is highly dependent on the local molecular structure and bonding between matrix/nano sizes reinforcing material. Moreover, due to small size, the lattice structure is not continuous so the conventional micromechanical approach is not preferred for predicting the bulk behavior of nanocomposites. Thus, authors suggested the application of equivalent continuum modeling methods to effective fiber. It is appropriate to note that the local polymer near the nanotube and the interface is named as an effective continuum fiber. The effective continuum fiber retained the continuity of structure so it is used for predicting the bulk mechanical properties of nano composites with various nanotube sizes and orientations.

Based on above review for effective continuum method, following conclusions can be drawn:

(i) Effective continuum approach for connecting atomistic models to continuum models uses relevant input from the atomistic simulations and carries forward the critical information to represent the continuum with the intrinsic nanoscale features incorporated into the model.
(ii) Continuum mechanical parameters, such as Young's modulus or stress, are classically defined with the assumption that the material is a mathematical continuum. However, a set of atoms in a molecular modeling simulation, which possess a structure that is in thermodynamic equilibrium, clearly does not resemble a mathematical continuum, but a discrete lattice structure. Therefore, the direct application of continuum mechanics analyses for molecular models is problematic unless steps are taken to secure their equivalency. Establishing an effective-continuum model for a discrete structure is the only way to reliably describe the behavior of the discrete structure in terms of continuum mechanics-based parameters.
(iii) This modeling technique takes into account the discrete nature of the atomic interactions at the nanometer length scale and the interfacial characteristics of the nanotube and the surrounding polymer matrix.
(iv) For many nanostructured materials, the trade-offs between structure and property must be established before the material can be optimized for any given application. The effective continuum method provides a means for parametrically exploring these structure–property relationships. This method is applicable to a wide range of problems that require the accuracy of atomistic-level descriptions coupled with the general applicability of continuum-level models.

2.4.3 Molecular Dynamics

Cornwell and Wille [68] applied quenched MD simulations to investigate the elastic behavior of open-ended, free-standing SWCNTs. Tersoff–Brenner potential has been used for describing the interatomic interaction. The tubes response to axial compression is examined and typical failure modes as well as stress–strain curves for a number of tube radii are plotted. The tubes could be compressed to some critical value of strain before they either kinked or buckled. The authors calculated the Young's modulus of all the tubes by calculating the force on the end rings and by using the cross-sectional area of the tube. The results obtained from the simulation approach are used for deriving the formula which is used for different radii CNTs to predict the Young's modulus in compression. The results obtained from the formula agreed to within 1% with the simulated values for CNTs with radii ranging from 5 to 22 Å. Schadler et al. [69] performed tension and compression tests on the MWCNT composites, indicating that the load transfer to the MWCNTs is higher in compression rather than in tension.

Simulations of individual CNTs using atomistic or MD models have provided abundant results for the understanding of mechanical and electrical behaviors of the CNTs [68, 70–77]. Continuum approaches based on continuum mechanics have also been applied successfully for simulating the mechanical response of individual or isolated CNTs which are treated as beams or thin shells or solids in cylindrical shapes [78–82]. These studies suggest that the continuum mechanics approach can be applied safely to investigate the mechanical behaviors of the CNTs when the lengths of the CNTs are about 100 nm and above.

Ruoff et al. [83] used molecular mechanics and MD to study the nature of load transfer in CNT bundles. One core tube is considered to be surrounded by number of tubes on the perimeter. The purpose of study involves predicting the effect of intertube interaction on the transfer of load in ropes. It is observed that two important factors such as surface tension and intertube corrugation are responsible for contributing the load transfer. Here, surface tension acting over the circumference of each tube and intertube corrugation is scaled linearly with respect to the contact surface area, and it is increased nonlinearly as the intertube distance decreased. It means that the contact length responsible for better load transfer depends on the intertube gap that is directly related to the corrugation effect. Twisting of CNT bundles significantly enhances the load transfer between the tubes as compared to parallel bundles.

Gao and Li [51] incorporated the atomistic structures of the capped SWCNTs in the molecular structure mechanics and render the Young's modulus of the effective cylindrical solids. Subsequently, the shear stress and the axial stress along the effective solid fiber embedded in the matrix was analyzed using shear-lag analysis.

Based on above review for MD method, following conclusions can be drawn:

(i) MD simulations have become a powerful tool for simulating complex physical phenomena. However, the length and time scales that can be probed using MD are still fairly limited.
(ii) For the study of problems, such as nanoscale coatings, ion-beam deposition, and nano indentation, models need to be on the scale of several microns, consisting of billions of atoms, which is too large for current MD simulations.
(iii) Efficiency and accuracy are the two most important criteria guiding the development of simulation methods. Both depend on the complexity of the interatomic potentials and the time-integration algorithms used in the simulation. With the computational power available today, a typical MD simulation domain contains several millions of atoms. Consequently, MD simulations have their own limitations.
(iv) Molecular mechanics methods have been used particularly to describe mechanical properties of nanotubes. However, little work has been carried out for nanofibers.

Table 2.2 shows material systems characterized by different modeling techniques. It also shows the simulation method, parametric study conducted, and some concluding remarks.

Table 2.2 Material systems characterized by different modeling techniques.

S. no.	Material system	Simulation method	Predicted properties	Conclusions/remarks
1.	Short fiber-reinforced composites Tucker and Liang [84]	Halpin–Tsai, Mori–Tanaka, Eshelby's model	Elastic modulus, damping	Different models are compared to find the best suitable model for predicting the properties of short fiber-reinforced composites
2.	Unidirectional multi-short fiber reinforced composites Fu et al. [85]	Cox model, Rosen model	Fiber axial stress, elastic modulus	Effect of volume fraction and *l*/*d* on fiber axial stress and modulus ratio is examined
3.	Unidirectional discontinuous fiber reinforced composites Hsueh [86]	Modified shear lag/Cox model	Elastic modulus	Young's modulus of unidirectional short fiber-reinforced composites is predicted and compared with FEM and other models
4.	Short fiber-reinforced/metal matrix composites Han et al. [87]	Cox model	Elastic modulus/FOD	Effect of fiber orientation distribution on elastic modulus of composites is predicted
5.	Carbon nanofiber/polypropylene Finegan et al. [46]	Halpin–Tsai, elastic–viscoelastic correspondence principle	Elastic storage modulus, damping	Effect of *l*/*d* and volume fraction on damping properties was determined
6.	CNT/polymer Odegard et al. [88]	Equivalent continuum modeling, effective fiber model	Effective elastic modulus	A new effective fiber model is used to model the properties at nanoscale
7.	Nanofiber/polymer composites Valavala and Odegard [89]	Computational chemistry, computational mechanics, and multiscale modeling	Elastic modulus	Comparison of modeling techniques for predicting the properties of polymer composites
8.	Silica particle/polymer composites Odegard et al. [54]	Molecular modeling, Mori–Tanaka, effective interface model	Elastic modulus	The moduli of functionalized and nonfunctionalized systems were determined and compared

Table 2.2 (Continued)

S. no.	Material system	Simulation method	Predicted properties	Conclusions/remarks
9.	CNT/polymer Papanikos and Tserpes [50]	FEM modeling of CNT as a 3D space frame structure	Elastic modulus	Effect of wall thickness, tube diameter, and chirality, of SWCNT, on elastic modulus is examined
10.	CNT/polymer composites Wang et al. [90]	Multiscale modeling	Elastic modulus	Effect of aspect ratio and volume fraction of CNT, the end gap between two coaxial nanotubes and the distance between two parallel nanotubes on the nanocomposites are investigated

2.5 Conclusion

Following conclusions can be drawn from different modeling methods studied for prediction of different properties of nanocomposites:

(i) The multiscale, computational materials modeling approach illustrates one of the primary challenges associated with hierarchical modeling of materials; namely, the accurate prediction of physical/chemical properties and behavior from nanoscale to macroscale without loss of intrinsic structural information. As one moves across a scale, overlaps on both time and length resolution occur, but the overall trend is consistent. It is recognized that at each level of homogenization or scale-up, the risk of losing the key structural information increases. The way to provide an accurate check and balance against these losses is to establish verification of analysis methods and validation of simulations at both the atomic and bulk scales.

(ii) Either molecular modeling or continuum modeling can be used for prediction of properties of nanocomposites. But for predicting exact bulk behavior of nanocomposites equivalent-continuum modeling and multiscale modeling methods are preferred. By using the above methods, the elastic properties of single fiber is predicted from molecular methods and then the properties are used in continuum models for predicting the overall behavior.

(iii) Most of the work on molecular modeling studies in nanocomposite materials relate to prediction of elastic behavior of CNTs. Effect of fiber–matrix interaction at the interface is accurately analyzed by the molecular modeling simulation methods with geometric parameters such as diameter, wall thickness, and length.

(iv) Continuum modeling methods studies in nanocomposite materials relates to prediction of effective elastic modulus and very less work on damping is available in literature.

Various factors that affect the elastic and dynamic properties are fiber orientation and their dispersion within the matrix, frequency, and stress dependence of damping. Different mathematical models are used for calculating the elastic modulus and damping properties of nanocomposites. In addition, FEM technique is also used for calculating the dynamic properties of the nanocomposites.

(v) Because of the complex interactions between constituent phases at the atomic level, a combination of modeling techniques is often required to simulate the bulk-level behavior of these composites. Computational Chemistry techniques assume the presence of a discrete molecular structure and are primarily used to predict the atomic structure of a material. Computational Mechanics techniques assume that the matter is composed of one or more continuous constituents and are used to predict the mechanical behavior of materials and structures. These two types of modeling techniques must be combined to an overall multiscale mode that is capable of predicting the structure and properties of polymer nanocomposites based on fundamental and scientific principles.

(vi) Most of the work has been done for continuous fibers. For aligned short/discontinuous aligned fibers, not much work has been carried out. In most of the cases, only longitudinal modulus and the corresponding loss factor has been evaluated and the other moduli have not been given much importance.

References

1 Chou, T.W., Thostenson, E.T., and Li, C. (2005). Nanocomposites in context. *Composites Science and Technology* 65: 491–516.

2 Alder, B.J. and Wainwright, T.E. (1957). Phase transition for a hard sphere system. *Journal of Chemical Physics* 27: 1208–1211.

3 Rahman, A. (1964). Correlations in the motion of atoms in liquid argon. *Physical Review* 136 (2A): 405–411.

4 Car, R. and Parrinello, M. (1985). Unified approach for molecular dynamics and density functional theory. *Physical Review Letters* 55: 2471.

5 Lopez, C.F. (2002). Computer simulation studies of bio membranes using a coarse grain model. *Computer Physics Communications* 147: 1–6.

6 Kremer, K. and Muller-Plathe, F. (2001). Multiscale problems in polymer science: simulation approaches. *Materials Resource Bulletin* 26: 205–210.

7 Binder, K. (1995). *Monte-Carlo and Molecular Dynamics Simulations in Polymer Science.* New York: Oxford University Press.

8 Raabe, D. (1998). *Computational Material Science.* Weinheim: Wiley-VCH.

9 Baschnagel, J. (2000). Bridging the gap between atomistic and coarse grain models of polymers: status and perspective. *Advances in Polymer Science* 152: 41–156.

10 Muller-Plathe, F. (2002). Coarse graining in polymer simulation: from the atomistic to the meso-scopic and back. *ChemPhysChem* 3 (9): 754–769.

11 Allen, M.P. and Tildesley, D.J. (1987). *Computer Simulation of Liquids.* New York: Oxford University Press.

12 Nemat-Nasser, S. and Hori, M. (1999). *Micromechanics: Overall Properties of Heterogeneous Materials*, 2e. Amsterdam: Elsevier.

13 Fung, Y.C. (1965). *Foundations of Solid Mechanics.* Englewood Cliffs, NJ: Prentice-Hall Inc.

14 Born, M. and Huang, K. (1954). *Dynamical Theory of Crystal Lattices*. London: Oxford University Press.

15 Kittel, C. (1971). *Introduction to Solid State Physics*, 4e. New York: Wiley.

16 Sun, C.T., Kim, B.J., and Bogdanoff, J.L. (1981). On the derivation of equivalent simple models for beam and plate like structures in dynamic analysis. In: *22nd AIAA/ASME/ASCE/AHS Structures, Structural Dynamics and Materials Conference*, vol. 81-0624, 523–532. Atlanta: American Institute of Aeronautics and Astronautics.

17 Noor, A.K. (1988). Continuum modeling for repetitive lattice structures. *Applied Mechanics Reviews* 41 (7): 285–296.

18 Lee, U. (1994). Equivalent continuum models of large plate like lattice structures. *International Journal of Solids and Structures* 31 (4): 457–467.

19 Eringen, A.C. (1966). Linear theory of micropolar elasticity. *Journal of Applied Mathematics and Mechanics* 15 (6): 909–923.

20 Tadmor, E.B., Ortiz, M., and Phillips, R. (1996). Mixed atomistic and continuum models of deformation in solids. *Langmuir* 12 (19): 4529–4534.

21 Shenoy, V.B. (1998). Quasi continuum models of interfacial structure and deformation. *Physical Review Letters* 80 (4): 742–745.

22 Nakano, A. (2001). Multi-scale simulation of nano-systems. *Computing in Science and Engineering* 3 (4): 56–66.

23 Abraham, F.F. (1998). Spanning the continuum to quantum length scales in dynamic simulation of brittle fracture. *Europhysics Letters* 44 (6): 783–787.

24 Broughton, J.Q. (1999). Concurrent coupling of length scales: methodology and application. *Physical Review B: Condensed Matter* 60 (4): 2391–2403.

25 Wagner, G.J. and Liu, W.K. (2003). Coupling of atomistic and continuum simulations using a bridging scale decomposition. *Journal of Computational Physics* 190 (1): 249–274.

26 Xiao, S.P. and Belytschko, T. (2004). A bridging domain method for coupling continua with molecular dynamics. *Computer Methods in Applied Mechanics and Engineering* 193 (17–20): 1645–1669.

27 Chen, X.L. and Liu, Y.J. (2004). Square representative volume elements for evaluating the effective material properties of carbon nanotube based composites. *Computational Materials Science* 29: 1–11.

28 Adams, R.D., Fox, M.A.O., Flood, R.J.L. et al. (1969). The dynamic properties of unidirectional carbon and glass fiber reinforced plastics in torsion and flexure. *Journal of Composite Materials* 3: 594–603.

29 Hashin, Z. (1970). Complex moduli of visco-elastic composites I. General theory and application to particulate composites. *International Journal of Solids and Structures* 6: 539–552.

30 McLean, D. and Read, B.E. (1975). Storage and loss moduli in discontinuous composites. *Journal of Materials Science* 10: 481–492.

31 Gibson, R.F. and Yau, A. (1980). Complex moduli of chopped fiber and continuous fiber composites: comparison of measurements with estimated bounds. *Journal of Composite Materials* 14: 155–167.

32 Sun, C.T., Chaturvedi, S.K., and Gibson, R.F. (1985). Internal damping of short fiber reinforced polymer matrix composites. *Computers and Structures* 20 (1–3): 391–400.

33 Suarez, S.A., Gibson, R.F., Sun, C.T., and Chaturvedi, S.K. (1986). The influence of fiber length and fiber orientation on damping and stiffness of polymer composite materials. *Experimental Mechanics* 26 (2): 175–184.

34 Gibson, R.F., Hwang, S.J., and Kwak, H. (1991). Micromechanical modeling of damping in composites including inter-phase effects. In: *Proceedings of 36th International SAMPE Symposium, Society for the Advancement of Material and Process Engineering, Covina*, vol. 1 (eds. J. Stinson, R. Adsit and F. Gordaninejad), 592–606. International Nuclear Information System.
35 Hwang, S.J. and Gibson, R.F. (1987). Micromechanical modeling of damping in discontinuous fiber composites using a strain energy/finite element approach. *Journal of Engineering Materials and Technology* 109: 47–52.
36 Zabaras, N. and Pervez, T. (1990). Viscous damping approximation of laminated anisotropic composite plates using finite element method. *Computer Methods in Applied Mechanics and Engineering* 81: 291–316.
37 Saravanos, D.A. and Chamis, C.C. (1990). An integrated methodology for optimizing the passive damping of composite structures. *Polymer Composites* 11 (6): 328–336.
38 Saravanos, D.A. and Chamis, C.C. (1990). Unified micromechanics of damping for unidirectional and off axis fiber composites. *Journal of Composites Technology and Research* 12: 31–40.
39 Ruoff, R.S. and Lorents, D.C. (1995). Mechanical and thermal properties of carbon nanotubes. *Carbon* 33: 925–930.
40 Ajayan, P.M., Schadler, L.S., and Giannaris, S.C. (1998). Load transfer in carbon nano-tube epoxy composites. *Applied Physics Letters* 73: 3842–3844.
41 Chandra, R., Singh, S.P., and Gupta, K. (1999). Damping studies in fiber reinforced composites – a review. *Composite Structures* 46: 41–51.
42 Chandra, R., Singh, S.P., and Gupta, K. (2002). Micromechanical damping models for fiber-reinforced composites: a comparative study. *Composites Part A Applied Science and Manufacturing* 33: 787–796.
43 Zengin, H., Zhou, W., Jin, J. et al. (2002). Carbon nanotube doped polyaniline. *Advanced Materials* 14 (20): 1480–1483.
44 Rajoria, H. and Jalili, N. (2004). Determination of strength and damping characteristics of carbon nanotube-epoxy composites. Proceedings of 2004 ASME International Mechanical Engineering Congress & Exposition. Dynamic Systems and Control, Parts A and B. Anaheim, California, USA. November 13–19, 2004. ASME.
45 Koratkar, N.A., Suhr, J., Ye, D., and Lu, T.-M. (2006). Damping properties of epoxy films with nanoscale fillers. *Journal of Intelligent Material Systems and Structures* 17: 255–260.
46 Finegan, I.C., Tibbetts, G.G., and Gibson, R.F. (2003). Modeling and characterization of damping in carbon nanofiber/polypropylene composites. *Composites Science and Technology* 63: 1629–1635.
47 Liu, Y.J. and Chen, X.L. (2003). Evaluations of the effective material properties of carbon nano-tube based composites using nano-scale representative volume element. *Mechanics of Materials* 35: 69–81.
48 Thostenson, E.T. and Chou, T.W. (2003). On the elastic properties of carbon nano-tube based composites: modeling and characterization. *Journal of Physics D: Applied Physics* 36 (5): 573–582.
49 Fisher, F.T., Bradshaw, R.D., and Brinson, L.C. (2003). Fiber waviness in nano-tube reinforced polymer composites-I: modulus predictions using effective nano-tube properties. *Composites Science and Technology* 63: 1689–1703.
50 Papanikos, P. and Tserpes, K.I. (2005). Finite element modeling of single-walled carbon nano-tubes. *Composites Part B Engineering* 36: 468–477.

51 Gao, X.L. and Li, K. (2005). A shear-lag model for carbon nanotube-reinforced polymer composites. *International Journal of Solids and Structures* 42: 1649–1667.

52 Schulte, K., Gojny, F.H., Wichmann, M.H.G., and Fiedler, B. (2005). Influence of different carbon nano-tubes on the mechanical properties of epoxy matrix composites – a comparative study. *Composites Science and Technology* 65: 2300–2323.

53 Srikanth, N., Zhong, X.L., and Gupta, M. (2005). Enhancing damping of pure magnesium using nano-size alumina particulates. *Materials Letters* 59: 3851–3855.

54 Odegard, G.M., Clancy, T.C., and Gates, T.S. (2005). Modeling of the mechanical properties of nano-particle/polymer composites. *Polymer* 46: 553–562.

55 Liu, Y., Nishimura, N., and Otani, Y. (2005). Large scale modeling of carbon-nanotube composites by a fast multipole boundary element method. *Computational Materials Science* 34: 173–187.

56 Haque, A. and Ramasetty, A. (2005). Theoretical study of stress transfer in carbon nano-tube reinforced polymer matrix composites. *Composite Structures* 71 (1): 68–77.

57 Youn, R.J. and Song, S.Y. (2006). Modeling of effective elastic properties for polymer based carbon nano-tube composites. *Polymer* 47: 1741–1748.

58 Shen, G.A., Namilae, S., and Chandra, N. (2006). Load transfer issues in the tensile and compressive behavior of multiwall carbon nano-tubes. *Materials Science and Engineering A* 429: 66–73.

59 Lagoudas, D.C. and Seidel, G.D. (2006). Micromechanical analysis of effective elastic properties of carbon nanotube reinforced composites. *Mechanics of Materials* 38: 884–907.

60 Hubert, P. and Ashrafi, B. (2006). Modeling the elastic properties of carbon nanotube array/polymer composites. *Composites Science and Technology* 66: 387–396.

61 Zalamea, L., Kim, H., and Pipes, R.B. (2007). Stress transfer in multi-walled carbon nanotubes. *Composites Science and Technology* 67: 3425–3433.

62 Guzman de Villoria, R. and Miravete, A. (2007). Mechanical model to evaluate the effect of the dispersion in nanocomposites. *Acta Materialia* 55: 3025–3031.

63 Basu, S., Bhattacharya, B., and Patel, R.K. (2007). Effect of inter-phase properties on the damping response of polymer nano-composites. *Mechanics Research Communications* 35 (1-2): 115–125.

64 Kanagaraj, S., Varanda, F.R., Zhil'tsova, T.V. et al. (2007). Mechanical properties of high density polyethylene/carbon nano-tube composites. *Composites Science and Technology* 67: 1–7.

65 Odegard, G.M., Gates, T.S., Nicholson, L.M., and Wise, K.E. (2001). Equivalent-continuum modeling of nano-structured materials. *Composites Science and Technology* 52: 1–30.

66 Odegard, G.M., Pipes, R.B., and Hubert, P. (2004). Comparisons of two models of SWCN polymer composites. *Composites Science and Technology* 64: 1011–1020.

67 Buryachenko, V.A. and Roy, A. (2005). Effective elastic moduli of nano-composites with prescribed random orientation of nano-fibers. *Composites Part B Engineering* 36: 405–416.

68 Cornwell, C.F. and Wille, L.T. (1997). Elastic properties of single-walled carbon nanotubes in compression. *Solid State Communications* 101 (8): 555–558.

69 Schadler, L.S., Giannaris, S.C., and Ajayan, P.M. (1998). Load transfer in carbon nanotube epoxy composites. *Applied Physics Letters* 73 (26): 3842–3844.

70 Han, J., Globus, A., Jaffe, R., and Deardorff, G. (1997). Molecular dynamics simulations of carbon nanotube-based gears. *Nanotechnology* 8 (3): 95–102.

71 Gao, G.H., Cagin, T., and Goddard, W.A. (1998). Energetics, structure, mechanical and vibrational properties of single-walled carbon nanotubes. *Nanotechnology* 9 (3): 184–191.

72 Halicioglu, T. (1998). Stress calculations for carbon nanotubes. *Thin Solid Films* 312 (1–2): 11–14.

73 Sinnott, S.B., Shenderova, O.A., White, C.T., and Brenner, D.W. (1998). Mechanical properties of nanotubule fibers and composites determined from theoretical calculations and simulations. *Carbon* 36 (1–2): 1–9.

74 Buongiorno Nardelli, M., Fattebert, J.L., Orlikowski, D., and Roland, C. (2000). Mechanical properties, defects and electronic behavior of carbon nanotubes. *Carbon* 38 (11–12): 1703–1711.

75 Kang, J.W. and Hwang, H.J. (2001). Mechanical deformation study of copper nanowire using atomistic simulation. *Nanotechnology* 12 (3): 295–300.

76 Macucci, M., Iannaccone, G., Greer, J., and Martorell, J. (2001). Status and perspectives of nanoscale device modelling. *Nanotechnology* 12 (2): 136–142.

77 Srivastava, D., Menon, M., and Cho, K. (2001). Computational nanotechnology with carbon nanotubes and fullerenes. *Computing in Science and Engineering* 3 (4): 42–55.

78 Wong, E.W., Sheehan, P.E., and Lieber, C.M. (1997). Nanobeam mechanics: elasticity, strength, and toughness of nanorods and nanotubes. *Science* 277 (5334): 1971–1975.

79 Sohlberg, K., Sumpter, B.G., Tuzun, R.E., and Noid, D.W. (1998). Continuum methods of mechanics as a simplified approach to structural engineering of nanostructures. *Nanotechnology* 9 (1): 30–36.

80 Govindjee, S. and Sackman, J.L. (1999). On the use of continuum mechanics to estimate the properties of nanotubes. *Solid State Communications* 110 (4): 227–230.

81 Ru, C.Q. (2000). Column buckling of multi-walled carbon nano-tubes with interlayer radial displacements. *Physical Review B: Condensed Matter* 62 (24): 16962–16967.

82 Qian, D., Wagner, G.J., Liu, W.K., and Yu, M.F. (2002). Mechanics of carbon nano-tubes. *Applied Mechanics Reviews* 55 (6): 495–533.

83 Ruoff, R.S., Liu, W.K., and Qian, D. (2003). Load transfer mechanism in carbon nanotube ropes. *Composites Science and Technology* 63: 1561–1569.

84 Tucker, C.L. and Liang, E. (1999). Stiffness predictions for unidirectional short-fiber composites: review and evaluation. *Composites Science and Technology* 59: 655–671.

85 Fu, S.Y., Yue, C.Y., Hu, X., and Mai, Y.W. (2000). On the elastic stress transfer and longitudinal modulus of unidirectional multi-short fiber composites. *Composites Science and Technology* 60: 3001–3012.

86 Hsueh, C.H. (2000). Young's modulus of unidirectional discontinuous fiber composites. *Composites Science and Technology* 60: 2671–2680.

87 Han, K.S., Nam, H.W., Kim, S.Y., and Jung, S.W. (2001). Measurements of fiber orientation and elastic modulus analysis in short fiber reinforced composites. *Composites Science and Technology* 61: 107–116.

88 Odegard, G.M., Gates, T.S., Wise, K.E. et al. (2003). Constitutive modeling of nanotube reinforced polymer composites. *Composites Science and Technology* 63: 1671–1687.

89 Valavala, P.K. and Odegard, G.M. (2005). Modeling techniques for determination of mechanical properties of polymer nanocomposites. *Advanced Material Science* 9: 34–44.

90 Wang, W.X., Luo, D., and Takao, Y. (2007). Effects of the distribution and geometry of carbon nanotubes on the macroscopic stiffness and microscopic stresses of nanocomposites. *Composites Science and Technology* 67: 2947–2958.

3

Modeling of Nanocomposites

Short-fiber composites have become increasingly popular in recent years because of their easy adaptability to conventional techniques of manufacturing and low-fabrication cost. The increasing number of applications of short-fiber composites makes it more important to understand and predict their mechanical and thermal properties, which depend strongly on fiber orientation and volume fraction. Methods of the analysis of relationships between microstructures and overall properties and strength of composites have been developed from the simplest relationships between averaged values (e.g. Voigt/Reuss estimations) to the complex multiscale numerical models, which take into account the nonlinear behavior of components, evolving microstructures, and real inhomogeneous phase arrangements. In this chapter, several approaches and methods of the analysis of the interrelations between the microstructures and the mechanical behavior and strength of materials in particular have been discussed. This chapter reviews and evaluates models that predict the stiffness of short-fiber composites. The polymer processing community has made substantial progress in modeling process-induced fiber orientation, particularly in injection molding, and these results are now routinely used to predict mechanical properties. Real injection-molded composites invariably have misoriented fibers of highly variable length, but aligned-fiber properties are always calculated as a prelude to modeling the more realistic situation. Hence, the focus here will be on composites having aligned fibers with uniform length and mechanical properties. In selecting models for consideration, the general requirements are that each model must include the effects of fiber and matrix properties and the fiber volume fraction, include the effect of fiber aspect ratio, and predict a complete set of elastic constants for the composite. Any model not meeting these criteria has been excluded from consideration. All of the models use the same basic assumptions:

(a) Fibers and the matrix are linearly elastic.
(b) Matrix is isotropic, and fibers are either isotropic or transversely isotropic.
(c) Fibers are axisymmetric, identical in shape and size, and can be characterized by an aspect ratio (l/d).
(d) Fibers and matrix are well bonded at their interface and remain that way during deformation.
(e) Interfacial slip, fiber/matrix debonding or matrix microcracking is not considered.

Mechanics of Particle- and Fiber-Reinforced Polymer Nanocomposites: From Nanoscale to Continuum Simulations, First Edition. Sumit Sharma.

3.1 Notation

Vectors will be denoted by lower-case Roman letters, second-order tensors by lower-case Greek letters, and fourth-order tensors by capital Roman letters. Whenever possible, vectors and tensors are written as boldface characters; indicial notation is used where necessary. A subscript or superscript f indicates a quantity associated with the fibers, and m denotes a matrix quantity. Thus, the fibers have Young's modulus E_f and Poisson ratio ν_f, while the corresponding matrix properties are E_m and ν_m. The symbol I represents the fourth-order unit tensor. **C** and **S** denote the stiffness and compliance tensors, respectively, and σ and ε are the total stress and infinitesimal strain tensors. Hence, the constitutive equations for the fiber and matrix materials are

$$\sigma_f = C_f \, \varepsilon_f \tag{3.1}$$

$$\sigma_m = C_m \, \varepsilon_m \tag{3.2}$$

Let x denote the position vector. When a composite material is loaded, the pointwise stress field $\sigma(x)$ and the corresponding strain field $\varepsilon(x)$ will be nonuniform on the microscale. The solution of these nonuniform fields is a formidable problem. However, many useful results can be obtained in terms of the average stress and strain. Consider a representative averaging volume V. Choose V large enough to contain many fibers, but small compared to any length scale over which the average loading or deformation of the composite varies. The volume-average stress $\overline{\sigma}$ is defined as the average of the pointwise stress $\sigma(x)$ over the volume V.

$$\overline{\sigma} \equiv \frac{1}{V}\int_V \sigma(x)\,dV \tag{3.3}$$

The average strain $\overline{\varepsilon}$ is defined similarly. It is also convenient to define volume-average stresses and strains for the fiber and matrix phases. To obtain these, first partition the averaging volume V into the volume occupied by the fibers V_f and the volume occupied by the matrix V_m. We consider only two-phase composites, so that

$$V = V_f + V_d \tag{3.4}$$

The fiber and matrix volume fractions are simply, $v_f = V_f/V$ and $v_m = V_m/V$ and, since only fibers and matrix are present, $v_f + v_m = 1$. The average fiber and the matrix stresses are averages over the corresponding volumes:

$$\overline{\sigma}_f \equiv \frac{1}{V_f}\int_{V_f} \sigma(x)\,dV \text{ and } \overline{\sigma}_m \equiv \frac{1}{V_m}\int_{V_m} \sigma(x)\,dV \tag{3.5}$$

Similarly, average strains for the fiber and matrix are defined. The relationships between the fiber and matrix averages and the overall averages can be derived from the preceding definitions; they are:

$$\overline{\sigma} = v_f\,\overline{\sigma}_f + v_m\overline{\sigma}_m \tag{3.6}$$

$$\overline{\varepsilon} = v_f\,\overline{\varepsilon}_f + v_m\,\overline{\varepsilon}_m \tag{3.7}$$

An important related result is the average strain theorem. Let the averaging volume V be subjected to surface displacements $u^0(x)$ consistent with a uniform strain ε^0. Then the average strain within the region is

$$\overline{\varepsilon} = \varepsilon^0 \tag{3.8}$$

This theorem is proved by substituting the definition of the strain tensor ε in terms of the displacement vector u into the definition of average strain $\overline{\varepsilon}$, and applying Gauss's theorem. The result is

$$\overline{\varepsilon}_{ij} = \frac{1}{V}\int_S \left(u_i^0 n_j + n_i u_j^0\right) \mathrm{d}S \tag{3.9}$$

where S denotes the surface of V and n is a unit vector normal to dS.

The average strain within a volume V is completely determined by the displacements on the surface of the volume, so displacements consistent with a uniform strain must produce the identical value of average strain. A corollary of this principle is that if we define a perturbation strain $\varepsilon^{\mathrm{C}}(x)$ as the difference between the local strain and the average

$$\varepsilon^{\mathrm{C}}(x) \equiv \varepsilon(x) - \overline{\varepsilon} \tag{3.10}$$

then the volume-average of $\varepsilon^{\mathrm{C}}(x)$ must equal zero:

$$\overline{\varepsilon}^{\mathrm{C}} = \frac{1}{V}\int_V \varepsilon^{\mathrm{C}}(x)\,\mathrm{d}V = 0 \tag{3.11}$$

The corresponding theorem for average stress also holds. Thus, if surface tractions consistent with a σ^0 are exerted on S, then the average stress is

$$\overline{\sigma} = \sigma^0 \tag{3.12}$$

3.2 Average Properties

Goal of micromechanics models is to predict the average elastic properties of the composite, but even these need careful definition. Direct approach given by Hashin [1] is followed here. Subject the representative volume V to surface displacements consistent with a uniform strain ε^0; the average stiffness of the composite is the tensor C that maps this uniform strain to the average stress. Using Eq. (3.6) we have

$$\overline{\sigma} = C\,\overline{\varepsilon} \tag{3.13}$$

Average compliance S is defined in the same way, applying tractions consistent with a uniform stress σ^0 on the surface of the averaging volume. Then, using Eq. (3.10),

$$\overline{\varepsilon} = S\,\overline{\sigma} \tag{3.14}$$

It should be clear that $S = C^{-1}$. Many authors have defined the average stiffness and compliance through the integral of the strain energy over V; this is equivalent to the direct approach. An important related concept, first introduced by Hill [2], is the idea of strain- and stress-concentration tensors A and B. These are essentially the ratios between the average fiber strain (or stress) and the corresponding average in the composite. More precisely,

$$\overline{\varepsilon}_{\mathrm{f}} = A\,\overline{\varepsilon} \tag{3.15}$$

$$\overline{\sigma}_{\mathrm{f}} = B\,\overline{\sigma} \tag{3.16}$$

A and **B** are fourth-order tensors and, in general, they must be found from a solution of the microscopic stress or strain fields. Different micromechanics models provide different ways to approximate **A** or **B**. Note that **A** and **B** have both the minor symmetries of a stiffness or compliance tensor, but lack the major symmetry. That is,

$$A_{ijkl} = A_{jikl} = A_{ijlk} \tag{3.17}$$

But in general;

$$A_{ijkl} \neq A_{klij} \tag{3.18}$$

For later use, it will be convenient to have an alternate strain concentration tensor $\hat{A}$ that relates the average fiber strain to the average matrix strain:

$$\bar{\varepsilon}_{\mathrm{f}} = \hat{A}\,\bar{\varepsilon}_{\mathrm{m}} \tag{3.19}$$

This is related to **A** by

$$\mathbf{A} = \hat{A}\left[\left(1 - V_{\mathrm{f}}\right)\mathbf{I} + V_{\mathrm{f}}\hat{A}\right]^{-1} \tag{3.20}$$

So the two forms are easily interchanged. Using equations now in hand, one can express the average composite stiffness in terms of the strain-concentration tensor **A** and the fiber and matrix properties. Combining Eqs. (3.1), (3.2), (3.6), (3.7), (3.13), and (3.15), the resulting equation is

$$\mathbf{C} = \mathbf{C_m} + V_{\mathrm{f}}\left(\mathbf{C_f} - \mathbf{C_m}\right)\mathbf{A} \tag{3.21}$$

The dual equation for the compliance is

$$\mathbf{S} = \mathbf{S_m} + V_{\mathrm{f}}\left(\mathbf{S_f} - \mathbf{S_m}\right)\mathbf{B} \tag{3.22}$$

Equations (3.21) and (3.22) are not independent, since $S = C^{-1}$. Hence, the strain-concentration tensor **A** and the stress-concentration tensor **B** are not independent either. The choice of which one to use in any instance is a matter of convenience. To illustrate the use of the strain-concentration and stress-concentration tensors, it can be noted that the Voigt average corresponds to the assumption that the fiber and the matrix both experience the same, uniform strain. Then, $\bar{\varepsilon}_{\mathrm{f}} = \bar{\varepsilon}$, $\mathbf{A} = \mathbf{I}$, and from Eq. (3.21), the composite modulus is

$$\mathbf{C}^{\mathbf{Voigt}} = \mathbf{C_m} + V_{\mathrm{f}}\left(\mathbf{C_f} - \mathbf{C_m}\right) = V_{\mathrm{f}}\mathbf{C_f} + V_{\mathrm{m}}\mathbf{C_m} \tag{3.23}$$

Voigt average is an upper bound on the composite modulus. Reuss average assumes that the fiber and matrix both experience the same, uniform stress. This means that the stress-concentration tensor B equals the unit tensor I, and from Eq. (3.22) the compliance is

$$\mathbf{S}^{\mathbf{Reuss}} = \mathbf{S_m} + V_{\mathrm{f}}\left(\mathbf{S_f} - \mathbf{S_m}\right) = V_{\mathrm{f}}\mathbf{S_f} + V_{\mathrm{m}}\mathbf{S_m} \tag{3.24}$$

This represents a lower bound on the stiffness of the composite. In Section 3.3, various models that can be used for predicting the modulus and damping properties of the composite have been reviewed.

3.3 Theoretical Models

Some major micromechanical models which have been employed for this comparative study are discussed in below sections.

3.3.1 Cox Shear Lag Model

Shear lag model is widely used to analyze the fiber–matrix stress transfer in undamaged and damaged composites. This model was developed initially by Cox [3] and then expanded and modified by many authors. The so-called "shear lag" method is often used for analysis of stress transfer problems in composites. The term "shear lag" can be traced, prior to its use in composites to analysis of bending of I beams and T beams with wide flanges and to box beams. Simple beam theory predicts that the axial displacements in the flanges of such beams are only a function of the distance from the neutral axis and independent of the distance from the web. This simple theory also predicts zero shear stress and zero shear strain in the flange. In reality, the true axial displacements "lag" behind the beam theory predictions. This "lag" is caused by load diffusion which can be viewed (using equilibrium arguments) as a consequence of nonzero shear stresses in the flange; hence, the term "shear lag." In these beam analyses, "shear lag" is an effect and not an analysis method. Many possible analysis methods can evaluate the "shear lag" effect. These methods generally result in defining an effective flange width that is less than the actual flange width. If effective width is not considered when designing beams, the resulting beams can be seriously under designed. Assuming that the load transfer from matrix to fiber occurs via shear stresses on the interface between them, Cox considered the force balance in a section of the fiber, and derived the formula which relates the rate of change of the stress along the fiber length, and the interfacial shear stress (Figure 3.1):

$$\frac{d\sigma_x}{dx} = -\frac{\tau}{r} \tag{3.25}$$

where r is the radius of the fiber, τ is the interface shear stress, and x is the coordinate along the fiber length. This formula is referred to as the basic shear lag equation. After some manipulations, this model leads to the following second-order differential equation:

$$\frac{d^2\sigma}{dx^2} - \beta^2\sigma = -\beta^2\sigma\infty \tag{3.26}$$

where β is the so-called shear lag parameter, σ_∞ is the far-field fiber stress, and σ is fiber stress.

Solving this equation, one determines the stress distribution along the axis x of the fiber:

$$\sigma = E\varepsilon\ [1 - \cosh(\beta x)\ \text{sech}(\beta xr)] \tag{3.27}$$

where E is the Young modulus of the fiber and ε is the strain in the composite. Cox derived a formula for the shear lag parameter, β for the case of a cylindrical fiber of radius r, embedded

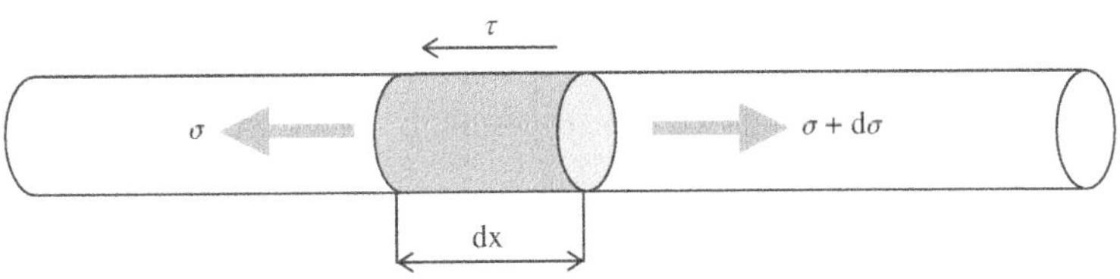

Figure 3.1 Cox shear lag model.

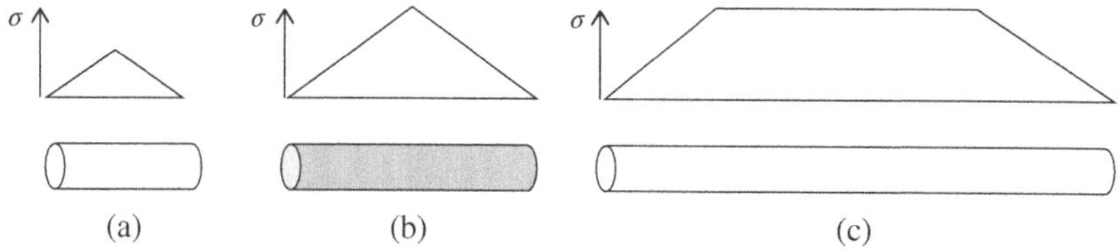

Figure 3.2 Stress distribution in the fiber (subject to the tensile loading) along the length: (a) short fiber ($l < l_c$); (b) critical fiber length $l = l_c$; and (c) longer fiber $l > l_c$.

into a cylindrical layer of matrix:

$$\beta = -\frac{1}{r}\sqrt{\frac{2G_m}{E\ \ln\left(\frac{s}{r}\right)}} \tag{3.28}$$

where G_m is the matrix shear modulus, s is the average distance between the fiber axes, r is the fiber radius, and E is the axial modulus of the fiber. Using Eqs. (3.1)–(3.3) one can determine the critical fiber length l_{cr}, at which both the matrix and fiber fail at the same strain. Assuming constant shear stress τ in Eq. (3.1), Kelly and Tyson [4] derived the following formula for the critical fiber length:

$$l_c = \frac{\sigma_c r}{\tau_c} \tag{3.29}$$

where σ_c is the fiber failure stress and τ_c is the matrix fiber interfacial shear strength. Figure 3.2 shows the effect of the fiber length on the stress distribution in the fiber under tensile loading. If $l < l_c$ (the case of discontinuous reinforcement), the stress in the fiber is below the critical level, and the fiber is not utilized fully. If $l > l_c$ (longer fibers), a large part of the fiber is overloaded, and multiple cracking can be observed in the fiber.

If the fiber length is equal to the critical size $l = l_c$, both the matrix and fiber fail at the same load, ensuring the most efficient reinforcement of the composite. Kelly and Tyson used this model to explain the experimental observation that the fiber breaking strength is a linear function of the wire content in different composites. Nairn [5] carried out a rigorous theoretical analysis of a model, for which the shear assumptions are exact. Using axisymmetric elasticity equations, he demonstrated that the rigorous analysis leads to Eq. (3.2) as well. Nairn demonstrated further that the shear lag method gives reasonably good estimations of average axial stress in the fiber and total strain energy in the specimen, yet the method is not applicable for low-fiber volume fractions. For the case of a broken fiber embedded into the ductile matrix, the shear-lag model was generalized by Landis and McMeeking [6]. They derived the shear lag equation for this case and verified the model by comparing it with finite element (FE) analysis. The longitudinal modulus E_{11}, which depends on the fiber aspect ratio s, is obtained from Cox's shear-lag model as
where

$$E_{cox} = (1 - V_f)\, E_m + \alpha \left(1 - \frac{\tanh \beta}{\beta}\right) V_f E_f \tag{3.30}$$

where

$$\beta = \frac{1}{d}\sqrt{\frac{E_m}{(1 + \nu) E_f \cdot \ln\left(\frac{\pi}{4V_f}\right)}} \tag{3.31}$$

$\alpha = 1$, for aligned fibers.

For the in-plane shear modulus G_{12}, we used the Wilczynski equation [7]

$$G_{12} = G_{\mathrm{m}} \frac{G_{\mathrm{f}}\left(1+V_{\mathrm{f}}\right)+G_{\mathrm{m}}\left(1-V_{\mathrm{f}}\right)}{G_{\mathrm{f}}\left(1-V_{\mathrm{f}}\right)+G_{\mathrm{m}}\left(1+V_{\mathrm{f}}\right)} \tag{3.32}$$

where G_{f} and G_{m} are the shear modulus of the fiber and matrix, respectively. Historically, shear lag models were the first micromechanics models for short-fiber composites, as well as the first to examine behavior near the ends of broken fibers in a continuous-fiber composite. Despite some serious theoretical flaws, shear lag models have enjoyed enduring popularity, perhaps due to their algebraic simplicity and their physical appeal. Classical shear lag models only predict the longitudinal modulus E_{11}, so they do not meet the criterion of predicting a complete set of elastic constants. However, they have been included here because of their historical importance and their widespread use. One could obtain a complete stiffness model by using the shear lag prediction for E_{11} and some continuous-fiber model for the remaining elastic constants. If the fiber is anisotropic, then its axial modulus should be used in the shear lag equations. Following Cox, the shear lag analysis focuses on a single fiber of length l and radius r_{f}, which is encased in a concentric cylindrical shell of matrix having radius R. The fiber is aligned parallel to the z axis, i.e. 1-direction, as shown in Figure 3.3. Only the axial stress σ_{11} and axial strain ε_{11} are of interest, and Poisson effects are neglected, so that $\sigma_{11}^{\mathrm{f}} = E_{\mathrm{f}}\varepsilon_{11}^{\mathrm{f}}$. The outer cylindrical surface of the matrix is subjected to displacement boundary conditions consistent with an average axial strain and one solves for the fiber stress $\sigma_{11}^{\mathrm{f}}(z)$, i.e. the average stress over the fiber cross-section at z. Axial equilibrium of the fiber requires that

$$\mathrm{d}\sigma_{11}^{\mathrm{f}}/\mathrm{d}z = -2\left(\tau_{12}/r_{\mathrm{f}}\right) \tag{3.33}$$

where τ_{12} is the axial shear stress at the fiber surface. The key assumption of shear lag theory is that τ_{12} is proportional to the difference in displacement (w) between the fiber surface and the outer matrix surface:

$$\tau_{12} = \left(H/2\pi r_{\mathrm{f}}\right)\left[w(R,z) - w\left(r_{\mathrm{f}},z\right)\right] \tag{3.34}$$

where H is a constant that depends on matrix properties and fiber volume fraction. Solving Eq. (3.33) for $\sigma_{11}^{\mathrm{f}}(z)$ and applying boundary conditions of zero stress at the fiber ends gives an average fiber stress of

$$\overline{\sigma}_{11}^{\mathrm{f}} = E_{\mathrm{m}}\,\overline{\varepsilon}_{11}\left[1 - \frac{\tanh(\beta l/2)}{(\beta l/2)}\right] \tag{3.35}$$

where

$$\beta^2 = \frac{H}{\pi r_{\mathrm{f}}^2 E_{\mathrm{f}}} \tag{3.36}$$

The above expression may be rewritten for the average fiber strain as

$$\overline{\varepsilon}_{11}^{\mathrm{f}} = \eta_{\mathrm{l}}\,\overline{\varepsilon}_{11} \tag{3.37}$$

where η_{l} is a length-dependent "efficiency factor," given as

$$\eta_{\mathrm{l}} = \left[1 - \frac{\tanh\left(\frac{\beta l}{2}\right)}{\left(\frac{\beta l}{2}\right)}\right] \tag{3.38}$$

Note that η_{l} is a scalar analog of the strain-concentration tensor **A** defined in Eq. (3.15), and $1/\beta$ is a characteristic length for stress transfer between the fiber and the matrix. Cox found

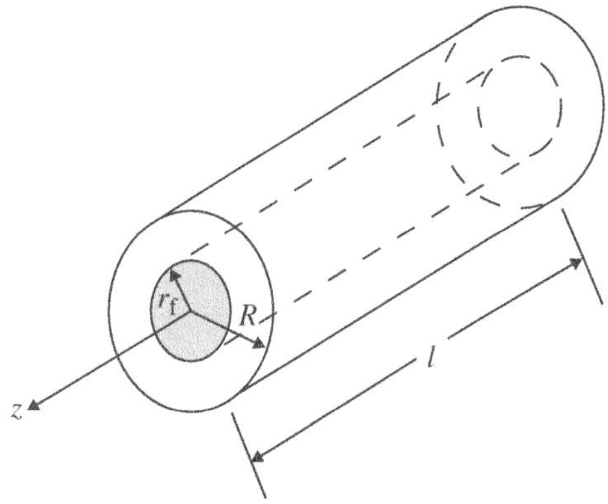

Figure 3.3 Idealized fiber and matrix geometry used in shear lag models.

the coefficient H by solving a second idealized problem. The concentric cylinder geometry is maintained, but the outer cylindrical surface of the matrix is held stationary and the inner cylinder, which is now rigid, is subjected to a uniform axial displacement. An elasticity solution for the matrix layer then gives

$$H = \frac{2\pi G_m}{\ln (R/r_f)} \quad (3.39)$$

Rosen [8] simplified this part of the problem by assuming that the matrix shell was thin compared to the fiber radius, $(R - r_f) \ll r_f$, obtaining

$$H = \frac{2\pi G_m}{\left(\frac{R}{r_f} - 1\right)} \quad (3.40)$$

Rosen's approximation gives much larger errors at lower volume fractions, so will not be considered further. It remains to choose the radius R of the matrix cylinder, and the exact choice is important. Several choices have been used, all of which can be written in the form:

$$\frac{R}{r_f} = \sqrt{\frac{K_R}{V_f}} \quad (3.41)$$

where K_R is a constant that depends on the assumption used to find R. Table 3.1 summarizes the choices for K_R. Cox assumed a hexagonal packing and chose R as the distance between centers of nearest-neighbor fibers as shown in Figure 3.4 (a). It seems more realistic to let R equal half of the distance between nearest neighbors (Figure 3.4b), a choice labeled "hexagonal" in Table 3.1. Rosen and later Carman and Reifsnider [9] chose $r_f^2/R^2 = V_f$, so that the concentric cylinder model in Figure 3.3 would have the same fiber volume fraction as the composite. This is the same R as the composite cylinders model of Hashin and Rosen [10]. Robinson and Robinson [11] assumed a square array of fibers and chose R as half the distance between centers of nearest neighbors as shown in Figure 3.4c. Each of these choices gives a somewhat different dependence of η_l on fiber volume fraction, with larger values of K_R producing lower values of E_{11}. Shear lag models are usually completed by combining the average fiber stress in Eq. (3.35) with an average matrix stress to produce a modified rule of mixtures for the axial modulus:

$$E_{11} = E_f V_f \eta_l + (1 - V_f) E_m = E_m + V_f (E_f - E_m) \eta_l \quad (3.42)$$

Table 3.1 Values of K_R used in shear lag models.

Fiber packing	K_R
Cox	$\frac{2\pi}{\sqrt{3}} = 3.628$
Composite cylinder	1.000
Hexagonal	$\frac{\pi}{2\sqrt{3}} = 0.907$
Square	$\frac{\pi}{4} = 0.785$

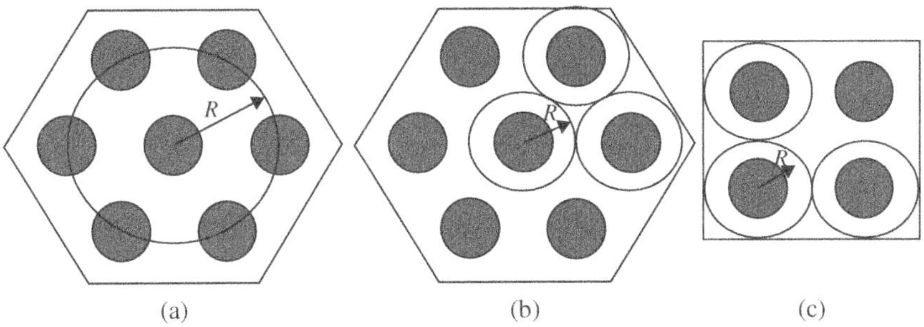

Figure 3.4 Fiber packing arrangements (a) cox, (b) hexagonal and (c) square, used to find R in shear lag models.

This equation is an exact scalar analog of the general tensorial stiffness formula, Eq. (3.21). A model by Fukuda and Kawata [12] for the axial stiffness of aligned short-fiber composites is closely related to shear lag theory. They begin with a 2-D elasticity solution for the shear stress around a single slender fiber in an infinite matrix. The usual shear lag relation, Eq. (3.33), is used to transform this into an equation for the fiber stress distribution, which is then approximated by a Fourier series. The coefficients of a truncated series are evaluated analytically using Galerkin's method. This is a dilute theory in which modulus varies linearly with fiber volume fraction. Like any shear lag theory, Fukuda and Kawata's theory predicts that E_{11} approaches the rule of mixtures result as the fiber aspect ratio approaches infinity. But for short fibers, Fukuda and Kawata's theory gives much lower E_{11} values than shear lag theory.

In Fukuda and Kawata's theory, the ratio of fiber strain to matrix strain is governed by the parameter $l/d(E_m/E_f)$. In contrast, for shear lag theory, Eq. (3.38), the governing parameter is $\beta l/2$, which is proportional to $l/d\sqrt{(E_m/E_f)}$. Thus, for high-modulus ratio and low-aspect ratio, Fukuda and Kawata's theory tends to under-predict E_{11}. For this reason, this theory is not pursued further.

3.3.2 Eshelby's Equivalent Inclusion

A fundamental result used in several different models is Eshelby's equivalent inclusion given by Eshelby [13]. Eshelby solved for the elastic stress field in and around an ellipsoidal particle in an infinite matrix. By letting the particle be a prolate ellipsoid of revolution, one can

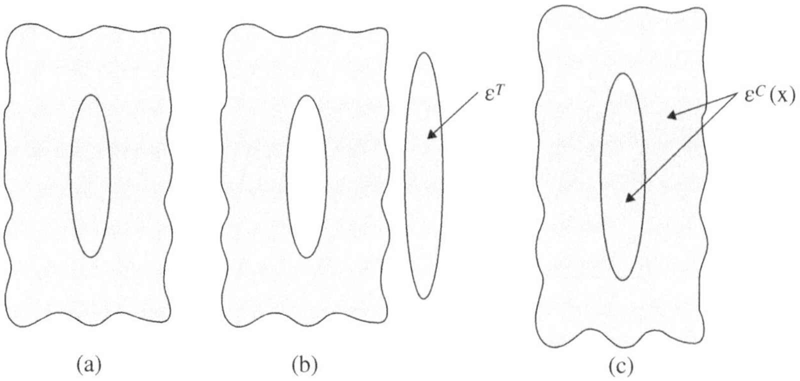

Figure 3.5 Eshelby's equivalent inclusion problem. Starting from the stress-free state (a), the inclusion undergoes a stress-free transformation strain ε^{T} (b). Fitting the inclusion and matrix back together (c) produces the strain state ε^{C} (x) in both the inclusion and the matrix.

use Eshelby's result to model the stress and strain fields around a cylindrical fiber. Eshelby first posed and solved a different problem, that of a homogeneous inclusion as shown in Figure 3.5. Consider an infinite solid body with stiffness $\mathbf{C_m}$ that is initially stress-free. All subsequent strains will be measured from this state. A particular small region of the body will be called the inclusion, and the rest of the body will be called the matrix. Suppose that the inclusion undergoes some type of transformation such that if it were a separate body, it would acquire a uniform strain ε^{T} with no surface traction or stress. ε^{T} is called the transformation strain, or the eigen strain. This strain might be acquired through a phase transformation or by a combination of a temperature change and a different thermal expansion coefficient in the inclusion. In fact the inclusion is bonded to the matrix, so when the transformation occurs, the whole body develops some complicated strain field $\boldsymbol{\varepsilon}^{\mathrm{C}}(x)$ relative to its shape before the transformation. Within the matrix, the stress $\boldsymbol{\sigma}_{\mathbf{m}}$ is simply the stiffness times this strain:

$$\boldsymbol{\sigma}_{\mathbf{m}}(x) = \mathbf{C}_{\mathbf{m}}\,\boldsymbol{\varepsilon}^{\mathrm{C}}(x) \tag{3.43}$$

But within the inclusion the transformation strain does not contribute to the stress, so the inclusion stress is

$$\sigma^{\mathrm{I}}(x) = \mathrm{C_m}\left(\varepsilon^{\mathrm{C}} - \varepsilon^{\mathrm{T}}\right) \tag{3.44}$$

The key result of Eshelby was to show that within an ellipsoidal inclusion, the strain ε^{C} is uniform and is related to the transformation strain by

$$\varepsilon^{\mathrm{C}} = \mathbf{E}\,\varepsilon^{\mathrm{T}} \tag{3.45}$$

where E is called Eshelby's tensor and it depends only on inclusion aspect ratio and the matrix elastic constants. A detailed derivation and application are given by Mura [14], and analytical expressions for Eshelby's tensor for an ellipsoid of revolution in an isotropic matrix appear in many papers. The strain field $\varepsilon^{\mathrm{C}}(x)$ in the matrix is highly nonuniform, but this more complicated part of the solution can often be ignored. Second step in Eshelby's approach is to demonstrate equivalence between the homogeneous inclusion problem and an inhomogeneous inclusion of the same shape. Consider two infinite bodies of matrix, as shown in Figure 3.6. One has a homogeneous inclusion with some transformation strain ε^{T}, and the

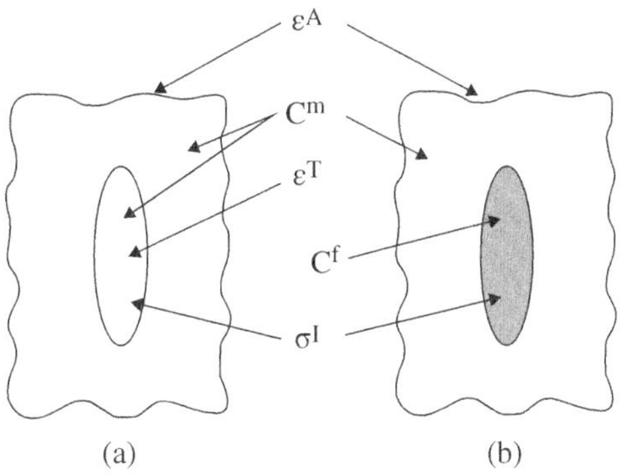

Figure 3.6 Eshelby's equivalent inclusion problem. The inclusion (a), with transformation strain ε^{T} has the same stress σ^{I} and strain as the in-homogeneity. (b). When both the bodies are subjected to a far field strain ε^{A}.

other has an inclusion with a different stiffness C_{f}, but no transformation strain. Both the bodies are subject to a uniform applied strain of ε^{A} at infinity. We wish to find the transformation strain ε^{T} that gives the two problems the same stress and strain distributions.

For the first problem the inclusion stress is Eq. (3.44) with the applied strain added:

$$\sigma^{\mathrm{I}} = C_{\mathrm{m}}\left(\varepsilon^{\mathrm{A}} + \varepsilon^{\mathrm{C}} - \varepsilon^{\mathrm{T}}\right) \tag{3.46}$$

While the second problem has no ε^{T}, but a different stiffness, giving a stress of

$$\sigma^{\mathrm{I}} = C_{\mathrm{f}}\left(\varepsilon^{\mathrm{A}} + \varepsilon^{\mathrm{C}}\right) \tag{3.47}$$

Equating these two expressions gives the transformation strain that makes two problems equivalent. Using Eq. (3.45) and some rearrangement, the result is

$$-\left[\mathbf{C_m} + \left(\mathbf{C_f} - \mathbf{C_m}\right)\mathbf{E}\right]\varepsilon^{\mathrm{T}} = \left(\mathbf{C_f} - \mathbf{C_m}\right)\varepsilon^{\mathrm{A}} \tag{3.48}$$

Note that ε^{T} is proportional to ε^{A}, which makes the stress in the equivalent in-homogeneity proportional to the applied strain.

3.3.3 Dilute Eshelby's Model

If the volume fraction of inclusions is very small, the interaction between reinforcing elements can be neglected. In this case, the dilute distribution model can be used to analyze the effective elastic properties of composites. The dilute distribution model, as well as many other micromechanical approaches, is based on the theory developed by Eshelby. Eshelby considered the stress and strain fields in a medium with an elliptical region which undergoes a transformation and changes its shape or size. He has shown that uniform stress and strain states are induced in the transformed elastic homogeneous inclusion ("elliptical region"), embedded into an infinite matrix subject to uniform strain. Eshelby introduced a so-called Eshelby tensor **E**, which relates the strains in an inclusion in the infinite elastic matrix $\varepsilon_{\mathrm{constr}}$ with the strain of the same inclusion, placed outside the matrix and free of the stresses imposed by the matrix $\varepsilon_{\mathrm{unconstr}}$. One can use Eshelby's result to find the stiffness of a composite with ellipsoidal fibers at dilute concentrations. From Eq. (3.21), we know that to find the stiffness one only has to find the strain-concentration tensor A. To do this, first note that for a dilute composite the average strain is identical to the applied strain, since this is the strain at infinity:

$$\bar{\varepsilon} = \varepsilon^{\mathrm{A}} \tag{3.49}$$

Also, from Eshelby, the fiber strain is uniform, and is given by

$$\bar{\varepsilon}_{\mathrm{f}} = \varepsilon^{\mathrm{A}} + \varepsilon^{\mathrm{C}} \tag{3.50}$$

where the right-hand side is evaluated within the fiber. Now writing the equivalence between the stresses in the homogeneous and the inhomogeneous inclusions, Eqs. (3.46) and (3.47):

$$C_{\mathrm{f}}\left(\varepsilon^{\mathrm{A}} + \varepsilon^{\mathrm{C}}\right) = C_{\mathrm{m}}\left(\varepsilon^{\mathrm{A}} + \varepsilon^{\mathrm{C}} - \varepsilon^{\mathrm{T}}\right) \tag{3.51}$$

and then using Eqs. (3.45), (3.49) and (3.50) to eliminate ε^{T}, ε^{A}, and ε^{C} from this equation, giving

$$\left[\mathbf{I} + \mathbf{E}\,\mathbf{S_m}\left(\mathbf{C_f} - \mathbf{C_m}\right)\right]\bar{\varepsilon}_{\mathrm{f}} = \bar{\varepsilon} \tag{3.52}$$

Comparing this equation with Eq. (3.15) shows that the strain concentration tensor for Eshelby's equivalent inclusion is

$$\mathbf{A}^{\mathrm{Eshelby}} = \left[\mathbf{I} + \mathbf{E}\,\mathbf{S_m}\left(\mathbf{C_f} - \mathbf{C_m}\right)\right]^{-1} \tag{3.53}$$

This can be used in Eq. (3.21) to predict the moduli of aligned-fiber composites, a result first developed by Russel [15]. Calculations using this model to explore the effects of particle aspect ratio on stiffness are presented by Chow [16]. While Eshelby's solution treats only ellipsoidal fibers, the fibers in most short-fiber composites are much better approximated as right circular cylinders. The relationship between ellipsoidal and cylindrical particles was considered by Steif and Hoysan [17], who developed a very accurate finite element technique for determining the stiffening effect of a single fiber of given shape. For very short particles, $l/d = 4$, they found reasonable agreement for E_{11} by letting the cylinder and the ellipsoid have the same l/d. The ellipsoidal particle gave a slightly stiffer composite, with the difference between the two results increasing as the modulus ratio, $E_{\mathrm{f}}/E_{\mathrm{m}}$ increased. Henceforth, the cylinder aspect ratio will be used in place of the ellipsoid aspect ratio in Eshelby-type models. Because Eshelby's solution only applies to a single particle surrounded by an infinite matrix, $\mathbf{A}^{\mathrm{Eshelby}}$ is independent of fiber volume fraction and the stiffness predicted by this model increases linearly with fiber volume fraction. Modulus predictions based on Eqs. (3.53) and (3.21) should be accurate only at low-volume fractions, say up to v_{f} of 1%. The more difficult problem is to find some way to include interactions between fibers in the model and to produce accurate results at higher-volume fractions. Section 3.3.4 deals with the approaches for doing that.

3.3.4 Mori–Tanaka Model

(a) In the approach suggested by Mori and Tanaka [18], each inclusion behaves as an isolated inclusion, subject to the averaged stress fields acting on it from all the other inclusions as shown in Figure 3.7. The stresses, acting on an inclusion and caused by the presence of other inclusions, are superimposed on the applied stress. The idea of Mori and Tanaka was to combine the Eshelby approach and the effective field concept. This is done by defining the strain concentration tensor, which relates the strain in the inclusion to the unknown strain in the matrix instead of the applied strain, as in the case of the dilute distribution model. Benveniste [19] expanded the relations suggested by Mori–Tanaka and provided a general method for determining the effective properties, based on this theory.

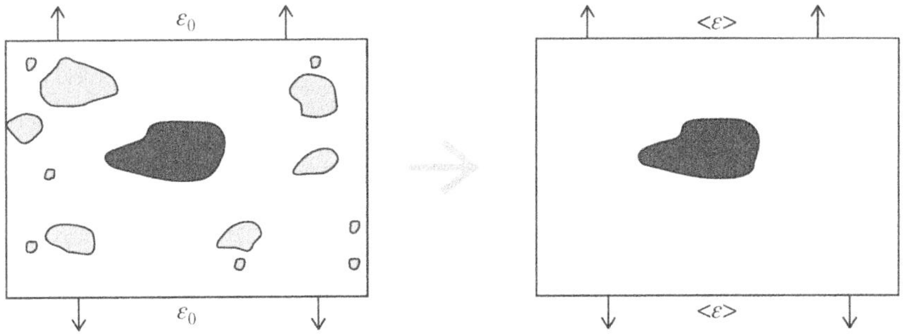

Figure 3.7 Mori–Tanaka model. The interaction between the inclusions is taken into account by averaging the strain fields acting on a given inclusion from all the other inclusions.

Mori and Tanaka approach belongs to the group of effective field methods. This model leads to the same formulae as the lower Hashin–Shtrikman bound (i.e. when the matrix is the softer phase) for spheres and many other inclusion shapes. A family of models for nondilute composite materials has evolved from a proposal originally made by Mori and Tanaka [18]. Benveniste [19] has provided a particularly simple and clear explanation of the Mori–Tanaka approach, which is used here to introduce the approach. The strain-concentration tensor has already been introduced in Eq. (3.15). Suppose that a composite is to be made of a certain type of reinforcing particle, and that, for a single particle in an infinite matrix, we know the dilute strain-concentration tensor $\mathbf{A}^{\text{Eshelby}}$:

$$\overline{\varepsilon}_{\text{f}} = \mathbf{A}^{\text{Eshelby}}\,\overline{\varepsilon} \tag{3.54}$$

Mori–Tanaka assumption is that when many identical particles are introduced in the composite, the average fiber strain is given by

$$\overline{\varepsilon}_{\text{f}} = \mathbf{A}^{\text{Eshelby}}\,\overline{\varepsilon}_{\text{m}} \tag{3.55}$$

that is, within a concentrated composite each particle "sees" a far-field strain equal to the average strain in the matrix. Using the alternate strain concentrator defined in the Eq. (3.19), the Mori–Tanaka assumption can be restated as follows:

$$\hat{\text{A}}^{\text{MT}} = \mathbf{A}^{\text{Eshelby}} \tag{3.56}$$

Equation (3.20) then gives the Mori–Tanaka strain concentrator as

$$\mathbf{A}^{\text{MT}} = \mathbf{A}^{\text{Eshelby}}\left[\left(1 - v_{\text{f}}\right)\boldsymbol{I} + v_{\text{f}}\,\mathbf{A}^{\text{Eshelby}}\right]^{-1} \tag{3.57}$$

This is the basic equation for implementing a Mori–Tanaka model. Mori–Tanaka approach for modeling composites was first introduced by Wakashima et al. [20] for modeling thermal expansions of composites with aligned ellipsoidal inclusions. (Mori and Tanaka's paper treats only the homogeneous inclusion problem and says nothing about composites). Mori–Tanaka predictions for the longitudinal modulus of a short-fiber composite were first developed by Taya and Mura [21] and Taya and Chou [22], whose work also included the effects of cracks and of a second type of reinforcement. Weng [23] generalized their method and Tandon and Weng [24] used the Mori–Tanaka approach to develop equations for the complete set of elastic constants of a short-fiber composite. Tandon and Weng's equations for the plane-strain bulk modulus k_{23} and the major Poisson ratio v_{12} must be solved iteratively. The usual development of the Mori–Tanaka model differs somewhat from Benveniste's explanation. For an average applied stress $\overline{\sigma}$, the reference strain ε^0 is defined as the strain in a homogeneous body of matrix at this stress:

$$\overline{\sigma} = \mathbf{C_m}\, \varepsilon^0 \tag{3.58}$$

Within the composite, the average matrix strain differs from the reference strain by some perturbation $\widetilde{\varepsilon^{\mathrm{m}}}$

$$\overline{\varepsilon}^{\mathrm{m}} = \varepsilon^0 + \widetilde{\varepsilon^{\mathrm{m}}} \tag{3.59}$$

A fiber in the composite will have an additional strain perturbation, $\widetilde{\varepsilon^{\mathrm{f}}}$, such that

$$\overline{\varepsilon}^{\mathrm{f}} = \varepsilon^0 + \widetilde{\varepsilon^{\mathrm{m}}} + \widetilde{\varepsilon^{\mathrm{f}}} \tag{3.60}$$

while the equivalent inclusion will have this strain plus the transformation strain ε^{T}. The stress equivalence between the inclusion and the fiber then becomes

$$\mathbf{C_f}\left(\varepsilon^0 + \widetilde{\varepsilon^{\mathrm{m}}} + \widetilde{\varepsilon^{\mathrm{f}}}\right) = \mathbf{C_m}\left(\varepsilon^0 + \widetilde{\varepsilon^{\mathrm{m}}} + \widetilde{\varepsilon^{\mathrm{f}}} - \varepsilon^{\mathrm{T}}\right) \tag{3.61}$$

comparing this to the dilute version, Eq. (3.51), and noting that ε^{A} in the dilute problem is equivalent to $(\varepsilon^0 + \widetilde{\varepsilon^{\mathrm{m}}})$ here. The development is completed by assuming that the extra-fiber perturbation is related to the transformation strain by Eshelby's tensor

$$\widetilde{\varepsilon^{\mathrm{f}}} = \mathbf{E}\, \varepsilon^{\mathrm{T}} \tag{3.62}$$

Combining this with Eqs. (3.59) and (3.60) reveals that Eq. (3.62) contains the essential Mori–Tanaka assumption: The fiber in a concentrated composite sees the average strain of the matrix.

(b) Tandon and Weng [24] derived explicit expressions for the elastic constants of a short-fiber composite using the Mori–Tanaka approach. Their formulae for the plane-strain bulk modulus k_{23} and the major Poisson ratio ν_{12} are coupled, and must be solved iteratively. Tandon and Weng calculations are given as

$$E_{11} = \frac{E_{\mathrm{m}}}{1 + \frac{V_{\mathrm{f}}\left(A_1 + 2V_{\mathrm{m}} A_2\right)}{A}} \tag{3.63}$$

$$E_{22} = \frac{E_{\mathrm{m}}}{1 + V_{\mathrm{f}}\left[\left(1 - V_{\mathrm{m}}\right) A_4 - 2V_{\mathrm{m}} A_3 + \left(1 + V_{\mathrm{m}}\right) A_5 A\right]/2A} \tag{3.64}$$

$$G_{12} = \frac{G_{\mathrm{m}}\left(1 + V_{\mathrm{f}}\right)}{2\left(1 - V_{\mathrm{f}}\right) S_{1212} + \frac{G_{\mathrm{m}}}{G_{\mathrm{f}} - G_{\mathrm{m}}}} \tag{3.65}$$

$$G_{23} \frac{G_{\mathrm{m}}\left(1 + V_{\mathrm{f}}\right)}{2\left(1 - V_{\mathrm{f}}\right) S_{2323} + \frac{G_{\mathrm{m}}}{G_{\mathrm{f}} - G_{\mathrm{m}}}} \tag{3.66}$$

$$\nu_{12} = \frac{v_{\mathrm{m}} A - V_{\mathrm{f}}\left(A_3 - v_{\mathrm{m}} A_4\right)}{A + V_{\mathrm{f}}\left(A_1 + 2v_{\mathrm{m}} A_2\right)} \tag{3.67}$$

$$\nu_{23} = \frac{E_{22}}{2G_{23}} - 1 \tag{3.68}$$

The equation for ν_{12} shown here was derived by Tucker and Liang [25] and provides a noniterative formula to the iterative equation of ν_{12} presented by Tandon and Weng. The constants are found using Eq. (3.69):

$$\begin{aligned}
A_1 &= D_1\left(B_4 + B_5\right) - 2B_2 \\
A_2 &= \left(1 + D_1\right)B_2 - \left(B_4 + B_5\right) \\
A_3 &= B_1 - D_1 B_3 \\
A_4 &= \left(1 + D_1\right)B_1 - 2B_3 \\
A_5 &= \left(1 - D_1\right) / \left(B_4 - B_5\right) \\
A &= 2B_2B_3 - B_1\left(B_4 + B_5\right)
\end{aligned} \tag{3.69}$$

The constants B_i and D_j are found from the following:

$$B_1 = V_f D_1 + D_2 + \left(1 - V_f\right)\left(D_1 S_{1111} + 2S_{2211}\right) \tag{3.70}$$

$$B_2 = V_f + D_3 + \left(1 - V_f\right)\left(D_1 S_{1122} + S_{2222} + S_{2233}\right) \tag{3.71}$$

$$B_3 = V_f + D_3 + \left(1 - V_f\right)\left(S_{1111} + \left(1 + D_1\right) S_{2211}\right) \tag{3.72}$$

$$B_4 = V_f D_1 + D_2 + \left(1 - V_f\right)\left(S_{1122} + D_1 S_{2222} + S_{2233}\right) \tag{3.73}$$

$$B_5 = V_f + D_3 + \left(1 - V_f\right)\left(S_{1122} + S_{2222} + D_1 S_{2233}\right) \tag{3.74}$$

$$D_1 = 1 + 2\left(\mu_f - \mu_m\right) / \left(\lambda_f - \lambda_m\right) \tag{3.75}$$

$$D_2 = \left(\lambda_m + 2\mu_m\right) / \left(\lambda_f - \lambda_m\right) \tag{3.76}$$

where μ_m, μ_f, λ_m, and λ_f are Lame's constants for the matrix and fiber materials. Lame's constants are related to the Young's modulus E and Poisson's ratio ν by

$$\lambda = E\nu / (1 + \nu)(1 - 2\nu) \tag{3.77}$$

$$\mu = E/2(1 + \nu) \tag{3.78}$$

The S_{ijkl} in Eqs. (3.70) to (3.74) are the Eshelby tensor components for a spheriodal inclusion defined as

$$S_{1111} = \frac{1}{2\left(1 - \nu_m\right)}\left\{\left(1 - 2\nu_m\right) + \frac{3\alpha^2 - 1}{\alpha^2 - 1} - \left[1 - 2\nu_m + \frac{3\alpha^2}{\alpha^2 - 1}\right] g\right\} \tag{3.79}$$

$$S_{2222} = S_{3333} = \frac{3}{8\left(1 - \nu_m\right)}\frac{\alpha^2}{\alpha^2 - 1} + \frac{1}{4\left(1 - \nu_m\right)}\left[1 - 2\nu_m - \frac{9}{4\left(\alpha^2 - 1\right)}\right] g \tag{3.80}$$

$$S_{2233} = S_{3322} = \frac{1}{4\left(1 - \nu_m\right)}\left\{\frac{\alpha^2}{2\left(\alpha^2 - 1\right)} - \left[1 - 2\nu_m + \frac{3}{4\left(\alpha^2 - 1\right)}\right] g\right\} \tag{3.81}$$

$$S_{2211} = S_{3311} = -\frac{1}{2\left(1 - \nu_m\right)}\frac{\alpha^2}{\alpha^2 - 1} + \frac{1}{4\left(1 - \nu_m\right)}\left\{\frac{3\alpha^2}{\alpha^2 - 1} - \left(1 - 2\nu_m\right)\right\} g \tag{3.82}$$

$$S_{1122} = S_{1133} = -\frac{1}{2\left(1-\nu_{\mathrm{m}}\right)}\left[1-2\nu_{\mathrm{m}}+\frac{1}{\alpha^{2}-1}\right]+\frac{1}{2\left(1-\nu_{\mathrm{m}}\right)}\left[1-2\nu_{\mathrm{m}}+\frac{3}{2\left(\alpha^{2}-1\right)}\right]g \quad (3.83)$$

$$S_{2323} = S_{3232} = \frac{1}{4\left(1-\nu_{\mathrm{m}}\right)}\left\{\frac{\alpha^{2}}{2\left(\alpha^{2}-1\right)}+\left[1-2\nu_{\mathrm{m}}-\frac{3}{4\left(\alpha^{2}-1\right)}\right]g\right\} \quad (3.84)$$

$$S_{1212} = S_{1313} = \frac{1}{4\left(1-\nu_{\mathrm{m}}\right)}\left\{\left(1-2\nu_{\mathrm{m}}\right)-\frac{\alpha^{2}+1}{\alpha^{2}-1}-\frac{1}{2}\left[1-2\nu_{\mathrm{m}}-\frac{3\left(\alpha^{2}+1\right)}{\alpha^{2}-1}\right]g\right\} \quad (3.85)$$

where α is the aspect ratio of the fiber, and g is a parameter defined as

$$g = \frac{\alpha}{\left(\alpha^{2}-1\right)^{3/2}}\left\{\alpha\left(\sqrt{\alpha^{2}-1}-\cosh^{-1}(\alpha)\right\}\right. \quad (3.86)$$

Using Eqs. (3.79)–(3.86), Eshelby's tensor **E** is evaluated and further used in Eq. (3.53) to find the strain concentration tensor **A**. This Eshelby's tensor **E** has been used in all the models to find the strain concentration tensor. The strain concentration tensor is then used in Eq. (3.21) to find the stiffness and damping of the composite.

3.3.5 Chow Model

Some other micromechanics models are equivalent to the Mori–Tanaka approach, though this equivalence has not always been recognized. Chow [26] considered Eshelby's inclusion problem and concluded that in a concentrated composite, the inclusion strain would be the sum of two terms: the dilute result given by Eshelby and the average strain in the matrix:

$$\left(\varepsilon^{\mathrm{C}}\right)_{\mathrm{f}} = \mathbf{E}\,\varepsilon^{\mathrm{T}} + \left(\widetilde{\varepsilon}^{\mathrm{C}}\right)_{\mathrm{m}} \quad (3.87)$$

This can be combined with the definition of the average strain from Eq. (3.7) to relate the inclusion strain $(\varepsilon^{\mathrm{C}})_{\mathrm{f}}$ to the transformation strain ε^{T}

$$\left(\varepsilon^{\mathrm{C}}\right)_{\mathrm{f}} = \left(1-V_{\mathrm{f}}\right)\mathbf{E}\,\varepsilon^{\mathrm{T}} \quad (3.88)$$

Chow then extended this result to an in-homogeneity following the usual arguments, Eqs. (3.46) to (3.53), which produces a strain-concentration tensor:

$$\mathbf{A}^{\mathrm{Chow}} = \left[\boldsymbol{I}+\left(1-V_{\mathrm{f}}\right)\boldsymbol{E}\,\boldsymbol{S}_{\mathbf{m}}\left(\boldsymbol{C}_{\mathbf{f}}-\boldsymbol{C}_{\mathbf{m}}\right)\right]^{-1} \quad (3.89)$$

which is equivalent to the Mori–Tanaka result (Eq. (3.57)). Chow was apparently unaware of the connection between his approach and the Mori–Tanaka scheme, but he seems to have been the first to apply the Mori–Tanaka approach to predict the stiffness of short-fiber composites. A more recent development is the equivalent polyinclusion model of Ferrari [27]. Rather than using the strain-concentration tensor A, Ferrari used an effective Eshelby tensor Ê, defined as the tensor that relates inclusion strain to transformation strain at finite volume fraction:

$$\left(\varepsilon^{\mathrm{C}}\right)_{\mathrm{f}} = \hat{E}\,\varepsilon^{\mathrm{T}} \quad (3.90)$$

Once $\hat{E}$ has been defined, it is straightforward to derive a strain concentration tensor **A** and a composite modulus. Ferrari considered admissible forms for $\hat{E}$, given the requirements that $\hat{E}$ must (i) produce a symmetric stiffness tensor C, (ii) approach Eshelby's tensor **E** as volume fraction approaches zero, and (iii) give a composite stiffness that is independent of the matrix stiffness as volume fraction approaches unity. He proposed a simple form that satisfies these criteria:

$$\hat{E} = \left(1 - V_{\mathrm{f}}\right) \mathbf{E} \tag{3.91}$$

Combination of Eqs. (3.90) and (3.91) is identical to Chow's assumption as in Eq. (3.88) and for aligned fibers of uniform length, Ferrari's equivalent polyinclusion model, Chow's model, and the Mori–Tanaka model are identical. Important differences between the equivalent polyinclusion model and the Mori–Tanaka model arise when the fibers are misoriented or have different lengths.

3.3.6 Modified Halpin–Tsai or Finegan model

Dynamic properties of carbon nanofiber/polypropylene composites can be expressed analytically from the theory of short-fiber composites. The elastic-viscoelastic correspondence principle – "A linear elastostatic analysis can be converted to a vibratory linear visco-elastic analysis by replacing the elastic moduli with the corresponding complex moduli," is used to define the dynamic properties. The complex modulus has a real and an imaginary part. The damping loss factor can be expressed as the ratio of imaginary to the real part of the complex modulus:

$$\eta = \frac{C''}{C'} \tag{3.92}$$

where

$C^* = C' + iC''$ = complex modulus of the composite
C' = storage modulus of the composite
C'' = loss modulus of the composite
H = damping loss factor of the composite

A composite modulus that takes into account the fiber orientation in the direction of injection has been expressed by Advani and Tucker [28] by using tensorial notation. The most general description of fiber orientation is the probability distribution function for orientation Ψ. Some of the assumptions made are as follows:

(a) Fibers are rigid cylinders, uniform in length and diameter.
(b) Number of fibers per unit volume is uniform, though the orientation of these fibers may not be uniform.

With these assumptions, the orientation of a single fiber can be described by the angles (θ, Φ) as shown in Figure 3.8. The figure presents the definition of the θ, Φ angles used in expressing the orientation of a single fiber by using three Cartesian components (p_1, p_2, p_3) of a unit vector oriented along the fiber axis:

$$\begin{aligned} p_1 &= \sin\theta \cos\Phi \\ p_2 &= \sin\theta \sin\Phi \\ p_3 &= \cos\theta \end{aligned} \tag{3.93}$$

a_{ij} and a_{ijkl} give information on how well the fibers are aligned to 1, 2, 3 axes (shown in Figure 3.8).

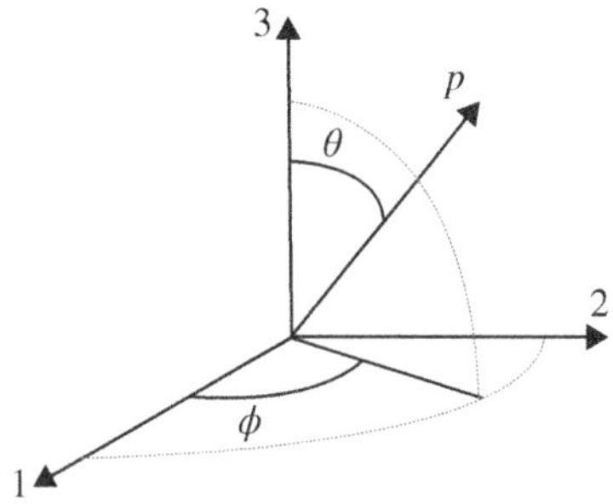

Figure 3.8 Coordinate system and definitions of θ, Φ, p.

The orientation state at a point in space can be described by a probability distribution function Ψ (θ, Φ). This function defines the probability of finding a fiber between angles θ_1 and $(\theta_1 + \mathrm{d}\theta)$ and Φ_1 and $(\Phi_1 + \mathrm{d}\Phi_1)$ in the way as follows:

$$P\left(\theta_1 \leq \theta \leq \theta_1 + \mathrm{d}\theta, \Phi_1 \leq \Phi \leq \Phi_1 + d\Phi\right) = \Psi\left(\theta_1, \Phi_1\right) \sin\,\theta_1\,\mathrm{d}\theta\, d\Phi \tag{3.94}$$

Another way is to associate a unit vector p with the fiber. Since the length of the vector is fixed; hence,

$$p_i = 1 \tag{3.95}$$

The set of all possible directions of p corresponds to unit sphere. The integral over the surface of the unit sphere is denoted by

$$\oint Dp = \int_{\Phi=0}^{2\pi}\int_{\theta=0}^{\pi} \sin\theta\,\mathrm{d}\theta\, d\Phi \tag{3.96}$$

The function Ψ must satisfy the following conditions:

(a) A fiber oriented at an angle (θ, Φ) is indistinguishable from a fiber oriented at angles $(\pi - \theta, \Phi + \pi)$, so Ψ should be periodic. Therefore,

$$\Psi\,(\theta, \Phi) = \Psi\,(\pi - \theta, \Phi + \pi) \tag{3.97}$$

or $\Psi(p) = \Psi(-p)$

(b) Since every fiber has some orientation, so Ψ must be normalized, that is,

$$\oint \Psi\,(p)\,\mathrm{d}p = \int_{\theta=0}^{\pi}\int_{\Phi=0}^{2\pi} \Psi\,(\theta, \Phi) \sin\,\theta\,\mathrm{d}\theta\,\mathrm{d}\Phi = 1 \tag{3.98}$$

(c) The third condition is continuity condition which describes the change in Ψ with time when fibers are changing orientation. If fibers move with bulk motion of fluid, then Ψ may be regarded as a convected quantity. One set of orientation tensors can be defined by forming products of vector **p** and integrating the product of these vectors with distribution function over all possible directions. Therefore,

$$\frac{D\Psi}{\mathrm{D}t} = -\frac{\alpha}{\alpha p}.(\Psi\,\dot{p}) \tag{3.99}$$

where $\frac{\alpha}{\alpha p}$ = gradient operator on the surface of unit sphere.

Because $\Psi(p) = \Psi(-p)$, the odd order integrals are zero. Thus, only even-order tensors are of interest. The second- and fourth-order orientation tensors are

$$a_{ij} = \oint p_i p_j \, \Psi(p) \, \mathrm{d}p$$

$$a_{ijkl} = \oint p_i p_j \, p_k p_l \, \Psi(p) \, \mathrm{d}p \tag{3.100}$$

One can also form another version of orientation tensors

$$b_{ij} = a_{ij} - \frac{1}{3}\delta_{ij} \tag{3.101}$$

$$b_{ijkl} = a_{ijkl} - \frac{1}{7}\left(\delta_{ij}\, a_{kl} + \delta_{ik}\, a_{jl} + \delta_{il}\, a_{jk} + \delta_{jk}\, a_{il} + \delta_{jl}\, a_{ik} + \delta_{kl}\, a_{ij}\right)$$
$$+ \frac{1}{35}\left(\delta_{ij}\,\delta_{kl} + \delta_{ik}\delta_{jl} + \delta_{il}\delta_{jk}\right) \tag{3.102}$$

where δ_{ij} is the unit tensor.

If we define tensor basis functions of p as follows:

$$f_{ij}(p) = p_i p_j - \frac{1}{3}\delta_{ij}$$

$$f_{ijkl}(p) = p_i p_j p_k p_l - \frac{1}{7}\left(\delta_{ij} p_k p_l + \delta_{ik} p_j p_l + \delta_{il} p_j p_k + \delta_{jk} p_i p_l + \delta_{jl} p_i p_k + \delta_{kl} p_i p_j\right)$$
$$+ \frac{1}{35}\left(\delta_{ij}\delta_{kl} + \delta_{ik}\delta_{jl} + \delta_{il}\delta_{jk}\right) \tag{3.103}$$

then the distribution function is

$$\Psi(p) = \frac{1}{4\pi} + \frac{15}{8\pi} b_{ij} f_{ij}(p) + \frac{315}{32\pi} b_{ijkl} f_{ijkl}(p) + \cdots \tag{3.104}$$

A planar orientation state is one in which all the fibers lie in a single plane. Taking 1–2 plane as the plane of orientation:

$$\Psi(\theta, \Phi) = \delta\left(\theta - \frac{\pi}{2}\right)\Psi(\Phi) \tag{3.105}$$

where

δ = Dirac delta function

$\Psi(\Phi)$ = Planar distribution function which satisfies the normalization condition;

$$\int_0^{2\pi} \Psi(\Phi)\, \mathrm{d}\Phi = 1 \tag{3.106}$$

Therefore, planar orientation tensors are

$$a_{ij} = \int_0^{2\pi} \Psi(\Phi)\, p_i p_j d\Phi$$

$$a_{ijkl} = \int_0^{2\pi} \Psi(\Phi)\, p_i p_j p_k p_l d\Phi$$

$$b'_{ij}(p) = a_{ij} - \frac{1}{2}\delta_{ij}$$

$$b'_{ijkl} = a_{ijkl} - \frac{1}{6}\left(\delta_{ij}a_{kl} + \delta_{ik}a_{jl} + \delta_{il}a_{jk} + \delta_{jk}a_{il} + \delta_{jl}a_{ik} + \delta_{kl}a_{ij}\right)$$
$$+ \frac{1}{24}\left(\delta_{ij}a_{kl} + \delta_{ik}a_{jl} + \delta_{il}a_{jk}\right) \tag{3.107}$$

where i and j can take values = 1 and 2.

Basis functions are

$$f'_{ij}(p) = p_i p_j - \frac{1}{2}\delta_{ij}$$

$$f'_{ijkl}(p) = p_i p_j p_k p_l - \frac{1}{6}\left(\delta_{ij}p_k p_l + \delta_{ik}p_j p_l + \delta_{il}p_j p_k + \delta_{jk}p_i p_l + \delta_{jl}p_i p_k + \delta_{kl}p_i p_j\right)$$
$$+ \frac{1}{24}\left(\delta_{ij}\delta_{kl} + \delta_{ik}\delta_{jl} + \delta_{il}\delta_{jk}\right) \tag{3.108}$$

therefore,

$$\Psi(\Phi) = \frac{1}{2\pi} + \frac{2}{\pi}b'_{ij}f'_{ij}(p) + \frac{8}{\pi}b'_{ijkl}f'_{ijkl}(p) + \ldots \tag{3.109}$$

Now, considering a tensor property $T(p)$ associated with a unidirectional microstructure aligned in the direction of p, for p to be a complete description of direction of T, T must be unchanged by any coordinate rotation that either preserves or reverses the direction of p, i.e. T must be transversely isotropic, with p as its axis of symmetry. For a second-order tensor to be transversely isotropic, it must have the form:

$$T_{ij}(p) = A_1 p_i p_j + A_2 \delta_{ij} \tag{3.110}$$

where A_1 and A_2 are scalar constants. If we take orientation average, then

$$<T>_{ij} = A_1 a_{ij} + A_2 \delta_{ij} \tag{3.111}$$

i.e. orientation average of second-order tensor is completely determined by the second order orientation tensor and by the unidirectional property tensor. Similarly, for a fourth-order tensor

$$T_{ijkl}(p) = B_1\left(p_i p_j p_k p_l\right) + B_2\left(p_i p_j \delta_{kl} + p_k p_l \delta_{ij}\right)$$
$$+ B_3\left(p_i p_k \delta_{jl} + p_i p_l \delta_{jk} + p_j p_l \delta_{ik} + p_j p_k \delta_{il}\right) + B_4\left(\delta_{ij}\delta_{kl}\right) + B_5\left(\delta_{ik}\delta_{jl} + \delta_{il}\delta_{jk}\right) \tag{3.112}$$

These B's are five scalar constants related to five independent components of a transversely isotropic elasticity tensor. Therefore,

$$\langle T\rangle_{ijkl} = B_1\left(a_{ijkl}\right) + B_2\left(a_{ij}\delta_{kl} + a_{kl}\delta_{ij}\right)$$
$$+ B_3\left(a_{ik}\delta_{jl} + a_{il}\delta_{jk} + a_{jl}\delta_{ik} + a_{jk}\delta_{il}\right) + B_4\left(\delta_{il}\delta_{kl}\right) + B_5\left(\delta_{ik}\delta_{jl} + \delta_{il}\delta_{jk}\right) \tag{3.113}$$

For longitudinal direction,

$$a_{11} = \int_{\Phi=0}^{2\pi}\int_{\theta=0}^{\pi} p_1(\theta,\Phi)\, p_1(\theta,\Phi)\, \Psi(\theta)\sin\theta\, d\theta\, d\Phi \tag{3.114}$$

$$a_{1111} = \int_{\Phi=0}^{2\pi}\int_{\theta=0}^{\pi} p_1(\theta,\Phi)\, p_1(\theta,\Phi)\, p_1(\theta,\Phi)\, p_1(\theta,\Phi)\, \Psi(\theta)\sin\theta\, d\theta\, d\Phi$$

where $p_1(\theta, \Phi)$ is given by components of p and weighting factor $\Psi(\theta)$ is determined experimentally as follows:

$$\Psi(\theta) = I(\theta) \times \text{Norm}$$

$$I(\theta) = I_1\left[\exp\left\{-\theta/\theta_{\text{fwhm}}\right\}^2\right] + I_0$$

$$I_1 = 44.3, I_0 = 37.9, \theta_{\text{fwhm}} = 50.4°$$

$$\text{Norm} = \left[\int_0^{2\pi}\int_0^{\pi} I(\theta)\sin\theta\, d\theta\, d\Phi\right] - 1 \tag{3.115}$$

For carbon nanofiber/polypropylene composite

$$a_{11} = 0.354$$
$$a_{1111} = 0.22 \tag{3.116}$$

Advani and Tucker [28] gave the stiffness of the composite as follows:

$$C^*_{ijkl} = B^*_1\left(a_{ijkl}\right) + B^*_2\left(a_{ij}\delta_{kl} + a_{kl}\delta_{ij}\right) + B^*_3\left(a_{ik}\delta_{jl} + a_{il}\delta_{jk} + a_{jl}\delta_{ik} + a_{jk}\delta_{il}\right)$$
$$+ B^*_4\left(\delta_{ij}\delta_{kl}\right) + B^*_5\left(\delta_{ik}\delta_{jl} + \delta_{il}\delta_{jk}\right) \tag{3.117}$$

Composite elastic modulus is replaced by the corresponding complex modulus according to the elastic-viscoelastic correspondence principle and composite damping can be calculated as shown in the Eq. (3.92). For the longitudinal direction, we get

$$C^*_{1111} = 0.22B^*_1 + 0.708B^*_2 + 1.416B_3 + B^*_4 + 2B^*_5 = E_{\text{longitudinal}} = E'_c + iE''_c \tag{3.118}$$

where $B^*_1, B^*_2, B^*_3, B^*_4, B^*_5$ are constants defined in terms of the stiffness matrix of the unidirectional aligned composite in complex form as follows:

$$B^*_1 = C^*_{11} + C^*_{22} - 2C^*_{12} - 4C^*_{66}$$
$$B^*_2 = C^*_{12} - C^*_{23}$$
$$B^*_3 = C^*_{66} + \frac{1}{2}\left(C^*_{23} - C^*_{22}\right)$$
$$B^*_4 = C^*_{23}$$
$$B^*_5 = \frac{1}{2}\left(C^*_{22} - C^*_{23}\right) \tag{3.119}$$

Components of the stiffness matrix, without taking into consideration the orientation tensors, C^*_{ij}, are calculated from the Halpin–Tsai equations for the unidirectional aligned composite in the complex form. By combining the self-consistent approach of Hill with the solutions of Hermans and making a few additional assumptions, Halpin and Tsai provide a simpler analytical form for predicting material properties of fiber composites. Halpin–Tsai equations need only one equation to find all the composite moduli, and the longitudinal Poisson's ratio is simply found from the rule of mixtures. It should be noted that the Halpin–Tsai equations are partially empirical where one parameter ξ is found by fitting the equations to numerical results:

$$C^*_{11} = E^*_{11} = \frac{1 + 2\left(\frac{l}{d}\right)\eta^*_L V_f}{1 - \eta^*_L V_f} E^*_m \tag{3.120}$$

$$C^*_{22} = E^*_{22} = \frac{1 + 2\eta^*_T V_f}{1 - \eta^*_T V_f} E^*_m \tag{3.121}$$

$$C^*_{12} = C^*_{66} = E^*_{12} = E^*_{66} = \frac{1 + \eta^*_G V_f}{1 - \eta^*_G V_f} G^*_m \tag{3.122}$$

$$C^*_{23} = E^*_{23} = \frac{1 + \xi\,\eta^*_{TS} V_f}{1 - \eta^*_{TS} V_f} G^*_m \tag{3.123}$$

where

$$\eta^*_L = \frac{\left[\frac{E^*_f}{E^*_m} - 1\right]}{\left[\frac{E^*_f}{E^*_m} + 2\left(\frac{l}{d}\right)\right]} \tag{3.124a}$$

Table 3.2 Dynamic properties of VGCF and polypropylene.

Dynamic properties	VGCF	Polypropylene
Storage modulus (E') (GPa)	1000 (measured)	1.23 (measured)
Loss modulus (E'') (GPa)	1.5 (calculated)	0.0738 (calculated)
Loss factor (η)	0.0015 [29]	0.06 [29]
Fiber volume fraction (V_f)	0.03	

$$\eta_T^* = \frac{\left[\frac{E_f^*}{E_m^*} - 1\right]}{\left[\frac{E_f^*}{E_m^*} + 2\right]} \tag{3.124b}$$

$$\eta_G^* = \frac{\left[\frac{G_f^*}{G_m^*} - 1\right]}{\left[\frac{G_f^*}{G_m^*} + 1\right]} \tag{3.124c}$$

$$\eta_{TS}^* = \frac{\left[\frac{G_f^*}{G_m^*} - 1\right]}{\left[\frac{G_f^*}{G_m^*} + \xi^*\right]} \tag{3.125a}$$

$$\xi * = \frac{\frac{K_m^*}{G_m^*}}{\left[\frac{K_m^*}{G_m^*} + 2\right]} \tag{3.125b}$$

$$K_m^* = \frac{E_m^*}{\left[3\left(1 - 2\,\nu_m\right)\right]} \tag{3.125c}$$

where

$$V_f = \text{Fiber volume fraction}$$

$$\frac{l}{d} = \text{Fiber aspect ratio}$$

$$E_f^* = E_f' + iE_f'' = \text{Complex elastic modulus of the fiber}$$

$$E_m^* = E_m' + iE_m'' = \text{Complex elastic modulus of the matrix}$$

$$G_f^* = G_f' + iG_f'' = \text{Complex elastic shear modulus of the fiber}$$

$$G_m^* = G_m' + iG_m'' = \text{Complex elastic shear modulus of the matrix}$$

Dynamic properties of vapor-grown carbon fibers (VGCFs) and polypropylene matrix as given by Gibson et al. [29] are used as input and shown in Table 3.2. With the above data, Eqs. (3.118)–(3.125) are solved using a mathematical tool such as Matlab. By writing a program in Matlab, we can solve the equations and can find the effect of various parameters such as aspect ratio and fiber volume fraction on loss modulus, storage modulus, and damping loss factor.

3.3.7 Hashin–Shtrikman Model

A rather different approach to modeling stiffness is based on finding upper and lower bounds for the composite moduli. All bounding methods are based on assuming an approximate field for either the stress or the strain in the composite. The unknown field is then found through a variational principle by minimizing or maximizing some functional of the stress and strain. Resulting composite stiffness is not exact, but it can be guaranteed to be either greater than or less than the actual stiffness, depending on the variational principle. This rigorous bounding property is the attraction of bounding methods. Historically, the Voigt and Reuss averages were the first models to be recognized as providing rigorous upper and lower bounds. To derive the Voigt model, Eq. (3.23), one assumes that the fiber and matrix have the same uniform strain, and then minimizes the potential energy. Since the potential energy will have an absolute minimum when the entire composite is in equilibrium, the potential energy under the uniform strain assumption must be greater than or equal to the exact result, and the calculated stiffness will be an upper bound on the actual stiffness. The Reuss model, Eq. (3.24), is derived by assuming that the fiber and matrix have the same uniform stress and then maximizing the complementary energy. Since the complementary energy must be maximum at equilibrium, the model provides a lower bound on the composite stiffness.

Detailed derivations of these bounds are provided by Wu and McCullough [30]. Voigt and Reuss bounds provide isotropic results (provided the fiber and matrix are themselves isotropic), when in fact we expect aligned-fiber composites to be highly anisotropic. More importantly, when the fiber and matrix have substantially different stiffnesses, then Voigt and Reuss bounds are quite far apart, and provide little useful information about the actual composite stiffness. This latter point motivated Hashin and Shtrikman to develop a way to construct tighter bounds. Hashin and Shtrikman [31] developed an alternate variational principle for heterogeneous materials. Their method introduces a reference material and bases the subsequent development on the differences between this reference material and the actual composite. Rather than requiring two variational principles, like the Voigt and Reuss bounds, their single variational principle gives both the upper and lower bounds by making appropriate choices of the reference material. For an upper bound, the reference material must be as stiff or stiffer than any phase in the composite (fiber or matrix), and for a lower bound, the reference material must have a stiffness less than or equal to any phase. In most composites, the fiber is stiffer than the matrix, so choosing the fiber as the reference material gives an upper bound and choosing the matrix as the reference material gives a lower bound. If the matrix is stiffer than the fiber, the bounds are reversed. The resulting bounds are tighter than the Voigt and Reuss bounds, which can be obtained from the Hashin–Shtrikman theory by giving the reference material infinite or zero stiffness, respectively. Hashin–Shtrikman's original bounds apply to isotropic composites with isotropic constituents. Frequently, the bounds are regarded as applying to composites with spherical particles, though a fiber composite with 3-D random fiber orientation must also obey the bounds. Walpole [32] rederived Hashin–Shtrikman bounds using classical energy principles and extended them to anisotropic materials. Walpole also derived results for infinitely long fibers and infinitely thin disks in both aligned and 3-D random orientations. Hashin–Shtrikman–Walpole bounds were extended to short-fiber composites by Willis [33] and by Wu and McCullough [30]. These workers introduced a two-point correlation function into the bounding scheme, allowing aligned ellipsoidal particles to be treated. Based on these extensions, explicit formulae for aligned ellipsoids were developed by Weng [34] and by Eduljee et al. [35]. The general bounding formula, shown here in the format developed by Weng, gives the composite stiffness C as follows:

$$\mathbf{C} = \left[V_f \mathbf{C_f} \mathbf{Q_f} + V_m \mathbf{C_m} \mathbf{Q_m}\right] \left[V_f \mathbf{Q_f} + V_m \mathbf{Q_m}\right]^{-1} \tag{3.126}$$

where the tensors $\mathbf{Q_f}$ and $\mathbf{Q_m}$ are defined as follows:

$$\mathbf{Q_f} = \left[\mathbf{I} + \mathbf{E^0} \mathbf{S^0} \left(\mathbf{C_f} - \mathbf{C^0}\right)\right]^{-1}$$

$$\mathbf{Q}_{\mathbf{m}} = \left[\mathbf{I} + \mathbf{E}^{\mathbf{0}}\,\mathbf{S}^{\mathbf{0}}\,\left(\mathbf{C}_{\mathbf{m}} - \mathbf{C}^{\mathbf{0}}\right)\right]^{-1} \tag{3.127}$$

Here $\mathbf{E^0}$ is Eshelby's tensor associated with the properties of the reference material, which has stiffness $\mathbf{C^0}$ and compliance $\mathbf{S^0}$. When the matrix is chosen as the reference material, Eq. (3.126) gives a strain concentrator of

$$\hat{\mathbf{A}}^{\text{lower}} = \left[\mathbf{I} + \mathbf{E}_{\mathbf{m}}\,\mathbf{S}_{\mathbf{m}}\,\left(\mathbf{C}_{\mathbf{f}} - \mathbf{C}_{\mathbf{m}}\right)\right]^{-1} \tag{3.128}$$

This result is labeled here as the lower bound on the presumption that fiber is stiffer than the matrix. The composite stiffness is found by substituting $\hat{\mathbf{A}}^{\text{lower}}$ in Eqs. (3.20)–(3.21). Eduljee et al. [35] argue that the lower bound provides the most accurate estimate of composite properties and recommend it as a model. Note that this lower bound prediction is identical to the Mori–Tanaka model, Eq. (3.57). This correspondence lends theoretical support to the Mori–Tanaka approach and guarantees that it will always obey the bounds.

3.3.8 Lielens Model

At fiber volume fractions close to unity, the matrix stiffness strongly influences the composite stiffness for the lower-bound/Mori–Tanaka models, despite the tiny amount of it that is present. Packing considerations suggest that the only way to approach such high-volume fractions is for the fiber phase to become continuous, and Lielens et al. [36] suggested that at very high-fiber volume fractions, the composite stiffness should be much closer to the upper bound, or equivalent to the Mori–Tanaka prediction using the fiber as the continuous phase. This insight prompted Lielens and coworkers to propose a model that interpolates between the upper and lower bounds, such that the lower bound dominates at low-volume fractions and the upper bound dominates at high-volume fractions (again presuming the fiber is the stiffer phase). They performed this interpolation on the inverse of the strain-concentration tensor $\hat{\mathbf{A}}$, producing the predictive equation:

$$\hat{\mathbf{A}}^{\text{lielens}} = \left\{(1-f)\left[\hat{\mathbf{A}}^{\text{lower}}\right]^{-1} + f\left[\hat{\mathbf{A}}^{\text{upper}}\right]^{-1}\right\}^{-1} \tag{3.129}$$

The interpolating factor f depends on fiber volume fraction, and they propose

$$f = \frac{V_{\text{f}} + V_{\text{f}}^2}{2} \tag{3.130}$$

This theory reproduces the lower bound and Mori–Tanaka results at low-volume fractions, but is said to give improved results at reinforcement volume fractions in the 40–60% range.

3.3.9 Self-Consistent Model

Another approach to account for finite-fiber volume fraction is the self-consistent method. This approach is generally credited to Hill [37] and Budiansky [38], whose original work focused on spherical particles and continuous aligned fibers. The application to short-fiber composites was developed by Laws and McLaughlin [39] and by Chou et al. [40]. In self-consistent scheme, one finds the properties of a composite in which a single particle is embedded in an infinite matrix that has the average properties of the composite. For this reason, self-consistent models are also called embedding models. Building on Eshelby's result for an ellipsoidal particle, we can create a self-consistent version of Eq. (3.53) by replacing the matrix stiffness and compliance tensors by the corresponding properties of the composite. This gives the self-consistent strain-concentration tensor as follows:

$$\mathbf{A}^{SC} = \left[\mathbf{I} + \mathbf{ES}\ (\mathbf{C}_f - \mathbf{C})\right]^{-1} \tag{3.131}$$

Of course, the properties **C** and **S** of the embedding "matrix" are initially unknown. When the reinforcing particle is a sphere or an infinite cylinder, the equations can be manipulated algebraically to find explicit expressions for the overall properties. For short fibers this has not proved possible, but numerical solutions are easily obtained by an iterative scheme. One starts with an initial guess at the composite properties, evaluates **E** and then $\mathbf{A}^{SC}$ from Eq. (3.131), and by substituting the result into Eq. (3.21) to get an improved value for the composite stiffness. The procedure is repeated using this new value, and the iterations continue until the results for **C** converge. An additional, but less obvious, change is that Eshelby's tensor E depends on the "matrix" properties, which are now transversely isotropic. Expressions for Eshelby's tensor for an ellipsoid of revolution in a transversely isotropic matrix are given by Lin and Mura [41] and by Chou et al. [42]. With these expressions in hand, one can use Eq. (3.131) together with Eq. (3.21) to find the stiffness of the composite. This is the self-consistent approach used for short-fiber composites. A closely related approach called "generalized self-consistent model" also uses an embedding approach. However, in these models, the embedded object comprises both fiber and matrix material. When the composite has spherical reinforcing particles, the embedded object is a sphere of the reinforcement encased in a concentric spherical shell of matrix; this is in turn surrounded by an infinite body with the average composite properties. The generalized self-consistent model is sometimes referred to as a "double embedding" approach. For continuous fibers, the embedded object is a cylindrical fiber surrounded by a cylindrical shell of matrix. The first generalized self-consistent models were developed for spherical particles by Kerner [43] and for cylindrical fibers by Hermans [44]. While the generalized self-consistent model is widely regarded as superior to the original self-consistent approach, no such model has been developed for short fibers.

Based on the above discussion, some conclusions which can be inferred from the study of different models are listed below:

(i) Cox model gives good approximation of average fiber axial stress and the total strain energy in the composite, yet this model is not applicable for low-fiber volume fractions. This model gives good results for longer fibers but fall below for short fibers. Cox model treats the fiber as a slender body. The true axial displacements lag behind the beam theory predictions.

(ii) Eshelby model is used when the volume fraction of fibers is very small. This model does not consider interfiber interaction and assumes that the strain is uniform in the fiber and highly nonuniform in the matrix. The stiffness predicted by this model increases linearly with fiber volume fraction.

(iii) Mori–Tanaka model assumes that the fiber in a concentrated composite sees the average strain of the matrix, i.e. the interaction between the fibers is taken into account by averaging the strain fields acting on a given fiber from all the other fibers. Most of other models reproduce the results of Mori–Tanaka. Hence, this model is widely used to predict stiffness and damping properties of short-fiber composites at different fiber volume fractions and aspect ratios of fiber.

(iv) Chow model is equivalent to the Mori–Tanaka approach. Chow concluded that at high-fiber volume fractions, the fiber strain would be the sum of the result given by Eshelby and the average strain in the matrix.

(v) Modified Halpin–Tsai or Finegan model uses the elastic-viscoelastic correspondence principle. This model gives satisfactory results for short fibers, but falls below for longer fibers. It contains no dependence on aspect ratio for E'_{22}, G'_{23} and G'_{12}. This model is semiempirical where one parameter $\xi = 2\, l/d$ is found by fitting the equations to numerical results.

(vi) Hashin–Shtrikman model provides upper and lower bounds on the composite moduli. Choosing the fiber as reference material gives an upper bound and choosing matrix as reference material gives a lower bound on the composite moduli.

(vii) Lielens model interpolates between the upper and lower bounds, such that lower bound dominates at low-fiber volume fractions and upper bound dominates at high-volume fractions. This theory reproduces the lower bound and Mori–Tanaka results at low-volume fractions.

(viii) In self-consistent model, we find the properties of the composite in which a single fiber is embedded in an infinite matrix that has the average properties of the composite. This model usually gives higher results as it starts with making an assumption of the composite properties.

3.3.10 Finite Element Modeling (FEM)

3.3.10.1 Introduction

Finite element method (FEM) (its practical application often known as finite element analysis [FEA]) is a numerical technique for finding approximate solutions of partial differential equations (PDEs) as well as of integral equations. The solution approach is based either on eliminating the differential equation completely (steady-state problems), or rendering the PDE into an approximating system of ordinary differential equations, which are then numerically integrated using standard techniques such as Euler's method, Runge–Kutta, etc. In solving PDEs, the primary challenge is to create an equation that approximates the equation to be studied, but is numerically stable, meaning that errors in the input and intermediate calculations do not accumulate and cause the resulting output to be meaningless. There are many ways of doing this, all with advantages and disadvantages. FEM is a good choice for solving PDEs over complicated domains (like cars and oil pipelines), when the domain changes (as during a solid-state reaction with a moving boundary), when the desired precision varies over the entire domain, or when the solution lacks smoothness. The FEM originated from the need for solving complex elasticity and structural analysis problems in civil and aeronautical engineering. Its development can be traced back to the work by Alexander Hrennikoff [45] and Richard Courant [46]. While the approaches used by these pioneers are dramatically different, they share one essential characteristic: Mesh discretization of a continuous domain into a set of discrete subdomains, usually called elements. Hrennikoff's work discretizes the domain by using a lattice analogy, while Courant's approach divides the domain into finite triangular subregions for solution of second-order elliptic PDEs that arise from the problem of torsion of a cylinder. Courant's contribution was evolutionary, drawing on a large body of earlier results for PDEs developed by Rayleigh, Ritz, and Galerkin. A variety of specializations under the umbrella of the mechanical engineering discipline (such as aeronautical, biomechanical, and automotive industries) commonly use integrated FEM in design and development of their products. Several modern FEM packages include specific components such as thermal, electromagnetic, fluid, and structural working environments. In a structural simulation, FEM helps tremendously in producing stiffness and strength visualizations and also in minimizing weight, materials, and costs. FEM allows detailed visualization of where structures bend or twist and indicates the distribution of stresses and displacements. FEM software provides a wide range of simulation options for controlling the complexity of both modeling and analysis of a system. Similarly, the desired level of accuracy required and associated computational time requirements can be managed simultaneously to address most engineering applications. FEM allows entire designs to be constructed, refined, and optimized before the design is manufactured. This powerful design tool has significantly improved both the standard of engineering designs and the methodology of the design process in many industrial applications. The introduction of FEM has substantially decreased the time to take products from concept to the production line. It is primarily through improved initial prototype designs using FEM that testing and development have been accelerated. In summary, benefits of FEM include increased accuracy, enhanced design, and better insight into critical design parameters, virtual prototyping, fewer hardware prototypes, a faster and less-expensive design cycle, increased productivity, and increased revenue.

Damping is the inherent material property that dissipates energy during cyclic loading. FEM can be used for numerical computation of bulk properties based on the geometry, properties, and volume fraction of constituent phases. FEM involves discretization of a material representative volume element (RVE) into elements for which the elastic solution leads to determination of stress and strain field. The coarseness of the discretization determines the accuracy of the solution. Nano-scale RVEs of different geometric shapes can be chosen for simulation of mechanical properties. Finite element modeling is performed to predict the response of a structure under prescribed boundary conditions and time-independent applied loads, when linear response behavior can be assumed with reasonable accuracy. The desired response quantities are generally, displacements, stresses, strains, reactions, and energy. In general, applied loads include prescribed forces/moments at nodes, uniform or nonuniform pressure on the faces of elements as well as gravity or centrifugal forces (body force) and loading due to thermal expansion/contraction. The boundary conditions are specified displacement values (zero or nonzero) at prescribed nodes, or they may include multipoint constraint equations, coupled displacements, or rigid links. The basic equation for linear static analysis may be written in the form:

$$[K]\ \{u\} = \{p\} \tag{3.132}$$

where $[K]$ is the linear stiffness matrix for the structure (known), $\{u\}$ is the nodal displacement vector (unknown), and $\{p\}$ is the load vector (known).

3.3.10.2 Representative Volume Elements (RVEs)

Several square RVE models for VGCF in a matrix material are studied using the FEM in order to evaluate material properties of the VGCF-based nanocomposites. One of the arrangements of the fibers within the RVE is shown in Figure 3.9. The same arrangement of the fibers in the RVE is drawn on NISA as shown in the Figure 3.10. Another type of arrangement of the fibers, where two centrally aligned short fibers are used for modeling on NISA software, is also shown in the Figure 3.11. The models are drawn for different fiber volume fractions and fiber aspect ratios. Each time, when we consider a new fiber volume fraction or a new fiber aspect ratio, a new model has to be drawn as we cannot use the same model for different fiber volume fractions and aspect ratios. In this study, the Z-direction represents the fiber direction, i.e. (1) direction and x- and y-directions are transverse directions which refer to (3) and (2), respectively. With these two types of arrangements, modeling is done on NISA. Different cases are considered in this modeling. These are listed below:

(i) In first case, length of the fibers is kept constant and diameter is varied.
(ii) In second case, diameter of the fibers is kept constant and length is varied. Since, length of the fibers is varying gap between the fibers is also varying.

Number of elements considered is 12 312. Number of elements in the fiber is 4104 and the remaining are the elements in the matrix material. Hexahedron elements are used for meshing. Strain energy method is used to predict the properties of the VGCF/pp. nanocomposite.

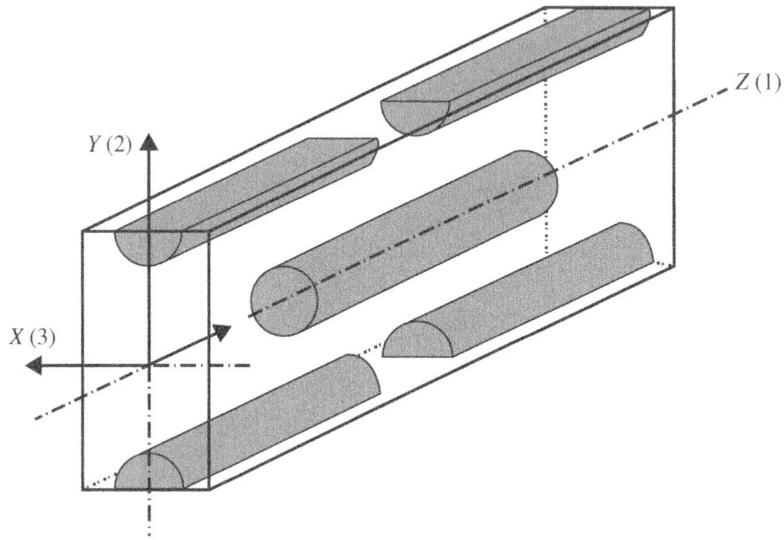

Figure 3.9 Unit cell for an array of straight fibers.

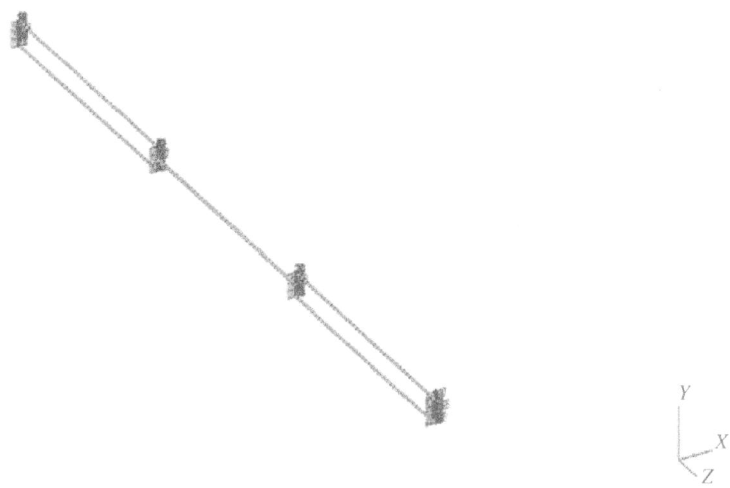

Figure 3.10 Arrangement of fibers while modeling on NISA software.

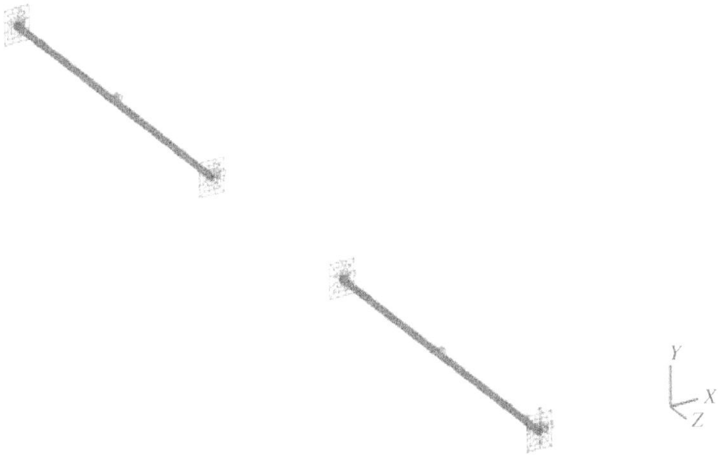

Figure 3.11 Arrangement of fibers while modeling on NISA software.

From 0.02 to 0.16 volume fractions, the models are created on NISA software at different fiber aspect ratios. Results of strain energy are taken from the NISA output file. The boundary conditions for different types of loading are shown in the Table 3.3. For the case of longitudinal loading, boundary conditions are as shown in the Figure 3.12.

For longitudinal loading, a full 3-D model is used. The RVE is fixed at one end, and the opposite end is given a small longitudinal displacement, $U_z = 0.001$, in the z-direction, which is the fiber direction. Similarly, for the case of transverse loading, the bottom of the RVE is fixed, and the top face is given a small displacement, $U_y = 0.001$, in the y-direction, which is the direction perpendicular to the fibers. The boundary conditions for the case of transverse loading are as shown in Figure 3.13. In order to determine the storage moduli and the corresponding loss factors for shear loading, such as, η_{23}, η_{31}, η_{12}, the boundary conditions are as shown in the Figures 3.14–3.16. For finding η_{21}, the bottom of the RVE is constrained in all directions, i.e. all the nodes in the bottom face of the RVE have all degrees of freedom equal to zero, and the nodes on the top face are given a small displacement, $U_z = 0.001$, in the z-direction, which is the fiber direction or 1 direction. For finding η_{23}, the bottom face of the RVE is fixed, i.e. all the nodes in the bottom face have all degrees of freedom equal to zero, while the top face is given a small nodal displacement, $U_x = 0.001$, in the x-direction. Similarly, for finding η_{31}, one side face of the RVE is fixed in such a way that all the nodes in that face have all degrees of freedom equal to zero, and the other side face is given a small nodal displacement, $U_z = 0.001$, in the z-direction. Fibers in the RVE of Figure 3.11 can be arranged in different ways. Two of these are as shown in the Figure 3.17. While making the models on NISA, several arrangements are used. These are explained as below:

(i) *Type 1*: In this type of model, the RVE consists of one full central fiber and the other fibers, on the top and bottom faces of the RVE are half of the full fiber, i.e. they are semicircles. On the top face, there are two semicircular fibers and similarly on the bottom face, there are two other semicircular fibers. This type may contain further subtypes as shown in the Figure 3.18.

(ii) *Type 2*: In this type of model, there are two central fibers separated by some interfiber spacing as shown in the Figure 3.19.

Table 3.3 Boundary conditions for different types of loading.

		U_x	U_y	U_z	T_x	T_y	T_z
Longitudinal loading (11)	Constraint			0	0	0	0
	Force/displacement			0.001	0	0	0
Transverse loading (22)	Constraint		0		0	0	0
	Force/displacement		0.001		0	0	0
Transverse shear loading (23) (Bottom face fixed and top face given displacement in *x*-direction)	Constraint	0	0	0	0	0	0
	Force/displacement	0.001	0	0	0	0	0
Longitudinal shear loading (31) (left face fixed and right face given displacement in *z*-direction)	Constraint	0	0	0	0	0	0
	Force/displacement	0	0	0.001	0	0	0
Longitudinal shear loading (12) (Bottom face fixed and top face given displacement in *z*-direction)	Constraint	0	0	0	0	0	0
	Force/displacement	0	0	0.001	0	0	0

Figure 3.12 Boundary conditions for longitudinal loading (11). Source: Based on Robinson and Robinson [11].

(iii) *Type 3*: In this type, there are two central fibers separated by some interfiber spacing as well as there is some distance between the first fiber and the front face of the RVE and between the second fiber and the back face of the RVE as shown in the Figure 3.20.

3.3.10.3 Modeling for E_{11}

Boundary conditions for longitudinal loading are shown in the Table 3.3. Strain energy plot in case of longitudinal loading (11) for Type 1 RVE, with $V_f = 0.03$ and $l/d = 10$, is shown in the Figure 3.21. RVE is fixed on back face and a small displacement (0.001) is applied to the

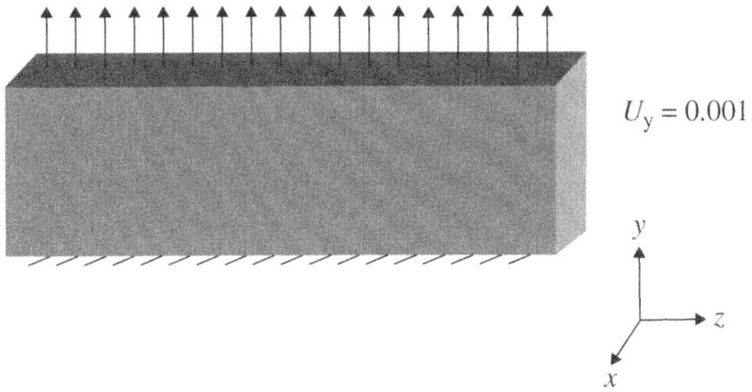

Figure 3.13 Boundary conditions for transverse loading (22). Source: Based on Taya and Chou [22].

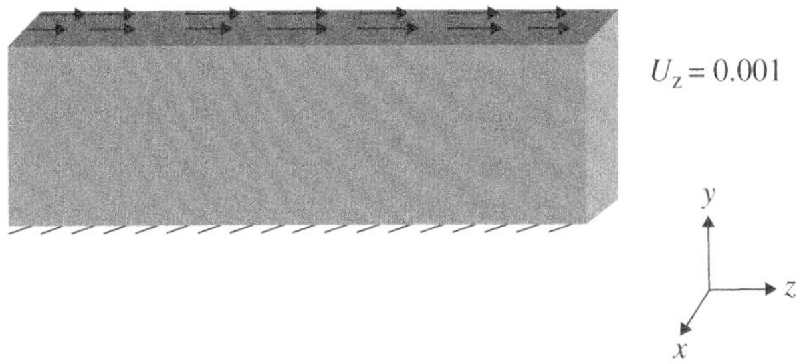

Figure 3.14 Boundary conditions for longitudinal shear loading (12). Source: Based on Fukuda and Kawata [13].

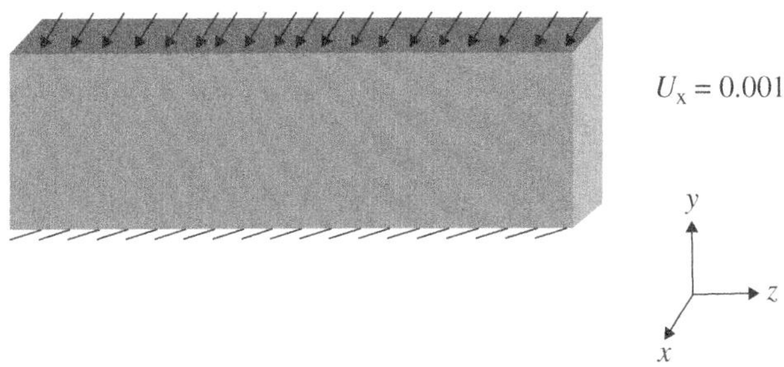

Figure 3.15 Boundary conditions for transverse shear loading (23). Source: Based on Weng [23].

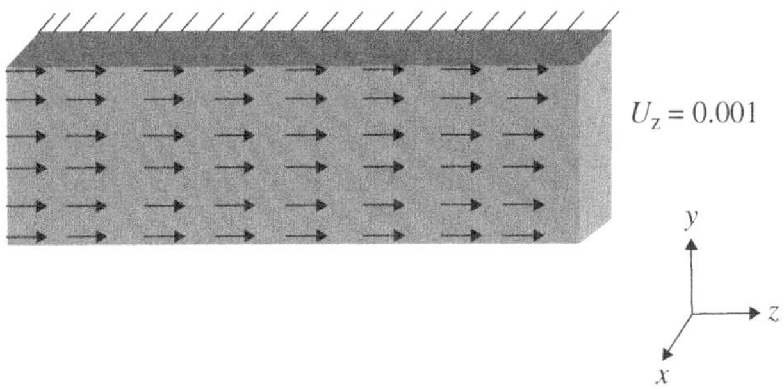

Figure 3.16 Boundary conditions for shear loading (13). Source: Based on Eshelby [13].

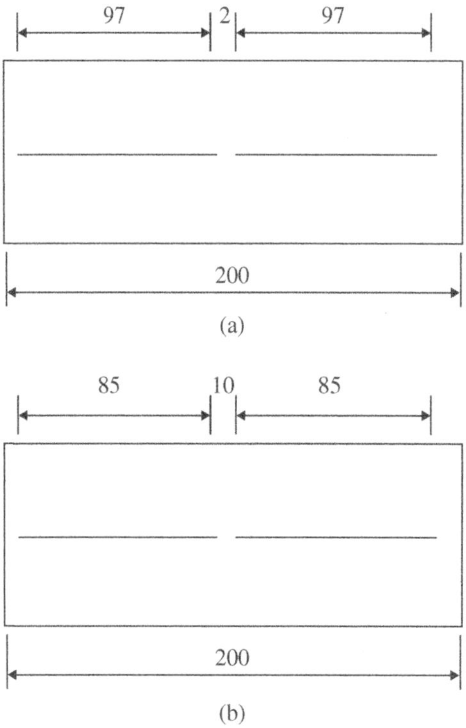

Figure 3.17 Different ways in which fibers can be arranged in the RVE. (All dimensions in nm).

front face. RVE consists of 4104 elements in the fiber and 8208 elements in the matrix. Thus, total number of elements in the model is 12 312. From this strain energy plot, it is observed that the strain energy in fiber (W_f) is less than the strain energy in the matrix (W_m). For the above model, strain energies in the fiber and matrix are

$$W_f = 3.29 \times 10^{-7}$$

$$W_m = 8.01 \times 10^{-6}$$

Total volume of the RVE, $V = 1.47894 \times 10^5$

Storage moduli and the corresponding loss factors can be determined using Eqs. (3.133–3.135). Figure 3.22 shows displaced shape for the above model for longitudinal loading. Figure 3.23 shows another case of longitudinal loading for Type 2 RVE, for fiber aspect ratio (l/d) of 97, and fiber volume fraction $V_f = 0.03$. There is an interfiber spacing of 6 nm in this case. Here, Type 2 RVE consists of 2160 elements in the fiber and 3672 elements in the matrix. Thus, the total number of elements in the model is 5832. The number of nodes in this model is 6403. Hexahedron elements are used for meshing both the fiber and the matrix. Strain energy shared by fiber and the matrix, obtained from the model are shown below:

$$W_f = 4.35 \times 10^{-7}$$

$$W_m = 5.82 \times 10^{-7}$$

Total volume of the RVE, $V = 7.688 \times 10^3$

Figure 3.24 shows another case, i.e. Type 3 RVE with longitudinal loading (11), where the fiber aspect ratio (l/d) is 97 and fiber volume fraction is 0.03. Between the front face and the

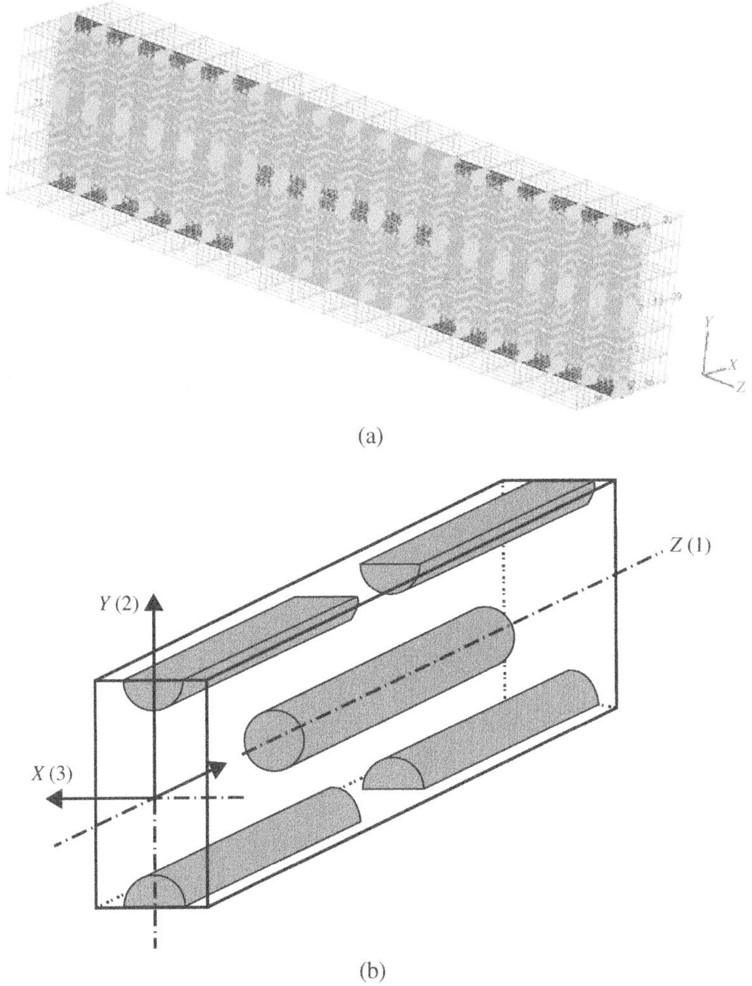

Figure 3.18 Arrangement of fibers in RVE of Type 1.

first fiber, there is a spacing of 2 nm. Then, there is a spacing of 2 nm between the first and the second fibers followed by a further spacing of 2 nm between the second fiber and the back face of the RVE. The effect of increase in this interfiber spacing on the modulus and the loss factor can be estimated. Here, the Type 3 RVE consists of 2160 elements in the fiber and the 7560 elements in the matrix. Total number of elements in the model is 9720 and the total number of nodes is 10 447. Strain energy shared by fiber and the matrix are as follows:

$$W_f = 2.72 \times 10^{-7}$$

$$W_m = 5.54 \times 10^{-7}$$

Total volume of the RVE, $V = 7.688 \times 10^3$

Figure 3.25 shows the strain energy plot for a fiber volume fraction (V_f) of 0.04 and an aspect ratio (l/d) of 97, for Type 2 RVE. This Type 2 RVE has 49 205 nodes and the total number of elements in the model is 13 824. Here, there are 5120 elements in the fiber and 8704

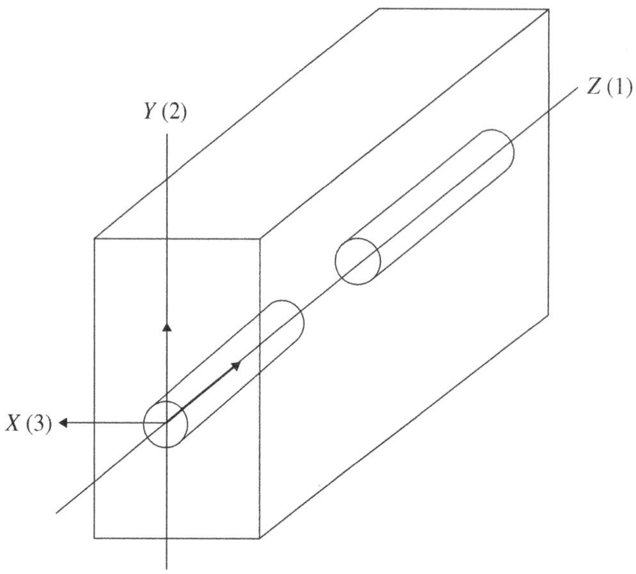

Figure 3.19 Arrangement of fibers in RVE of Type 2. The two fibers are touching the front and rear faces of RVE.

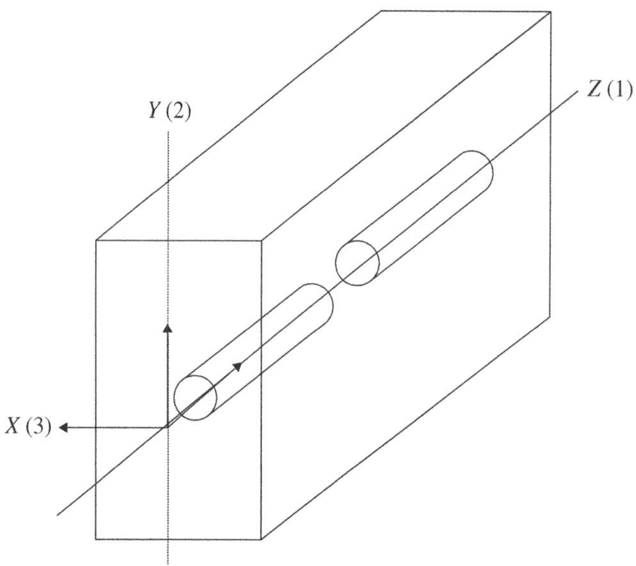

Figure 3.20 Arrangement of fibers in RVE of Type 3. The two fibers are not touching the front and rear faces of RVE, hence, fibers are shown as hidden.

elements in the matrix. The strain energy shared by fiber and the matrix are shown below:

$$W_f = 2.64 \times 10^{-7}$$

$$W_m = 4.98 \times 10^{7}$$

Total volume of the RVE, $V = 3.801\,92 \times 10^{3}$

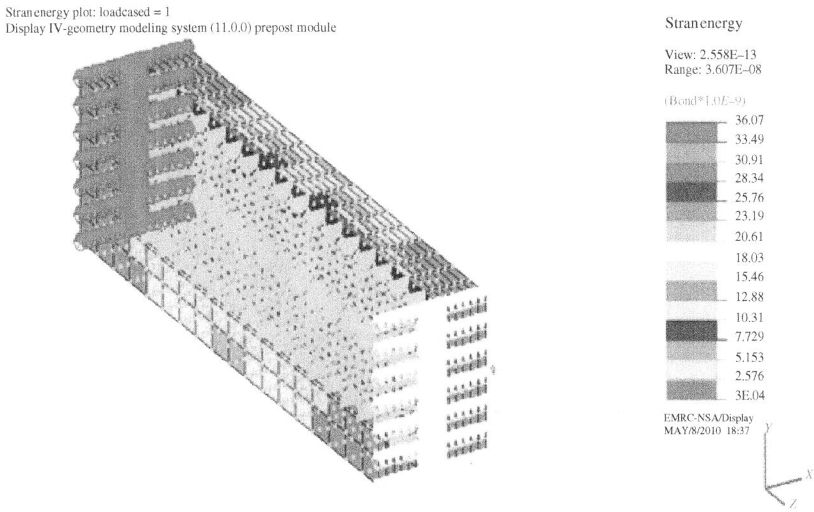

Figure 3.21 Strain energy plot for longitudinal loading (11) for Type 1 RVE with $V_f = 0.03$ and $(l/d) = 10$. Source: Based on Robinson and Robinson [11].

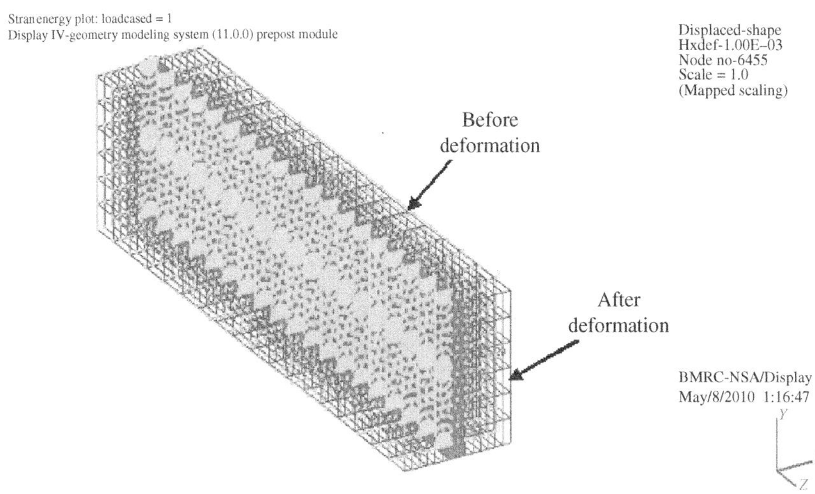

Figure 3.22 Displaced shape for longitudinal loading for Type 1 RVE with $V_f = 0.03$ and fiber aspect ratio $(l/d) = 10$.

3.3.10.4 Modeling for E_{22}

Boundary conditions for the case of transverse loading (22) are shown in the Table 3.3. Figure 3.26 shows the boundary conditions on NISA software. Bottom face of the RVE is fixed, and the top face is given a small displacement of 0.001 in y-direction. In Figure 3.26, red color shows the fixed position and the yellow color shows that a small displacement is given to the top face. Figure 3.27 shows the strain energy plot for transverse loading (22), for a fiber volume fraction of 0.03 and an aspect ratio of 10, for Type 1 RVE. Here, there are 12 312 elements in the model. Number of elements in the fiber and matrix, respectively, are 4104 and 8208. Displaced shape for the case of transverse loading for Figure 3.27 is shown

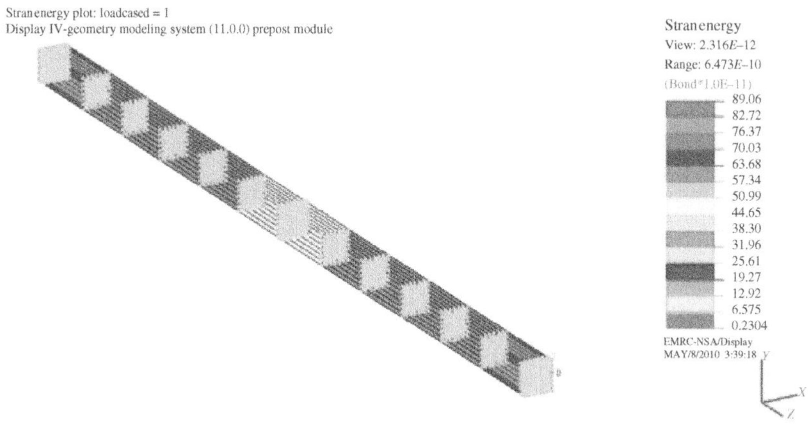

Figure 3.23 Strain energy plot for longitudinal loading for Type 2 RVE with $V_f = 0.03$ and $(l/d) = 97$ with interfiber spacing of 6 nm.

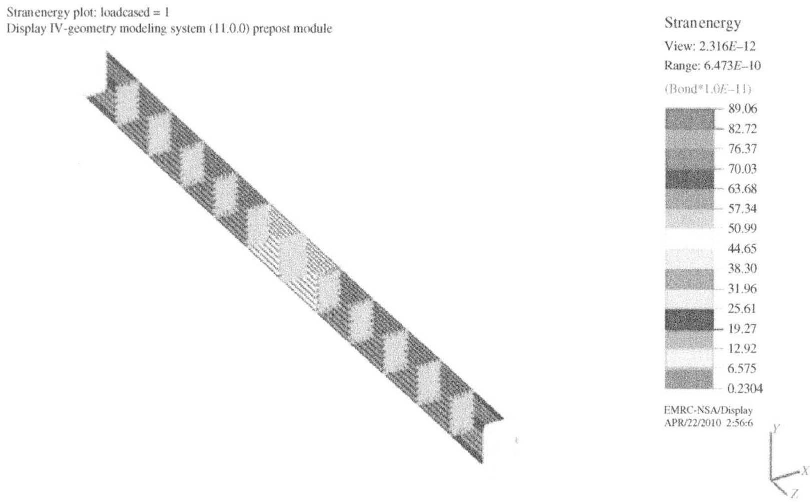

Figure 3.24 Strain energy plot for longitudinal loading for $V_f = 0.03$ and fiber aspect ratio $(l/d) = 97$ with interfiber spacing of Type 3.

in the Figure 3.28. Strain energy shared by fiber and the matrix are as shown below:

$$W_f = 7.51 \times 10^{-8}$$

$$W_m = 9.79 \times 10^{-5}$$

Total volume of the RVE, $V = 1.478\,94 \times 10^5$

Figures 3.29–3.30 show the strain energy plots for $V_f = 0.03$, but having different aspect ratios of $l/d = 20$ and $l/d = 50$, respectively. The number of elements in the Figures 3.29–3.30 is 12 312, and the number of nodes is 13 699. The volume of the RVE and the strain energy shared by fiber and the matrix in Figure 3.29 are as shown below:

$$W_f = 2.28 \times 10^{-7}$$

$$W_m = 1.00 \times 10^{-4}$$

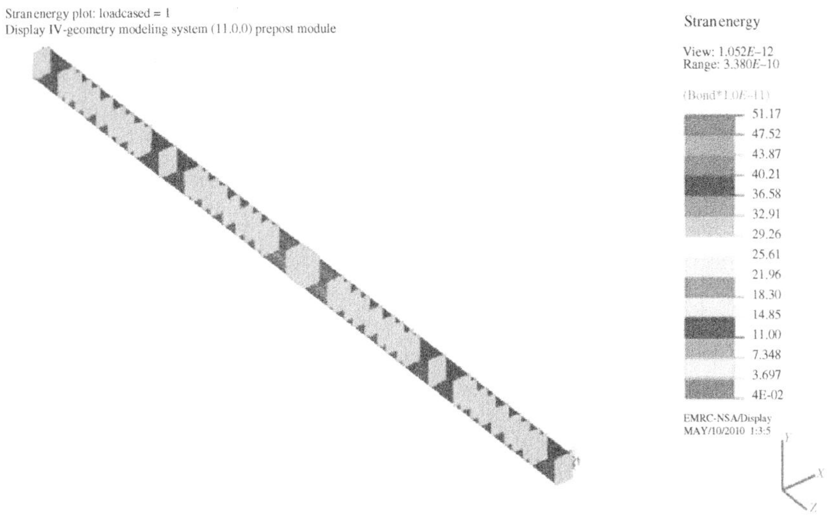

Figure 3.25 Strain energy plot for longitudinal loading for Type 2 RVE with $V_f = 0.04$ and $(l/d) = 97$ with interfiber spacing of 6 nm.

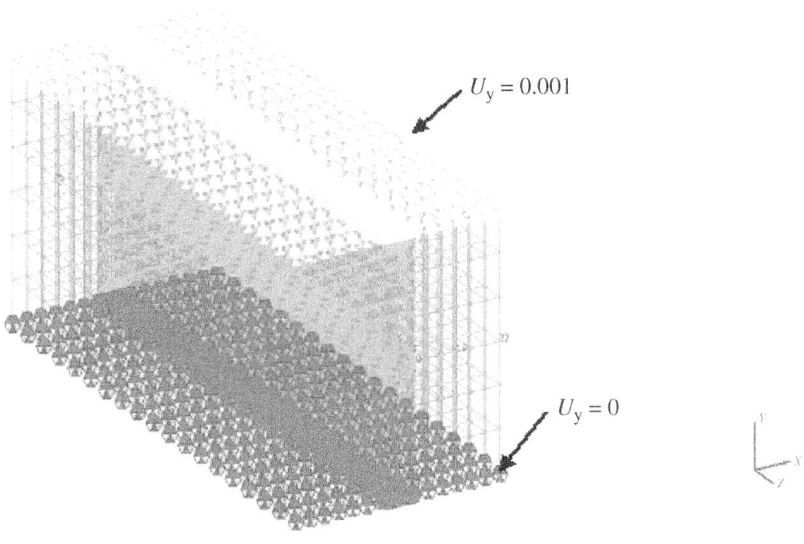

Figure 3.26 Boundary conditions for transverse loading (22). Source: Based on Taya and Chou [22].

$$V = 3.650\,40 \times 10^4$$

Volume of the RVE and the strain energy shared by fiber and the matrix in Figure 3.30 are as shown below:

$$W_f = 4.38 \times 10^{-7}$$

$$W_m = 1.02 \times 10^{-4}$$

$$V = 5.878\,14 \times 10^3$$

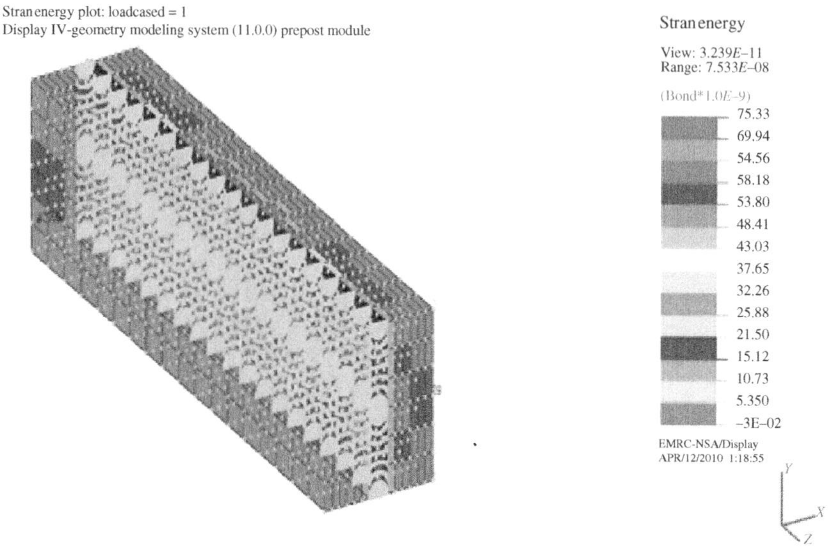

Figure 3.27 Strain energy plot for transverse loading for Type 1 RVE with $V_f = 0.03$ and $(l/d) = 10$.

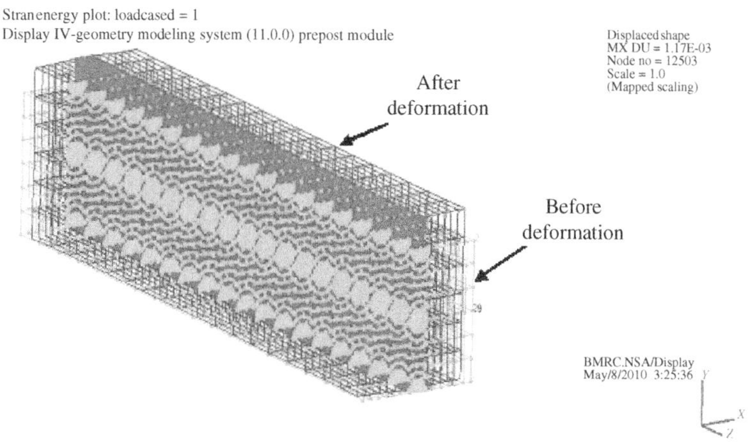

Figure 3.28 Displaced shape for transverse loading for $V_f = 0.03$ and $(l/d) = 10$.

Figure 3.31 shows the strain energy plot for Type 2 RVE having a fiber volume fraction of 0.03 and an aspect ratio of 97, with an interfiber spacing of 6 nm. Here, there are 5832 elements in the model and the number of nodes is 6403. The RVE consists of 2160 elements in the fiber and 3672 elements in the matrix. The strain energy shared by fiber and the matrix for transverse loading are shown below:

$$W_f = 6.59 \times 10^{-7}$$

$$W_m = 1.37 \times 10^{-4}$$

$$V = 7.688 \times 10^3$$

Figure 3.32 shows another case of transverse loading where the model belongs to the "Type 3." Between the front face and the first fiber, there is a spacing of 2 nm. Then, there is

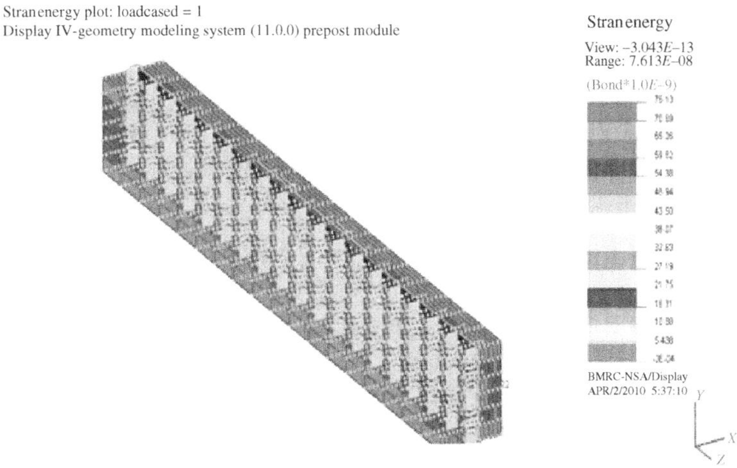

Figure 3.29 Strain energy plot for transverse loading for Type 1 RVE for $V_f = 0.03$ and $(l/d) = 20$.

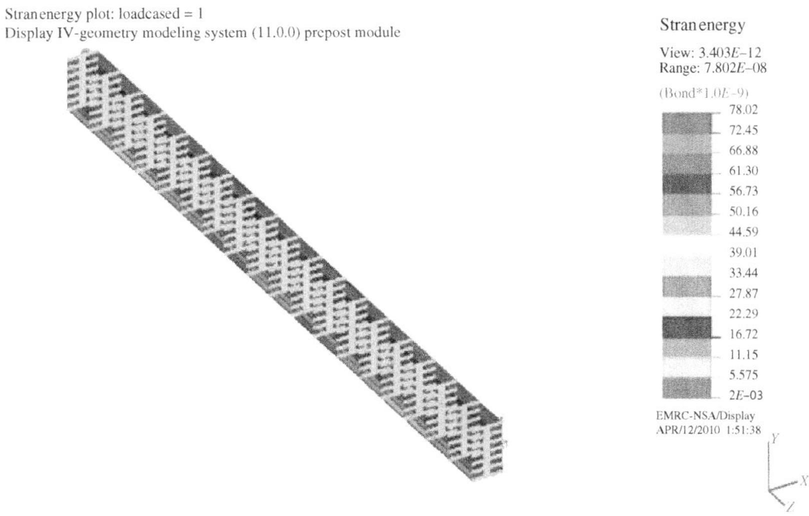

Figure 3.30 Strain energy plot for transverse loading for Type 1 RVE for $V_f = 0.03$ and $(l/d) = 50$.

a spacing of 2 nm between the first and the second fibers followed by a further spacing of 2 nm between the second fiber and the back face of the RVE. The effect of increase in this interfiber spacing on the modulus and the loss factor can be estimated. RVE consists of 2160 elements in the fiber and 7560 elements in the matrix. The total number of elements in the model is 9720 and the total number of nodes is 10 447. The strain energy shared by fiber and the matrix are as shown below:

$$W_f = 6.66 \times 10^{-7}$$

$$W_m = 1.37 \times 10^{-4}$$

Total volume of the RVE, $V = 7.688 \times 10^3$

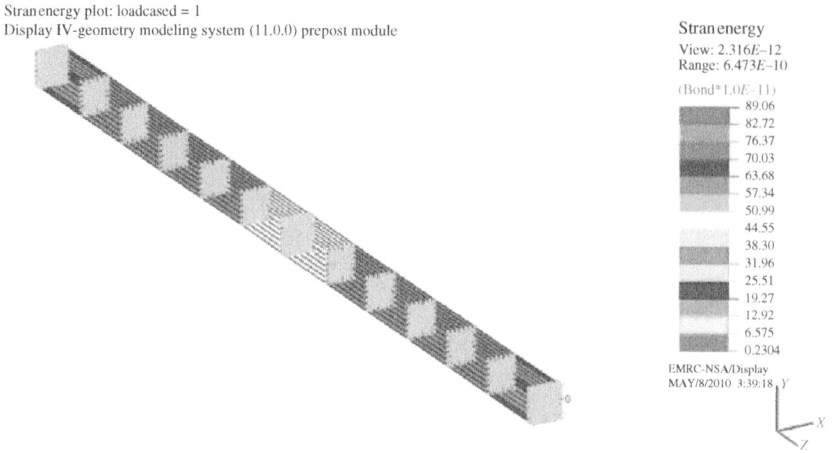

Figure 3.31 Strain energy plot for transverse loading for Type 2 RVE for $V_f = 0.03$ and $(l/d) = 97$ with interfiber spacing of 6 nm.

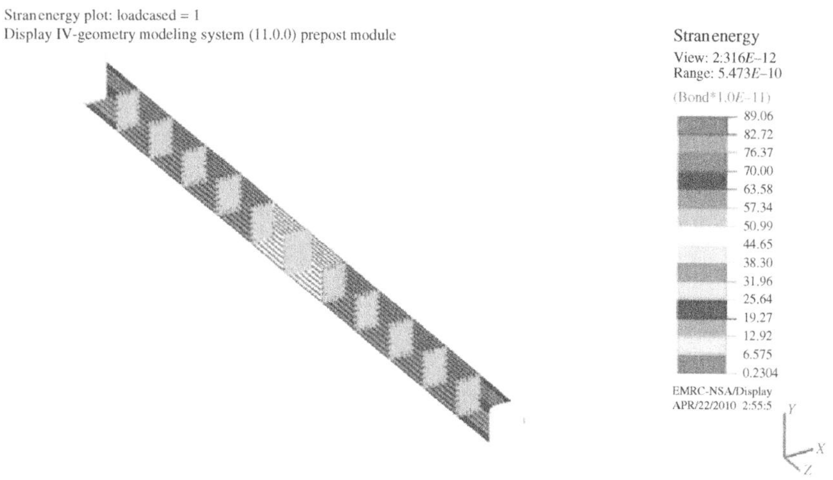

Figure 3.32 Strain energy plot for transverse loading for $V_f = 0.03$ and fiber aspect ratio $(l/d) = 97$ with interfiber spacing of Type 3.

Figure 3.33 shows the strain energy plot for transverse loading (22), for Type 2 RVE, for $V_f = 0.04$ and fiber aspect ratio (l/d) =97, with interfiber spacing of 6 nm. The number of nodes in the model is 49 205 and the total number of elements in the model is 13 824. RVE has 5120 elements in the fiber and 8704 elements in the matrix. The strain energy shared by fiber and the matrix are as shown below:

$$W_f = 4.01 \times 10^{-7}$$

$$W_m = 1.42 \times 10^{-4}$$

Total volume of the RVE, $V = 3.801\,92 \times 10^3$

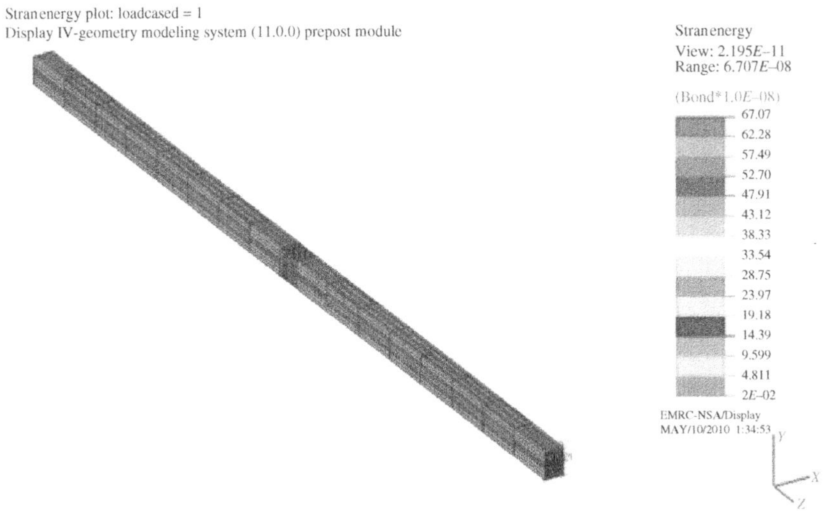

Figure 3.33 Strain energy plot for transverse loading for Type 2 RVE with $V_f = 0.04$ and $(l/d) = 97$ with interfiber spacing of 6 nm.

3.3.10.5 Modeling for G_{23}

Boundary conditions for the case of transverse shear loading (23) are shown in the Table 3.3. Figure 3.34 shows the boundary conditions on NISA. The bottom face of the RVE is fixed in this case and the top face is given a small displacement of 0.001 in the x-direction. In Figure 3.34, the red color shows the fixed position and yellow color shows that a small displacement is given to the top face. Figure 3.35 shows the strain energy plot for shear loading (23), for Type 1 RVE, for a fiber volume fraction of 0.03, and an aspect ratio of 10. Number of elements in the model is 12 312. Number of elements in the fiber and matrix, respectively, are 4104 and 8208. The displaced shape is shown in the Figure 3.36. Strain energy shared by fiber and the matrix are as shown below:

$$W_f = 1.03 \times 10^{-8}$$

$$W_m = 2.34 \times 10^{-5}$$

$$\text{Total volume of the RVE, } V = 1.478\,94 \times 10^5$$

Figures 3.37–3.38 show, respectively, the strain energy plot and the displaced shape for transverse shear loading (23), for Type 1 RVE having $V_f = 0.03$ and $l/d = 20$. The number of elements in the RVE is 12 312 and the number of nodes is 13 699. The strain energy shared by fiber and the matrix and volume of the RVE are as shown below:

$$W_f = 1.02 \times 10^{-8}$$

$$W_m = 2.35 \times 10^{-5}$$

$$V = 3.650\,40 \times 10^4$$

Figure 3.39 shows the strain energy plot for Type 1 RVE, having $V_f = 0.03$ and $l/d = 50$. The number of elements in the RVE is 12 312 and the number of nodes is 13 699. Figure 3.40 shows the strain energy plot for Type 2 RVE with an interfiber spacing of 6 nm and having a fiber

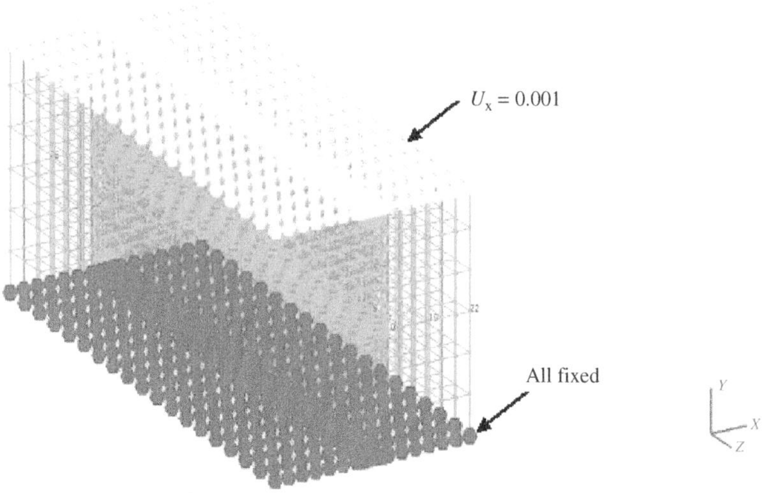

Figure 3.34 Boundary conditions for transverse shear loading (23). Source: Based on Weng [23].

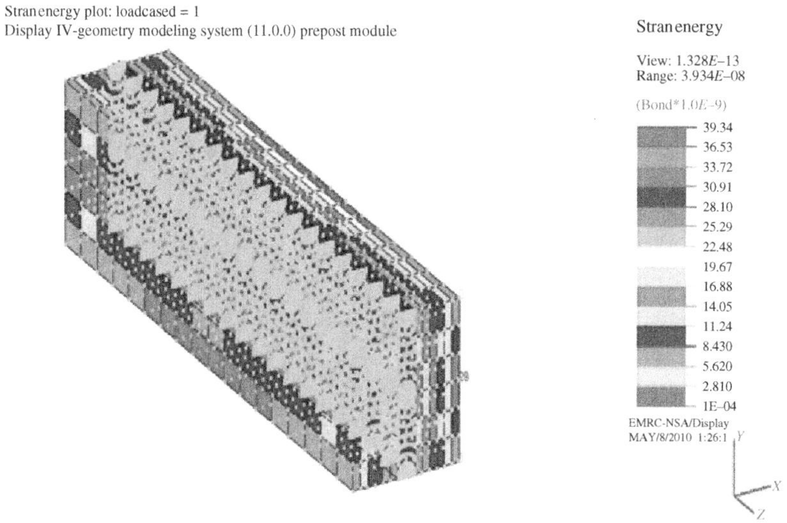

Figure 3.35 Strain energy plot for shear loading (23) for Type 1 RVE for $V_f = 0.03$ and $(l/d) = 10$. Source: Based on Weng [23].

volume fraction of 0.03 and an aspect ratio of 97. Here, the RVE consists of 5832 elements and 6403 nodes. There are 2160 elements in the fiber and 3672 elements in the matrix. Strain energy shared by fiber and the matrix for the case of shear loading (23) are shown below:

$$W_f = 3.27 \times 10^{-9}$$

$$W_m = 3.41 \times 10^{-5}$$

$$V = 7.688 \times 10^3$$

Figure 3.41 shows the strain energy plot for transverse shear loading (23), for a Type 3 RVE, having $V_f = 0.03$ and $l/d = 97$. Between the front face and the first fiber, there is spacing of

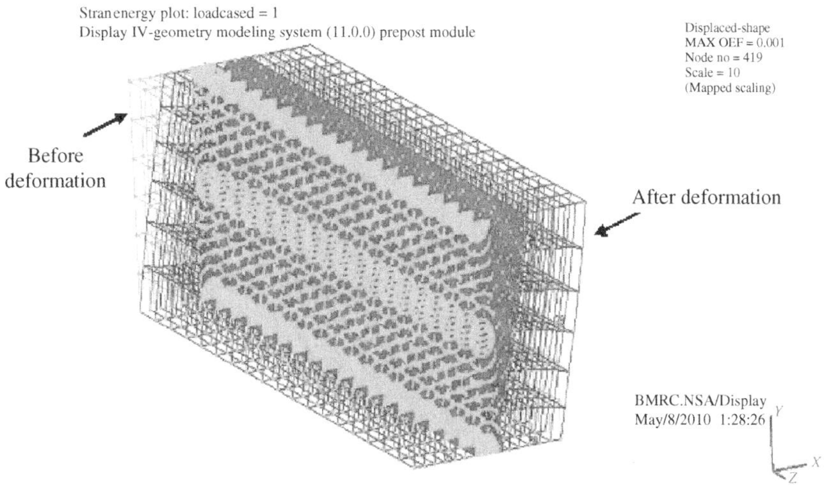

Figure 3.36 Displaced shape for shear loading (23) for $V_f = 0.03$ and $(l/d) = 10$. Source: Based on Weng [23].

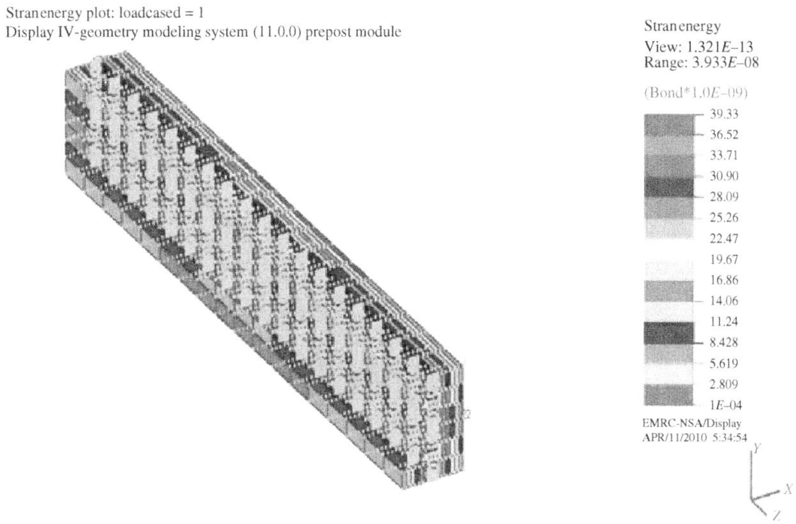

Figure 3.37 Strain energy plot for shear loading (23) for Type 1 RVE for $V_f = 0.03$ and $(l/d) = 20$. Source: Based on Weng [23].

2 nm. Then, there is a spacing of 2 nm between the first and the second fibers followed by a further spacing of 2 nm between the second fiber and the back face of the RVE. Here, the RVE consists of 2160 elements in the fiber and 7560 elements in the matrix. The total number of elements in the model is 9720 and the total number of nodes is 10 447. Strain energy shared by fiber and the matrix are shown below:

$$W_f = 3.28 \times 10^{-9}$$

$$W_m = 3.41 \times 10^{-5}$$

Total volume of the RVE, $V = 7.688 \times 10^3$

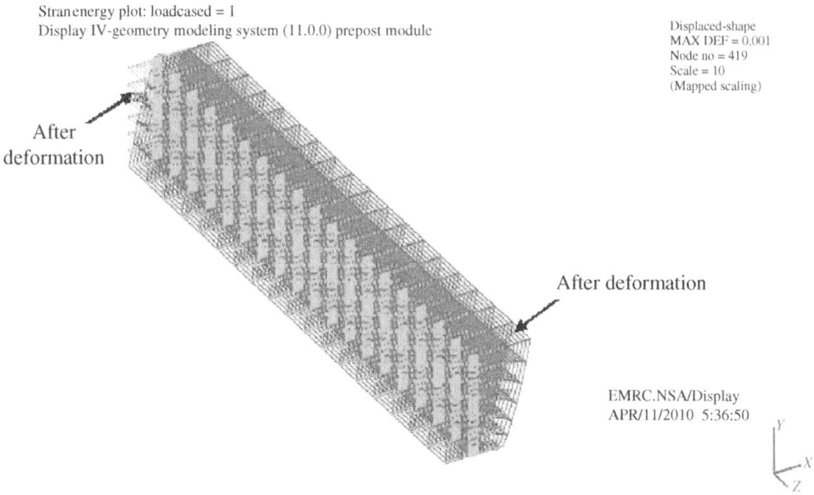

Figure 3.38 Displaced shape for shear loading (23) for Type 1 RVE for $V_f = 0.03$ and $(l/d) = 20$. Source: Based on Weng [23].

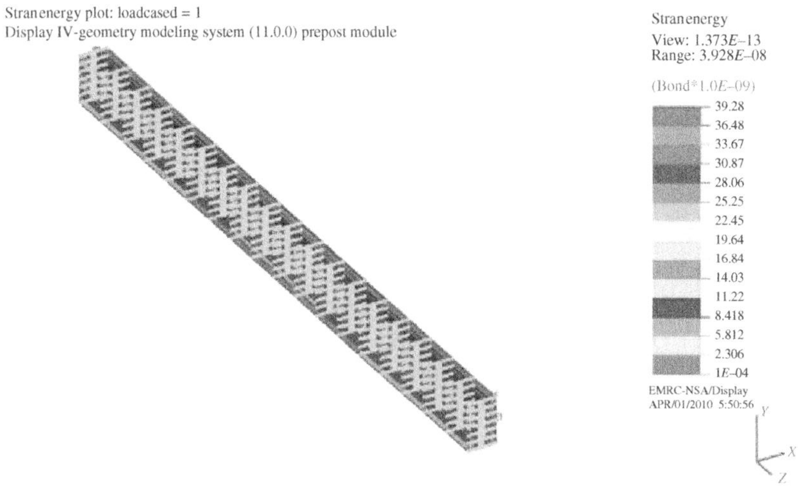

Figure 3.39 Strain energy plot for shear loading (23) for Type 1 RVE for $V_f = 0.03$ and $(l/d) = 50$. Source: Based on Weng [23].

Figure 3.42 shows the strain energy plot for shear loading (23), for Type 3 RVE, for $V_f = 0.04$ and fiber aspect ratio $(l/d) = 97$, with interfiber spacing of 6 nm. The number of nodes in the model is 49 205 and the total number of elements in the model is 13 824. The RVE consists of 5120 elements in the fiber and 8704 elements in the matrix. The strain energy shared by fiber and matrix are

$$W_f = 4.01 \times 10^{-7}$$

$$W_m = 1.42 \times 10^{-4}$$

Total volume of the RVE, $V = 3.801\,92 \times 10^3$

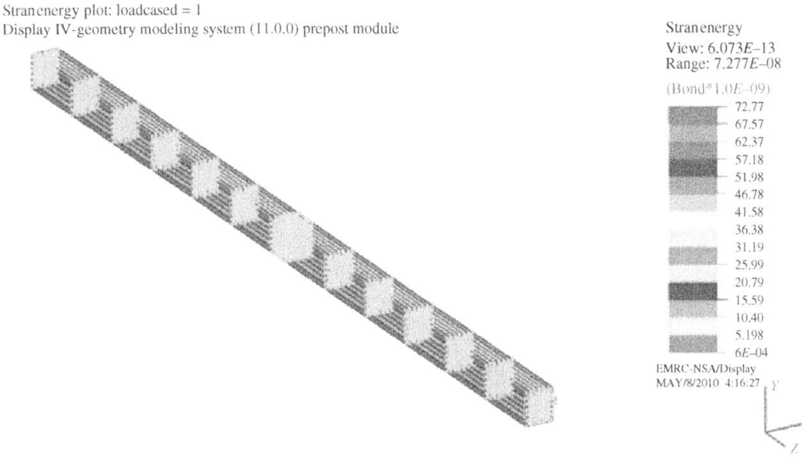

Figure 3.40 Strain energy plot for shear loading (23) for Type 2 RVE for $V_f = 0.03$ and $(l/d) = 97$ with interfiber spacing of 6 nm. Source: Based on Weng [23].

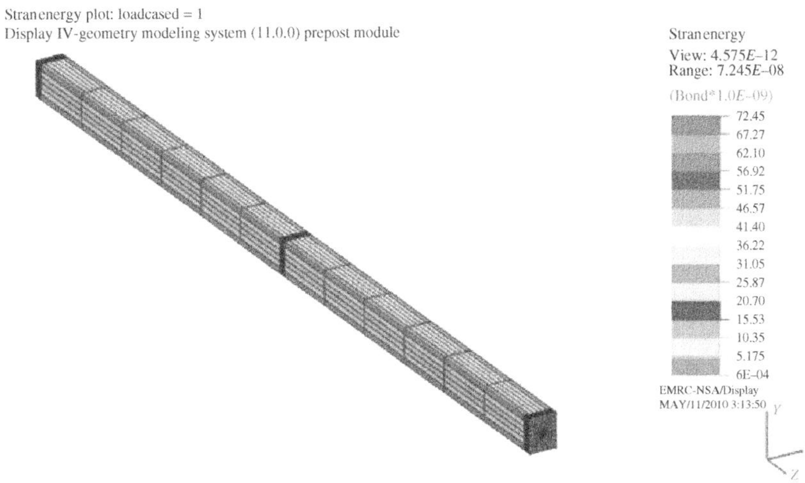

Figure 3.41 Strain energy plot for shear loading (23) for $V_f = 0.03$ and $(l/d) = 97$ for RVE of Type 3. Source: Based on Weng [23].

3.3.10.6 Modeling for G_{31}

Boundary conditions for the case of longitudinal shear loading (31) are shown in the Table 3.3. Figure 3.43 shows the boundary conditions on NISA. Left face of the RVE is fixed and the right face is given a small displacement (0.001) in the z-direction. Here, yellow colored dots represent fixed state and red-colored dots represent a small displacement in the z-direction. Figure 3.44 shows the strain energy plot for Type 1 RVE with a fiber volume fraction of 0.03 and an aspect ratio of 10. Here, the RVE consists of 12 312 elements and the number of elements in the fiber and matrix, respectively, is 4104 and 8208. The displaced shape for the case of shear loading (31) for the Figure 3.44 is shown in the Figure 3.45. Strain energy shared by fiber and the matrix are as shown below:

$$W_f = 4.42 \times 10^{-8}$$

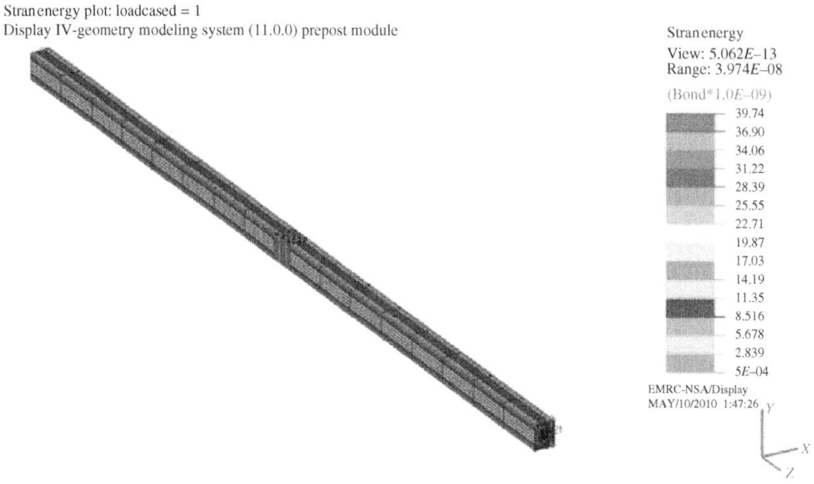

Figure 3.42 Strain energy plot for shear loading (23) for Type 2 RVE for $V_f = 0.04$ and $(l/d) = 97$ with interfiber spacing of 6 nm. Source: Based on Weng [23].

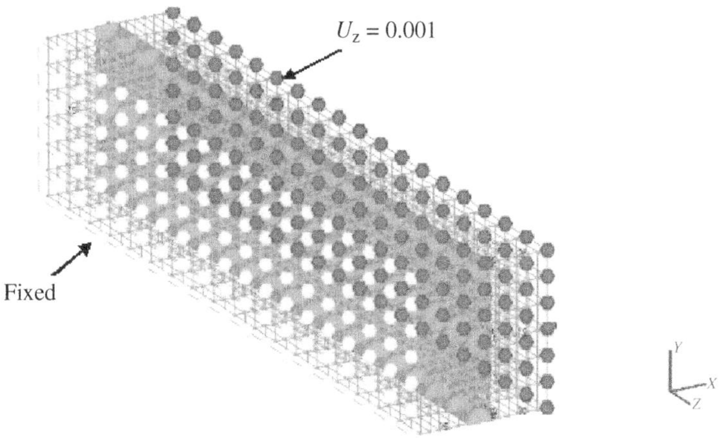

Figure 3.43 Boundary conditions for shear loading (31). Source: Based on Hashin and Shtrikman [31].

$$W_m = 1.65 \times 10^{-7}$$

$$\text{Total volume of the RVE, } V = 1.478\,94 \times 10^5$$

Figure 3.46 shows the strain energy plot for shear loading (31), for Type 1 RVE, having $V_f = 0.03$ and $l/d = 20$. The RVE consists of 12 312 elements and 13 699 nodes. The strain energy shared by fiber and matrix and the volume of the RVE are shown below:

$$W_f = 3.15 \times 10^{-9}$$

$$W_m = 1.09 \times 10^{-8}$$

$$V = 3.650\,40 \times 10^4$$

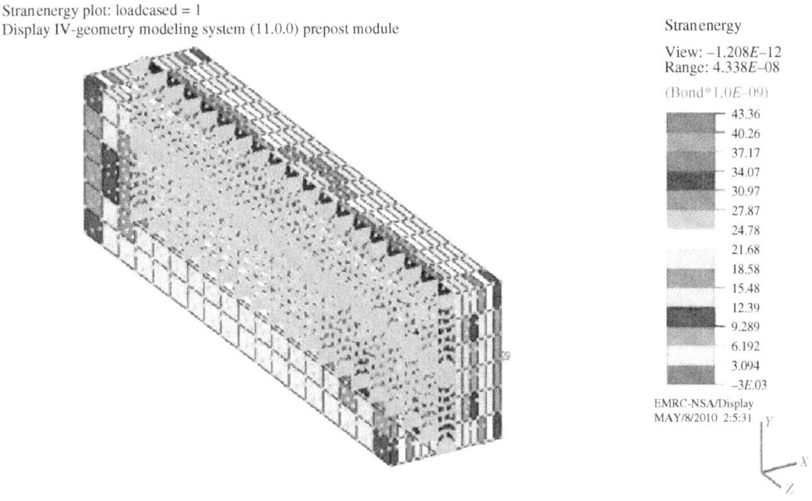

Figure 3.44 Strain energy plot for shear loading (31) for Type 1 RVE for $V_f = 0.03$ and $(l/d) = 10$. Source: Based on Hashin and Shtrikman [31].

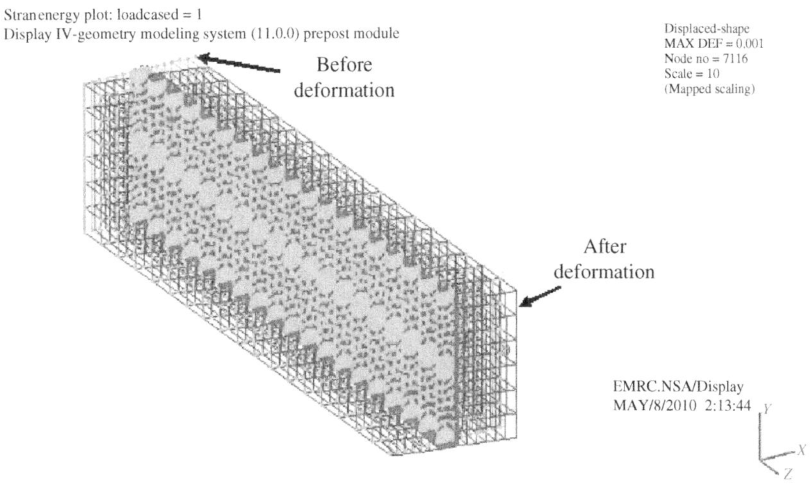

Figure 3.45 Displaced shape for shear loading (31) for $V_f = 0.03$ and $(l/d) = 10$. Source: Based on Hashin and Shtrikman [31].

Figure 3.47 shows the strain energy plot for shear loading (31), for Type 1 RVE, having $V_f = 0.03$ and $l/d = 50$. The RVE consists of 12 312 elements and 13 699 nodes. Figure 3.48 shows the strain energy plot, for shear loading (31), for Type 2 RVE, having an interfiber spacing of 6 nm, with fiber volume fraction of 0.03 and an aspect ratio of 97. The RVE consists of 5832 elements and 6403 nodes. There are 2160 elements in the fiber and 3672 elements in the matrix. The strain energy shared by fiber and the matrix for shear loading (31) are shown below:

$$W_f = 5.80 \times 10^{-9}$$

$$W_m = 4.92 \times 10^{-5}$$

$$V = 7.688 \times 10^{3}$$

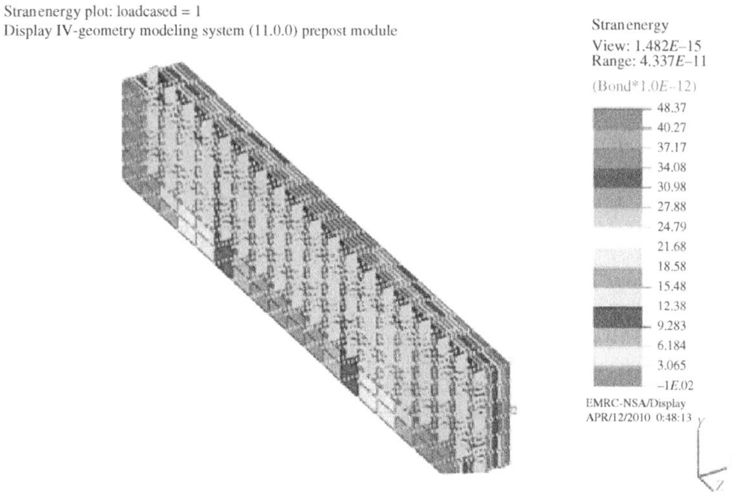

Figure 3.46 Strain energy plot for shear loading (31) for Type 1 RVE for $V_f = 0.03$ and $(l/d) = 20$. Source: Based on Hashin and Shtrikman [31].

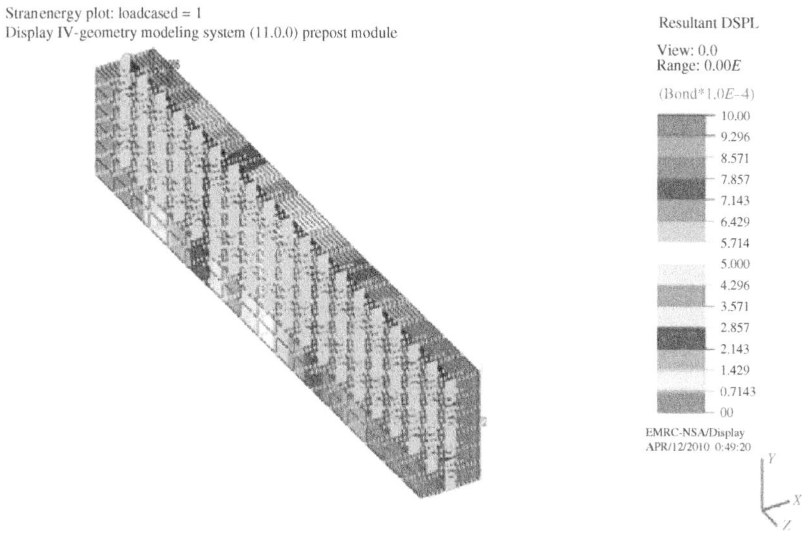

Figure 3.47 Strain energy plot for shear loading (31) for Type 1 RVE having $V_f = 0.03$ and $(l/d) = 50$. Source: Based on Hashin and Shtrikman [31].

Figure 3.49 shows another case of transverse shear loading (31) for Type 3 RVE having $V_f = 0.03$ and $l/d = 97$. Between the front face and the first fiber, there is a spacing of 2 nm. Then, there is a spacing of 2 nm between the first and the second fibers followed by a further spacing of 2 nm between the second fiber and the back face of the RVE. The RVE consists of 2160 elements in the fiber and 7560 elements in the matrix. The total number of elements in the model is 9720 and the total number of nodes is 10 447. The strain energy shared by fiber and the matrix are as shown below:

$$W_f = 5.89 \times 10^{-9}$$

$$W_m = 4.88 \times 10^{-5}$$

Total volume of the RVE, $V = 7.688 \times 10^3$

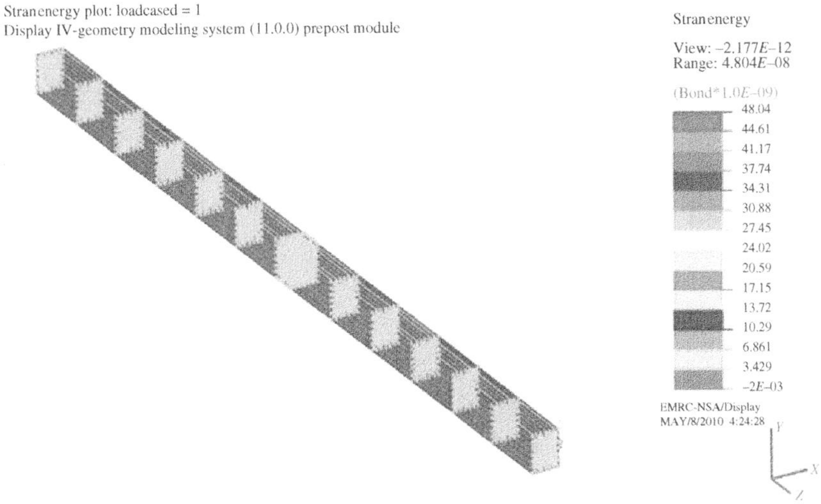

Figure 3.48 Strain energy plot for shear loading (31) for Type 2 RVE having $V_f = 0.03$ and $(l/d) = 97$ with interfiber spacing of 6 nm. Source: Based on Hashin and Shtrikman [31].

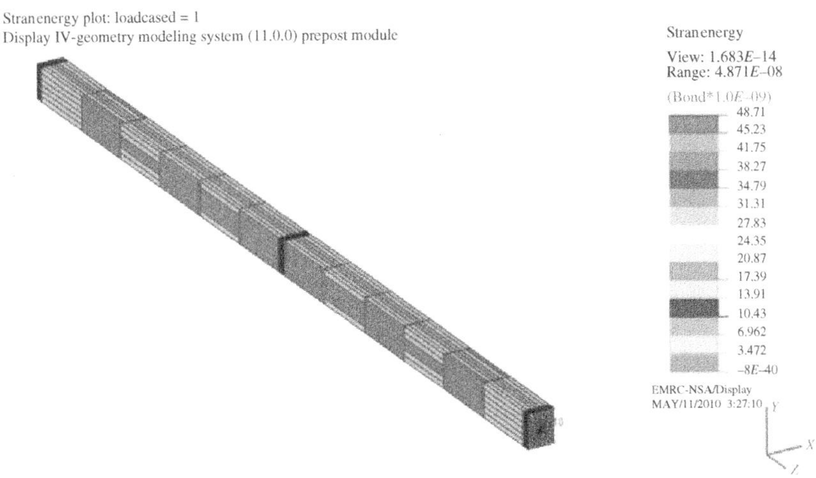

Figure 3.49 Strain energy plot for shear loading (31) for $V_f = 0.03$ and fiber aspect ratio $(l/d) = 97$ with interfiber spacing of Type 3. Source: Based on Hashin and Shtrikman [31].

Figure 3.50 shows the strain energy plot for shear loading (31) for Type 2 RVE, having $V_f = 0.04$ and fiber aspect ratio $(l/d) = 97$, with interfiber spacing of 6 nm. The RVE consists of 49 205 nodes and 13 824 elements. There are 5120 elements in the fiber and 8704 elements in the matrix. The strain energy shared by fiber and matrix are shown below:

$$W_f = 1.16 \times 10^{-8}$$

$$W_m = 5.12 \times 10^{-5}$$

$$\text{Total volume of the RVE, } V = 3.801\,92 \times 10^3$$

Shear loading (12) is similar to (31) as we have considered transverse isotropy.

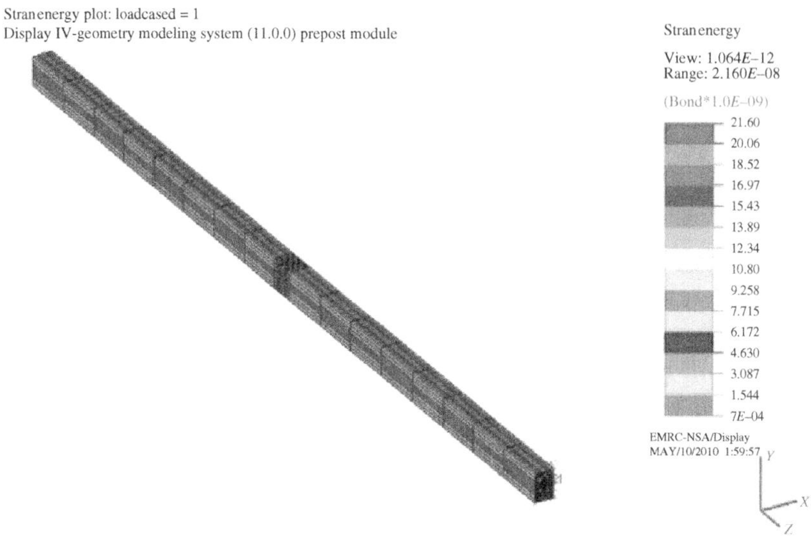

Figure 3.50 Strain energy plot for longitudinal shear loading (31) for $V_f = 0.04$ and fiber aspect ratio (l/d) = 97 with interfiber spacing of 6 nm. Source: Based on Hashin and Shtrikman [31].

3.3.10.7 Theoritical Formulation

The following theoretical formulation is used for predicting the properties of VGCF/pp nanocomposites:

$$\text{Storage modulus } (E') = \frac{2 \times W}{V \times \varepsilon^2} \tag{3.133}$$

where

W = Total strain energy
V = Total volume of the RVE
ε = strain (change in length per unit original length)

$$\text{Loss factor } (\eta) = \frac{\eta_f W_f + \eta_m W_m}{W} \tag{3.134}$$

where

η_f = Loss factor of fiber
η_m = Loss factor of matrix
W_f = Strain energy in the fiber
W_m = Strain energy in the matrix
W = Total strain energy

$$\text{Loss modulus } (E'') = \eta \times E' \tag{3.135}$$

Thus, the FEM results obtained through NISA software are used in Eqs. (3.133)–(3.135) for determining the moduli and corresponding loss factors.

3.3.10.8 Comparison of Results

Figures 3.51–3.66 show the comparison of results obtained from FEM for three different types of RVEs. These results are explained below:

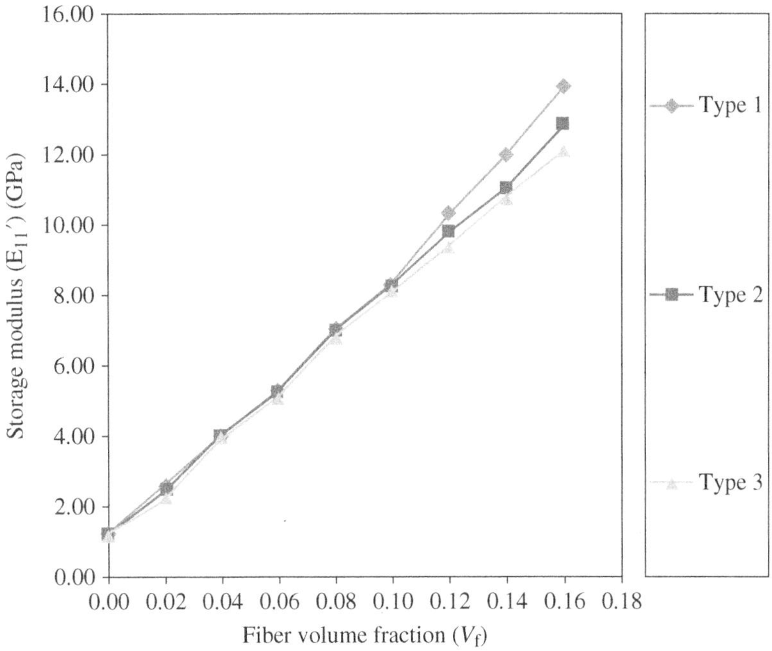

Figure 3.51 Variation of E'_{11} with V_f for $l/d = 19$ for three different types of RVEs.

(i) Storage moduli

(a) *Variation of* E'_{11}: Figure 3.51 shows the variation of E'_{11} with V_f for $l/d = 19$. It can be observed that E'_{11} increases with V_f for all types of RVE's. Upto $V_f = 0.10$, the three types of RVEs give almost same values of E'_{11} and thereafter the Type 1 RVE gives higher values as compared to Type 2 and Type 3. For Type 1 RVE, E'_{11} increases 332.5% from $V_f = 0$ to 0.06 and from $V_f = 0.06$ to 0.16, the increase in E'_{11} reduces to 162.4%. At $V_f = 0.06$, the difference between E'_{11} values for Type 1 and Type 2 RVE's is 1.879%. This difference increases to 7.87% at $V_f = 0.16$. Similarly, the difference between E'_{11} values for Type 1 and Type 3 RVEs at $V_f = 0.06$ is 3.759%, and this difference increases to 13.18% at $V_f = 0.16$. The longitudinal storage modulus (E'_{11}) increases with the fiber volume fraction (V_f), for a constant fiber aspect ratio ($l/d = 19$), as shown in the Figure 3.51. This is due to the fact that the fibers are aligned in the longitudinal direction. As V_f increases for a constant aspect ratio ($l/d = 19$), the volume of the fibers is increasing. Since, the fibers have high modulus in the longitudinal direction, the composite has improved properties in the said direction. Staggered array of fibers gives higher values of storage modulus as compared to regular array. The higher values probably occur because this type of RVE tends to create long chains of nearly touching fibers with a high degree of overlap. Figure 3.52 shows the variation of E'_{11} with fiber aspect ratio (l/d) for $V_f = 0.03$ for three different types of RVEs. It shows that the RVE of Type 1, which has a staggered array of fibers, gives higher modulus as compared to the other two arrangements, i.e. Type 2 and Type 3, which is due to better interaction between fiber–matrix and fiber–fiber interaction. Further, it can be observed that E'_{11} increases sharply upto $l/d = 100$, which is the same for all types of RVEs. However, for $l/d > 100$, E'_{11}

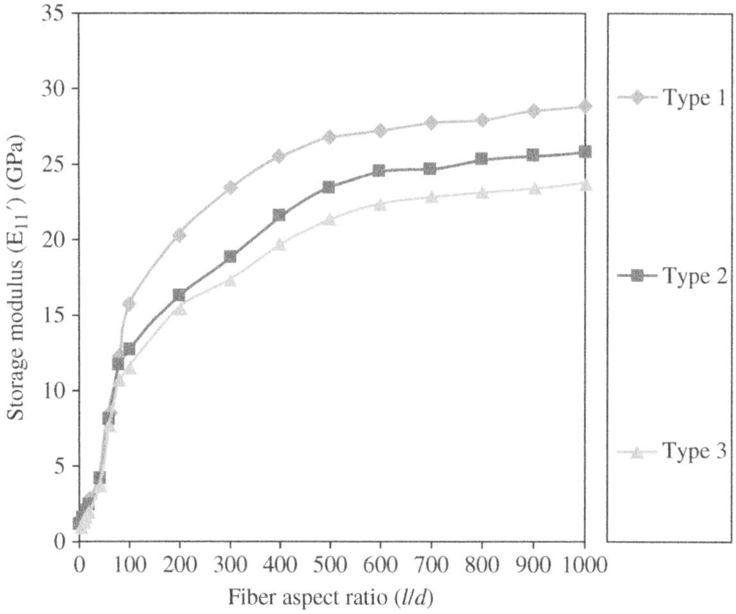

Figure 3.52 Variation of E'_{11} with l/d for $V_f = 0.03$ for three different types of RVEs.

increases slowly and is higher for the Type 1 RVE as compared to Type 2 and Type 3. It can be observed that at $l/d = 100$, the percentage difference between E'_{11} values for FEM models of Type 1 and Type 2, is 18.98%. The percentage difference between E'_{11} values for FEM models of Type 1 and Type 3 is 25.94%. At $l/d = 1000$, the differences reduce, respectively, to 10.38% and 17.64%. From Figure 3.52, it can also be observed that for l/d = 5 to 100, the percentage increase in storage modulus E'_{11} is 939% for Type 1 RVE, whereas the increase is only 82.9% from l/d = 100 to 1000.

(b) *Variation of* E'_{22}: Figure 3.53 shows the variation of E'_{22} with V_f for $l/d = 19$. The results again show that the RVE having staggered array of fibers gives higher modulus as compared to the other two arrangements. The percentage increase in E'_{22}, from $V_f = 0$ to 0.06, for Type 1 RVE is 38.21%, whereas from $V_f = 0.06$ to 0.16, the percentage increase in E'_{22} is only 18.90%. At $V_f = 0.06$, the difference between E'_{22} values for Type 1 and Type 2 RVEs is 1.897%. This difference increases to 3.17% at $V_f = 0.16$. Similarly, the difference between E'_{22} values for Type 1 and Type 3 RVEs at $V_f = 0.06$ is 5.6%, and this difference increases to 6.133% at $V_f = 0.16$. E'_{22} values are considerably lower than the corresponding values of E'_{11}. This is due to the fact that in transverse direction, the properties of the composite are matrix dominated and hence, lower values of E'_{22}.

Figure 3.54 shows the variation of E'_{22} with l/d for $V_f = 0.03$. The results show that the transverse storage modulus decreases rapidly with l/d till $l/d < 100$. Thereafter, the decrease in E'_{22} with increase in l/d, for $l/d > 100$, is very slow and become insensitive to l/d at large aspect ratios. At l/d = 100, the percentage difference between the E'_{22} values for FEM models of Type 1 and 2, is 0.003 58%, while at $l/d = 1000$, the difference reduces to 0.000 057 9%. Thus, it can be seen that at large l/d ratios, E'_{22} is insensitive to the variation in l/d. From l/d = 5 to 100, the percentage decrease

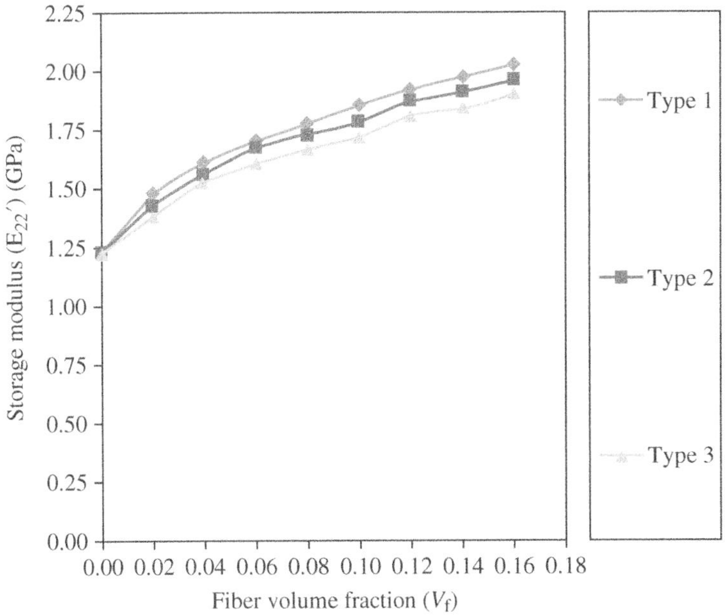

Figure 3.53 Variation of E'_{22} with V_f for $l/d = 19$ for three different types of RVEs.

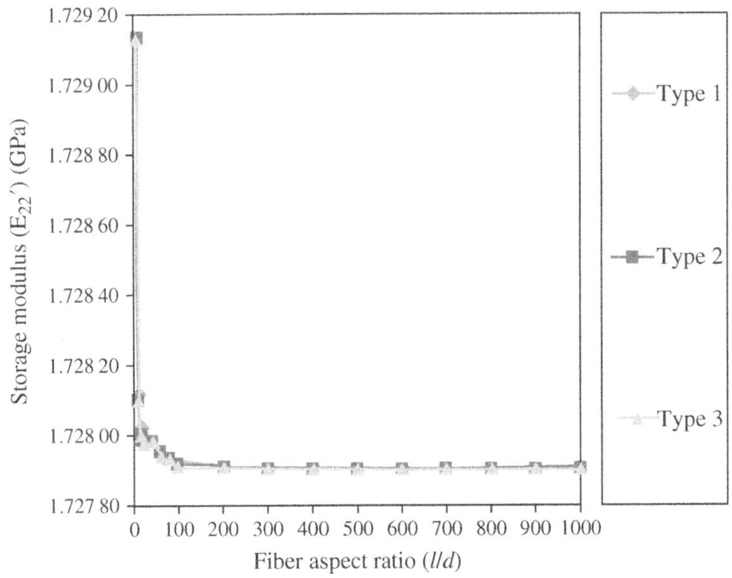

Figure 3.54 Variation of E'_{22} with l/d for $V_f = 0.03$ for three different types of RVEs.

in E'_{22} for Type 1 RVE is 0.069% and from $l/d = 100$ to 1000, percentage decrease in E'_{22} reduces to 0.0013%. A probable reason for rapid fall of storage moduli till $l/d = 100$ is increasing the fiber length decreases the number of fiber ends and thus the amount of matrix between the fiber ends. Since, the amount of matrix is fixed, it

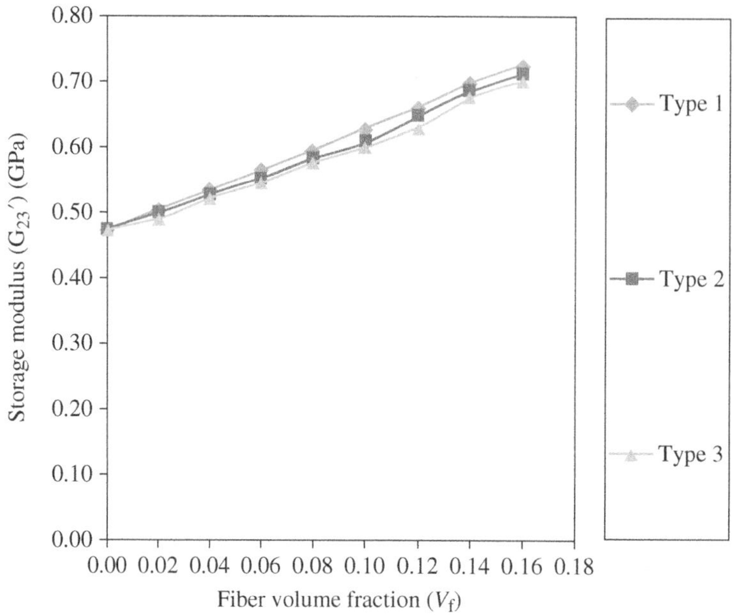

Figure 3.55 Variation of G'_{23} with V_f for $l/d = 19$ for three different types of RVEs.

moves toward the space between the fibers and increases the interfiber spacing and thus decreases the values of E'_{22}.

(c) *Variation of* G'_{23}: Figure 3.55 shows the variation of transverse storage shear modulus (G'_{23}) with V_f for $l/d = 19$. It shows that the shear modulus (G'_{23}) also increases with V_f. The increase is small in comparison to the increase in the longitudinal storage modulus values. Here also, the RVE having staggered array of fibers gives higher modulus as compared to the other two arrangements. The percentage increase in G'_{23}, for Type 1 RVE, from $V_f = 0$ to 0.06, is 19.2389%, whereas the percentage increase in G'_{23} from $V_f = 0.06$ to 0.16 is 28.43%. At $V_f = 0.06$, the difference between G'_{23} values for Type 1 and Type 2 RVEs is 2.496%. This difference increases to 2.58% at $V_f = 0.16$. Similarly, the difference between G'_{23} values for Type 1 and Type 3 RVEs at $V_f = 0.06$ is 3.49%, and this difference increases to 3.52% at $V_f = 0.16$. G'_{23} values are considerably lower than the corresponding values of E'_{11}. This is again due to the fact that in transverse shear, the properties of the composite are matrix dominated and hence, lower values of G'_{23}.

Figure 3.56 shows the variation of G'_{23} with l/d for constant $V_f = 0.03$. The results obtained show that the trend of variation of G'_{23} with l/d is same as that of variation of E'_{22} with l/d. G'_{23} also decreases with l/d till $l/d < 100$ and thereafter, almost become constant. At $l/d = 100$, the difference between the G'_{23} values for FEM models of Type 1 and 2 is 0.009 179%, whereas at $l/d = 1000$, the difference reduces to 0.0098%, again showing that l/d has little effect on shear storage modulus. The storage modulus G'_{23} decreases to 0.337% for an increase in aspect ratio from $l/d = 5$ to 100, whereas G'_{23} decreases to 0.002 12% for an aspect ratio increase from $l/d = 100$ to 1000. Reason for rapid fall in G'_{23} values till $l/d = 100$ is the same as that for E'_{22}.

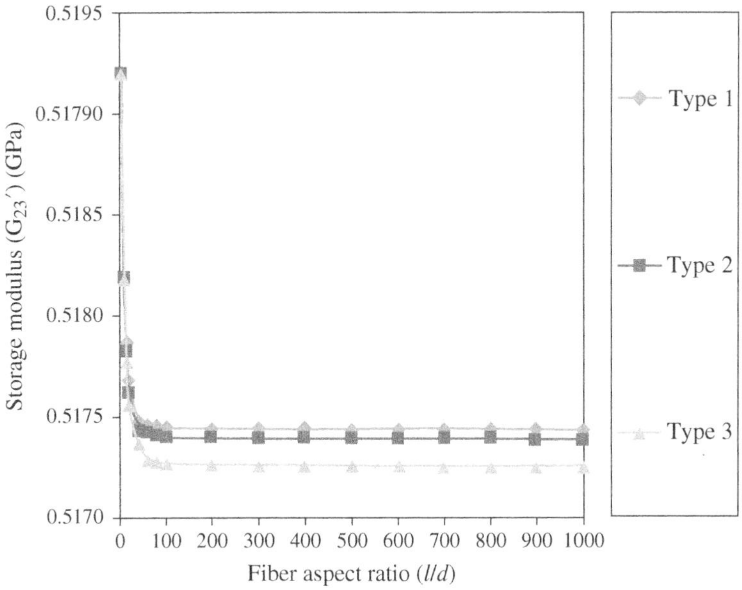

Figure 3.56 Variation of G'_{23} with l/d for $V_f = 0.03$ for three different types of RVEs.

(d) *Variation of* G'_{12}: Figure 3.57 shows the variation of G'_{12} with V_f for $l/d = 19$. The trend is similar to that of Figure 3.55. Storage modulus (G'_{12}) increases to 25.53% for Type 1 RVE, for an increase in fiber volume fraction from $V_f = 0$ to 0.06. Similarly, for an increase in fiber volume fraction from $V_f = 0.06$ to 0.16, G'_{12} increases to 35.59%. At $V_f = 0.06$, the difference between G'_{12} values for Type 1 and Type 2 RVEs is 3.62%. This difference increases to 3.75% at $V_f = 0.16$. Similarly, the difference between G'_{12} values for Type 1 and Type 3 RVE's at $V_f = 0.06$ is 6.20%, and this difference increases to 6.25% at $V_f = 0.16$.

Figure 3.58 shows the variation of G'_{12} with l/d for $V_f = 0.03$. Here also, G'_{12} decreases rapidly with l/d till $l/d < 100$ and then becomes almost constant. For Type 1 RVE, G'_{12} decreases to 0.609% for an increase in fiber aspect ratio from $l/d = 5$ to 100, whereas from $l/d = 100$ to 1000, the decrease in G'_{12} reduces to 0.001%. At $l/d = 100$, the difference between the G'_{12} values for FEM models of Type 1 and 2 is 0.0377%, whereas at $l/d = 1000$, the difference reduces to 0.0328%, again showing that l/d has little effect on shear storage modulus. Similarly, the difference between G'_{12} values for Type 1 and Type 3 RVEs at $l/d = 100$ is 0.094 28%, and this difference decreases to 0.094 23% at $l/d = 1000$.

(ii) Loss factors

(a) *Variation of* η_{11}: Figure 3.59 shows the variation of η_{11} with fiber volume fraction (V_f) for $l/d = 19$ for three different types of RVEs which indicates inverse relation vis-à-vis E'_{11}. The loss factor η_{11} decreases to 5.422% for Type 1 RVE, for an increase in fiber volume fraction from $V_f = 0$ to 0.06 and from $V_f = 0.06$ to 0.16, the decrease in η_{11} reduces to 1.8%. At $V_f = 0.06$, the percentage difference between η_{11} values for Type 1 and Type 2 RVEs is 0.266%. This difference increases to 0.6065% at $V_f = 0.16$. Similarly, at $V_f = 0.06$, the percentage difference between η_{11} values for Type 1 and

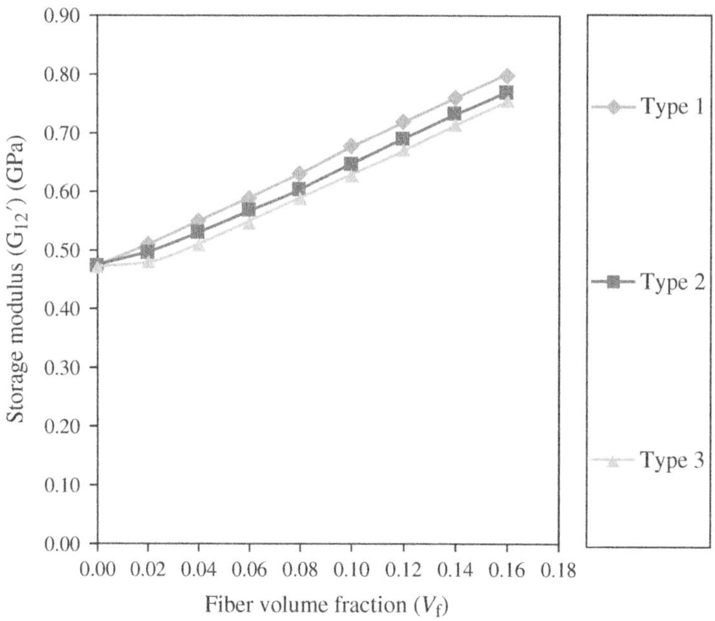

Figure 3.57 Variation of G'_{12} with V_f for $l/d = 19$ for three different types of RVEs.

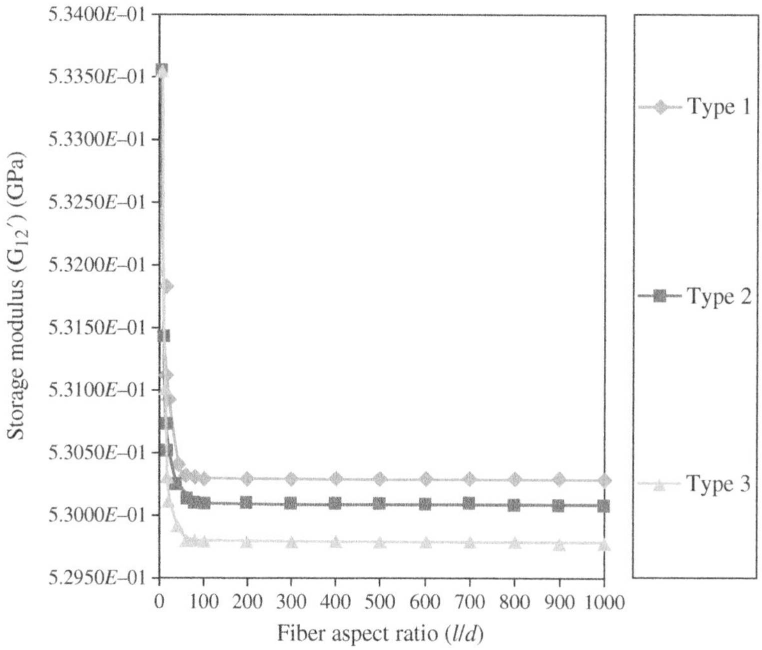

Figure 3.58 Variation of G'_{12} with l/d for $V_f = 0.03$ for three different types of RVEs.

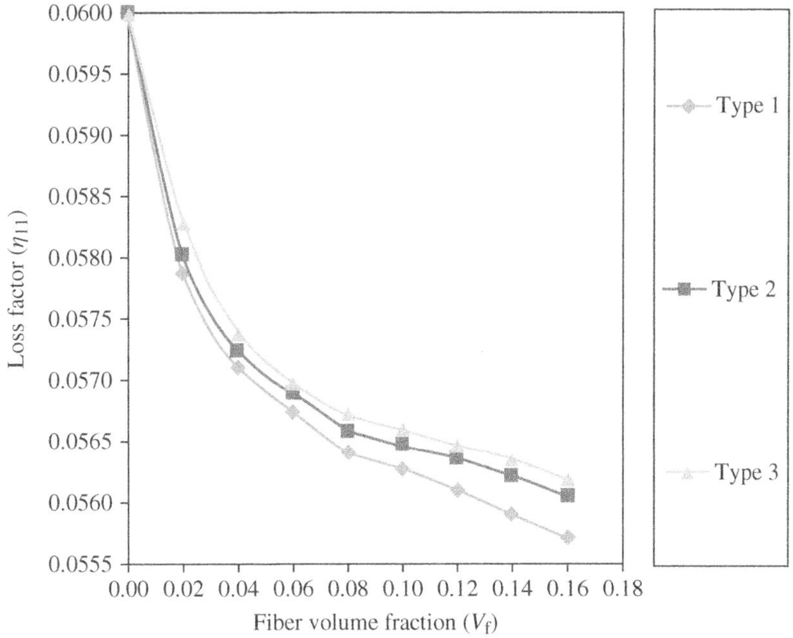

Figure 3.59 Variation of η_{11} with V_f for $l/d = 19$ for three different types of RVEs.

Type 3 RVE's is 0.415%. This difference increases to 0.897% at $V_f = 0.16$. The loss factor η_{11} is decreasing with V_f for a constant aspect ratio of $l/d = 19$.

Figure 3.60 shows the variation of η_{11} with fiber aspect ratio for $V_f = 0.03$ which also indicates inverse relation vis-à-vis E'_{11}. The loss factor η_{11} decreases sharply upto $l/d = 100$ and thereafter, it decreases slowly. The loss factor η_{11} predicted by all types of RVEs is almost the same. Loss factor η_{11} decreases 75.10% from l/d = 5 to 200, whereas the decrease from l/d = 200 to l/d = 1000 is 64.28%. It can be observed that at l/d = 100, the percentage difference between η_{11} values for FEM models of Type 1 and Type 2 is 0.4311%. The percentage difference between η_{11} values for FEM models of Type 1 and Type 3 is 1.8046%. At $l/d = 1000$, differences reduce to 0.245% and 0.174%, respectively.

(b) *Variation of* η_{22}: Figure 3.61 shows the variation of η_{22} with V_f for $l/d = 19$ which indicates inverse relation vis-à-vis E'_{22}. The percentage decrease in η_{22} from $V_f = 0$ to 0.06 for Type 1 RVE is 0.0149%, and from $V_f = 0.06$ to 0.16, the percentage decrease in η_{22} increases to 0.0298%. At $V_f = 0.06$, the percentage difference between η_{22} values for Type 1 and Type 2 RVEs is 0.001 16%. This difference increases to 0.001 86% at $V_f = 0.16$. Similarly, at $V_f = 0.06$, the percentage difference between η_{22} values for Type 1 and Type 3 RVEs is 0.002 17%. This difference increases to 0.003 53% at $V_f = 0.16$.

Figure 3.62 shows the variation of η_{22} with l/d for a constant fiber volume fraction (V_f) of 0.03. The results show that the transverse loss factor (η_{22}) increases rapidly with l/d till $l/d < 100$. Thereafter, the increase in η_{22} with increase in l/d, for $l/d > 100$, is very slow and become insensitive to l/d at large aspect ratios. At $l/d = 100$, the percentage difference between the η_{22} values for FEM models of Type 1 and 2 is 1.85×10^{-5} %, while at $l/d = 1000$, the difference reduces to 1.83×10^{-5}%.

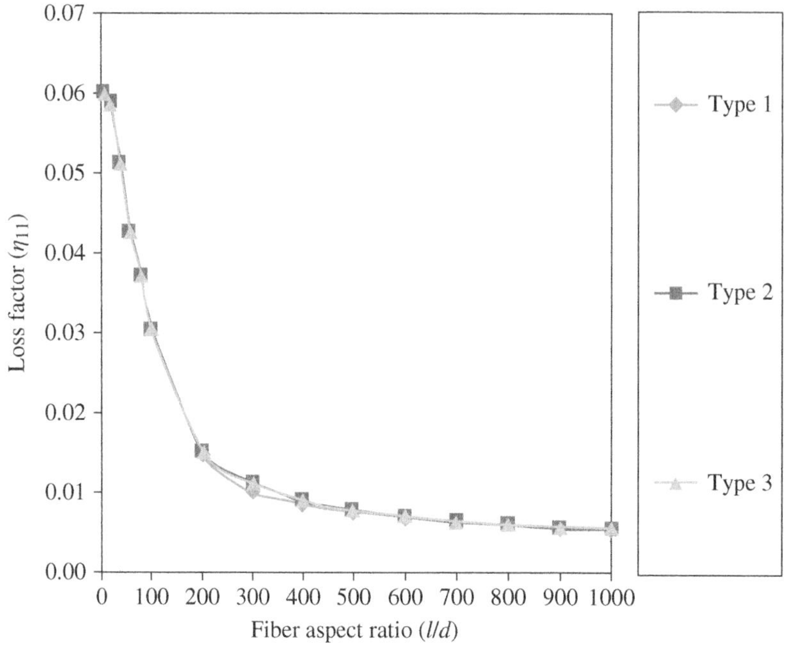

Figure 3.60 Variation of η_{11} with l/d for $V_f = 0.03$ for three different types of RVEs.

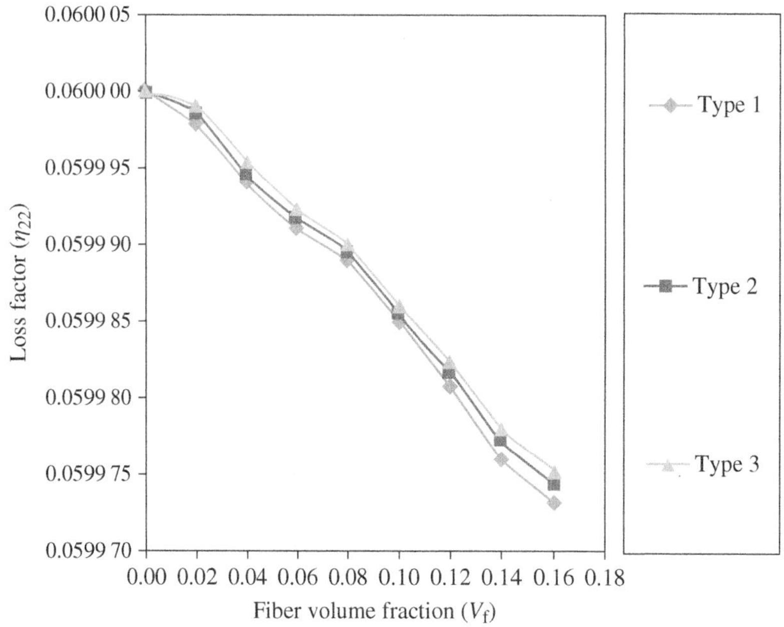

Figure 3.61 Variation of η_{22} with V_f for $l/d = 19$ for three different types of RVEs.

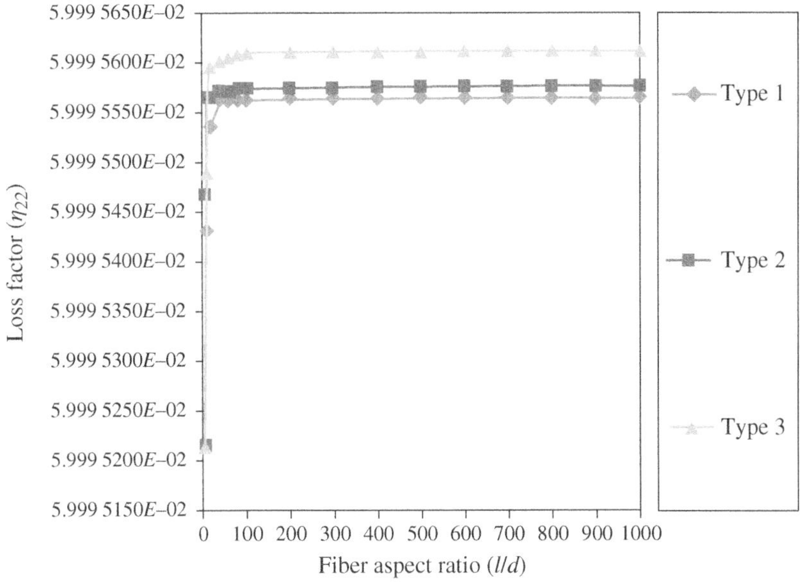

Figure 3.62 Variation of η_{22} with l/d for $V_f = 0.03$ for three different types of RVEs.

Similarly, at $l/d = 100$, the percentage difference between the η_{22} values for FEM models of Type 1 and 3 is $7.833 \times 10^{-5}\%$, while at $l/d = 1000$, the difference reduces to $7.832 \times 10^{-5}\%$. Thus, it can be seen that at large l/d ratios, E'_{22} is insensitive to the variation in l/d. From $l/d = 5$ to 100, the percentage decrease in E'_{22} for Type 1 RVE is 0.069%, and from $l/d = 100$ to 1000, the percentage decrease in E'_{22} reduces to 0.0013%.

(c) *Variation of* η_{23}: Figure 3.63 shows the variation of loss factor (η_{23}) with V_f for $l/d = 19$ which indicates inverse relation vis-à-vis G'_{23}. It shows that the shear loss factor (η_{23}) also decreases with V_f. The decrease is small in comparison to the decrease in the longitudinal loss factor (η_{11}) values. Since RVE having staggered array of fibers gives higher modulus as compared to the other two arrangements; hence, loss factor values will be lower for Type 1 arrangement in comparison to Type 2 and Type 3 RVEs. The percentage decrease in η_{23} for Type 1 RVE from $V_f = 0$ to 0.06 is 0.0633%, whereas the percentage decrease in η_{23} from $V_f = 0.06$ to 0.16 is 0.0867%. At $V_f = 0.06$, the difference between η_{23} values for Type 1 and Type 2 RVEs is 0.00696%. This difference increases to 0.00716% at $V_f = 0.16$. Similarly, the difference between η_{23} values for Type 1 and Type 3 RVEs at $V_f = 0.06$ is 0.0176% and this difference increases to 0.0184% at $V_f = 0.16$.

Figure 3.64 shows the variation of η_{23} with l/d for $V_f = 0.03$. For Type 1 RVE, loss factor η_{23} increases to 0.002 78% for an increase in aspect ratio from $l/d = 5$ to 100, whereas for an aspect ratio increase from $l/d = 100$ to 1000, the increase in η_{23} reduces to 0.0001%. The results show that the loss factor (η_{23}) increases rapidly with l/d till $l/d < 100$. Thereafter, the increase in η_{23} with increase in l/d, for $l/d > 100$, is very slow and become insensitive to l/d at large aspect ratios. At $l/d = 100$, the percentage difference between the η_{23} values for FEM models of Type 1 and 2 is $1.65 \times 10^{-4}\%$, while at $l/d = 1000$, the difference reduces to $1.286 \times 10^{-4}\%$. Similarly, at $l/d = 100$, the percentage difference between the η_{23} values for FEM

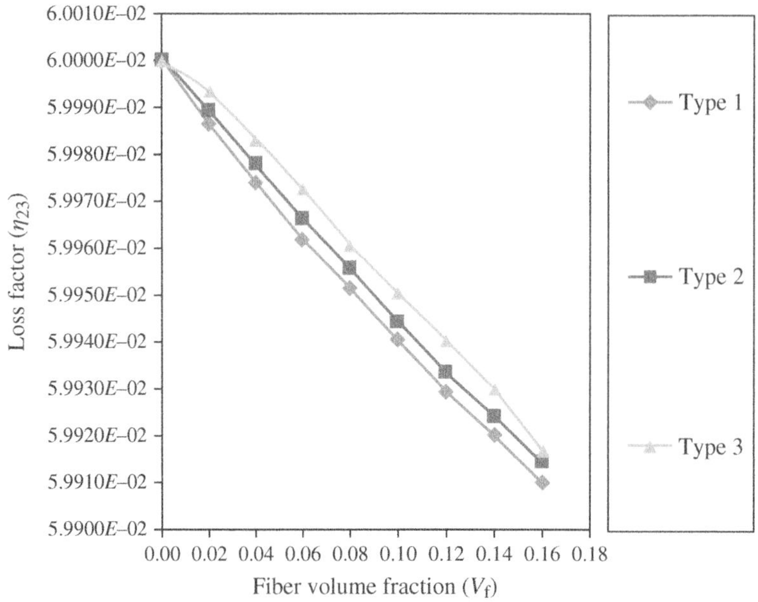

Figure 3.63 Variation of η_{23} with V_f for $l/d = 19$ for three different types of RVEs.

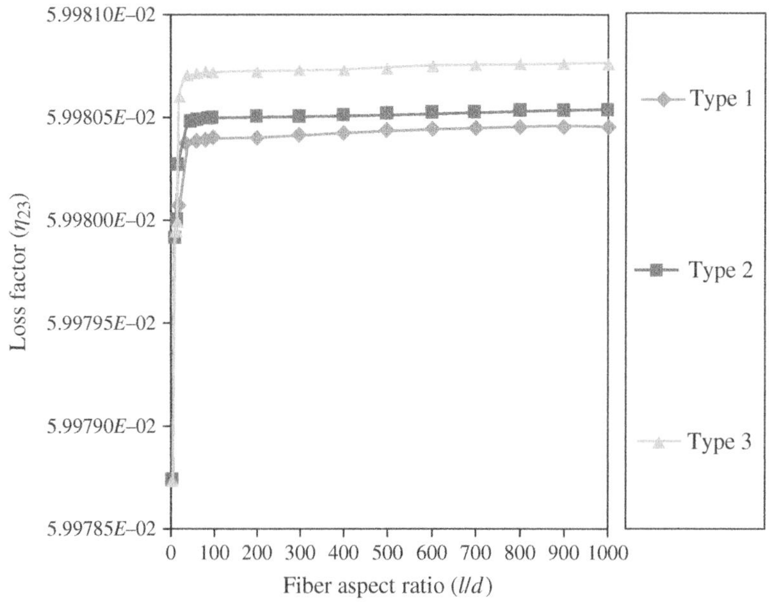

Figure 3.64 Variation of η_{23} with l/d for $V_f = 0.03$ for three different types of RVEs.

models of Type 1 and 3 is $5.404 \times 10^{-4}\%$, while at $l/d = 1000$, the difference reduces to $5.060 \times 10^{-4}\%$. The results obtained show that the trend of variation of η_{23} with l/d is same as that of variation of η_{22} with l/d.

(d) *Variation of η_{12}*: Figure 3.65 shows the variation of η_{12} with V_f for $l/d = 19$ which indicates inverse relation vis-à-vis G'_{12}. Loss factor (η_{12}) decreases to 0.10% for Type 1 RVE for an increase in fiber volume fraction from $V_f = 0$ to 0.06. Similarly, for an

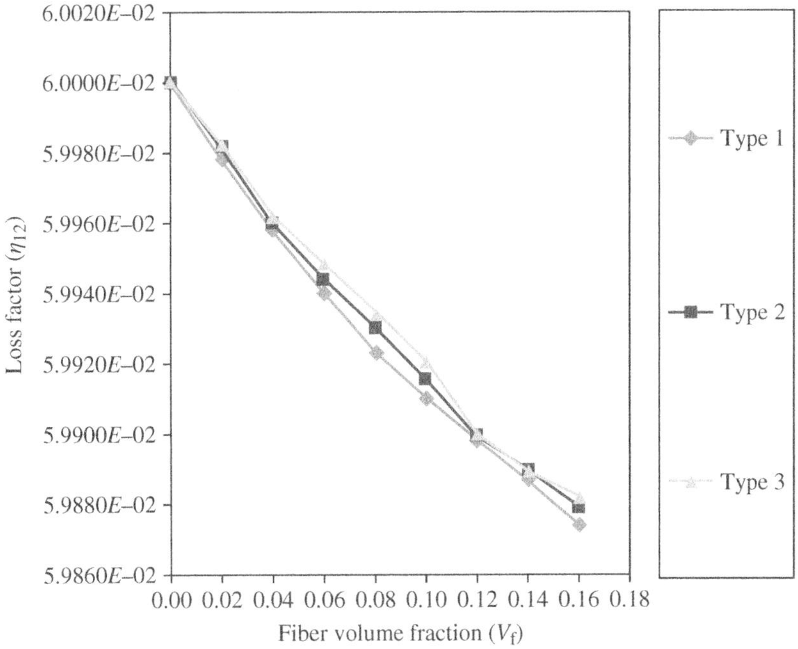

Figure 3.65 Variation of η_{12} with V_f for $l/d = 19$ for three different types of RVEs.

increase in fiber volume fraction from $V_f = 0.06$ to 0.16, η_{12} decreases to 0.1099% for Type 1 RVE. At $V_f = 0.06$, the difference between η_{12} values for Type 1 and Type 2 RVE's is 0.006 84%. This difference increases to 0.009 02% at $V_f = 0.16$. Similarly, the difference between η_{12} values for Type 1 and Type 3 RVEs at $V_f = 0.06$ is 0.0138%, and this difference increases to 0.0139% at $V_f = 0.16$.

Figure 3.66 shows the variation of η_{12} with l/d for $V_f = 0.03$. For Type 1 RVE, η_{12} increases to 0.006% for an increase in fiber aspect ratio from $l/d = 5$ to 100, whereas from $l/d = 100$ to 1000, η_{12} increases to 0.000 18% for Type 1 RVE. At $l/d = 100$, the percentage difference between the η_{12} values for FEM models of Type 1 and 2 is 6.168×10^{-4}%, while at $l/d = 1000$, the difference reduces to 5.921×10^{-4}%. Similarly, at $l/d = 100$, the percentage difference between the η_{12} values for FEM models of Type 1 and 3 is 9.5×10^{-4}%, while at $l/d = 1000$, the difference reduces to 9.25×10^{-4}%. The results obtained show that the trend of variation of η_{12} with l/d is same as that of variation of η_{22} with l/d.

3.4 Fast Fourier Transform Numerical Homogenization Methods

The properties of composites not only depend upon the constituent materials but also on their microstructures. Some paradigms are a diversity of biological materials which can assemble a few ingredients like carbon and minerals into a large number of structures featured with outstanding properties and multifunctionalities. The effective properties of such composites can be estimated by theories like Hashin–Shtrikman bounds and Budiansky bounds without the consideration of structural complexity. In order to design advanced materials,

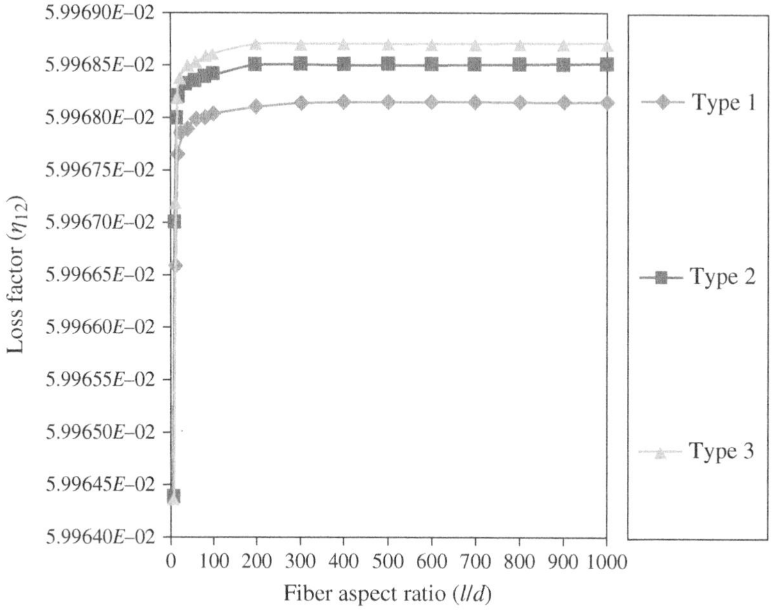

Figure 3.66 Variation of η_{12} with l/d for $V_f = 0.03$ for three different types of RVEs.

however, it is highly desirable to know their accurate properties after the microstructure is determined. Based on Eshelby's equivalent inclusion theory and combined with the average stress, some approaches such as self-consistent method, generalized self-consistent method, and Mori–Tanaka method have been proposed to predict the effective properties of composites. Initially, these methods are only applicable to composites with ellipsoid-shaped inclusions because this theory is based on an ellipsoid. With a step-by-step analysis, these methods were improved to be used for multiphase composite and graphene sheets dispersed in polymer nanocomposites [47, 48].

For composites in more complex shapes, their RVE (unit cell) was simulated by FEA with periodic boundary conditions. By averaging the local stress and the local strain within the unit cell, some mechanical properties can be obtained straightforward or in two steps. Virtual element method has also been developed to retrieve the effective properties for heterogeneous materials [49]. However, these FEA-based methods lack a rigorous mathematic framework to explicitly illuminate the essential relationships between the effective properties and material layouts. In 1978, a homogenization theory was proposed to evaluate the effective properties of composites by assuming the microdisplacement of their unit cell is governed by a characteristic equation [50, 51]. When the size of the unit cell tends to be infinitesimally small, the effective elastic tensor can be approximated by integrating the difference between test strains and microstrains. This method has been successfully used for porous materials, lattice, and fiber-reinforced composites. Because a well-defined formula is given, the sensitivity of the effective properties with respect to local density, namely the ratio of solid material occupation in an element, can be derived. Such a sensitivity analysis has been used in structural topology optimization to systematically design advanced materials with specified properties or extreme properties on Hashin–Shtrikman bounds. Very recently, a novel homogenization method based on enriched microinertial continuum was explored for resonant meta-materials to analyze their negative stiffness and negative effective mass density [52]. Meanwhile, the effective medium model of periodic composites was analyzed by a multiscale nonlocal homogenization method [53].

To investigate the effective properties of linear and nonlinear periodic materials as well as their local response, a homogenization method based on fast Fourier transform (FFT) was developed by Moulinec and Suquet in 1994 [54]. Later, the authors extended this method to complex microstructures under nonlinear plastic deformation by imposing macroscopic stress [55]. Though this method works extremely well for composites with small contrasts in Young's modulus of constitutive materials, it converges slowly for porous materials with infinite ratio of stiffness [56]. To remedy such an insufficiency, an augmented Lagrangian method with the consideration of local strain potential was developed [57]. A conjugate gradients method [58, 59] and a polarization FFT iterative scheme based on energy principle [60, 61] were also established to solve this problem. To improve the efficiency of FFT-based homogenization, an accelerated scheme based on a contraction mapping and nested uniform grids was proposed [62]. Compared to asymptotic homogenization, the method of FFT-based homogenization avoids the time-consuming FEA for calculate local displacement, but obtains the strain using the Green tensor operator in the frequency domain. Because of its versatility and high efficiency, FFT-based homogenization methods have been extended to retrieve the effective physical properties used in rigid-plastics preformation [63], viscose-plastic preformation [64], crack propagation [65], electrical engineering [66, 67], and optics [68].

In this section, we review the essential ideas and numerical implementation of FFT-based homogenization and asymptotic homogenization, compare their efficiency and applicability. Numerical examples show that the former can be applicable to composites with an extremely large contrast of Young's modulus and eliminates inappropriate values (e.g. negative entries) in the effective elastic tensor. We propose an algorithm to calculate the local displacement of unit cell under test strains in principal directions for FFT-based homogenization. We find the boundaries of unit cell are wave-like even under test strains in principal directions, while they are straight in asymptotic homogenization. Because such deformation patterns reflect the periodic boundary conditions more reasonably, FFT-based homogenization is more likely to correctly retrieve the effective properties of composites with unsymmetrical microstructures.

3.4.1 FFT-based Homogenization Method

The linear elastic composite considered in this section, without loss of generality, is constructed by self-repeated unit cells. Under an external distributed force $\mathbf{f}$, the stress $\boldsymbol{\sigma}$ in the unit cell is governed by the equation of elasticity:

$$\operatorname{div} \boldsymbol{\sigma} + \mathbf{f} = 0 \tag{3.136}$$

where the relation between the stress $\boldsymbol{\sigma}$ to strain $\boldsymbol{\varepsilon}$ is given as follows:

$$\boldsymbol{\sigma} = \mathbf{C} : \boldsymbol{\varepsilon} = \left(\mathbf{C} - \mathbf{C}^0\right) : \boldsymbol{\varepsilon} + \mathbf{C}^0 : \boldsymbol{\varepsilon} \tag{3.137}$$

If the composite is assumed as a homogeneous material with an elastic tensor $\mathbf{C}^0$, at any arbitrary location occupied by solid material with an elastic tensor $\mathbf{C}$ or a void material, the material difference will result in the polarization stress tensor τ, given as follows:

$$\tau = \left(\mathbf{C} - \mathbf{C}^0\right) : \varepsilon \tag{3.138}$$

Substituting Eq. (3.137) into Eq. (3.136) gives

$$\text{div}\ \left(\tau + \mathbf{C}^0 : \varepsilon\right) + \mathbf{f} = 0 \tag{3.139}$$

Extending Eq. (3.139) and writing it in the Einstein tensor form, we get

$$\mathbf{C}^0_{ijkl}\ \mathbf{u}_{k,lj} + f_i + \tau_{ij,j} = 0 \tag{3.140}$$

where the displacement **u** is subjected to periodic boundary condition. Based on the superposition law of a linear elastic problem, the displacement, namely the solution to Eq. (3.140), is the sum of two components, given by

$$\mathbf{u} = \mathbf{u^0} + \mathbf{u^1} \tag{3.141}$$

where $\mathbf{u^0}$ is the solution to

$$C^0_{ijkl} u^0_{k,lj} + f_i = 0 \tag{3.142}$$

If deleting $\mathbf{f_i}$ and considering $\tau_{ij,j}$ as a load in Eq. (3.140), $\mathbf{u^1}$ represents the displacement induced by the polarization stress and can be determined by solving:

$$C^0_{ijkl} u_{k,lj} + \tau_{ij,j} = 0 \tag{3.143}$$

According to Green's function theory, the solution of Eq. (3.142) is

$$u^0_j(\mathbf{x}) = \int_\Omega G_{ij}\left(\mathbf{x} - \mathbf{x}'\right) f_i\left(\mathbf{x}'\right) \mathrm{d}\mathbf{x}' \tag{3.144}$$

where the Green's function $G_{ij}(\mathbf{x} - \mathbf{x}')$ represents the displacement at point x in i-direction under a unit force at point x' in j-direction. Mathematically, it is the solution to the equation of elasticity under load $\delta(x - x')\delta_{im}$, given by

$$C^0_{imkl} G_{ij,lm}\left(\mathbf{x} - \mathbf{x}'\right) + \delta\left(\mathbf{x} - \mathbf{x}'\right)\delta_{kj} = 0 \tag{3.145}$$

where δ is the Kronecker function. In the composite, the Green's function only relies on the relative distance between two points. For simplicity, we set $\mathbf{x}' = (0{,}0{,}0)$ and obtain:

$$G_{ij}\left(\mathbf{x} - \mathbf{x}'\right) = G_{ij}(\mathbf{x}) \tag{3.146}$$

Substituting Eq. (3.146) into Eq. (3.145), we get

$$C^0_{imkl} G_{ij,lm}(\mathbf{x}) + \delta(\mathbf{x})\delta_{kj} = 0 \tag{3.147}$$

By applying Fourier transform on both sides of Eq. (3.145), we obtain

$$\frac{1}{(2\pi)^3}\int_{-\infty}^{\infty}\left(C^0_{imkl}\,\hat{G}_{ij,lm}(\xi) + \delta_{kj}\right)\mathrm{e}^{q\xi x}\mathrm{d}\xi = 0 \tag{3.148}$$

herein $q = (-1)^{1/2}$ is the unit imaginary number and $\boldsymbol{\xi}$ denotes the frequency vector. By using the derivative rule, we get

$$\frac{1}{(2\pi)^3}\int_{-\infty}^{\infty}\left(C^0_{imkl}\left(q\xi_1\right)\left(q\xi_m\right)\hat{G}_{ij}(\xi) + \delta_{kj}\right)\mathrm{e}^{q\xi x}\mathrm{d}\xi = 0 \tag{3.149}$$

Therefore, we obtain

$$\left(\left(\xi_1\right)\left(\xi_m\right)C^0_{imkl}\right)\hat{G}_{ij}(\xi) = \delta_{kj} \tag{3.150}$$

Rearranging Eq. (3.150), we get

$$\hat{G}_{ij}(\xi) = \left(\left(\xi_k\right)\left(\xi_l\right)C^0_{ijkl}\right)^{-1} \tag{3.151}$$

After obtaining the Green's function, the displacement $\mathbf{u}^0$ that resulted from f_i can be easily calculated in accordance with Eq. (3.144). Similarly, $\mathbf{u}^1$ can be determined as

$$u_j^1(\mathbf{x}) = \int_\Omega G_{ij}(\mathbf{x}-\mathbf{x}')\,\tau_{ij,j}(\mathbf{x}')\,d\mathbf{x}' \tag{3.152}$$

Therefore, Eq. (3.141) becomes

$$u_j(\mathbf{x}) = u_j^0(\mathbf{x}) - \int_\Omega h_{ijk}(\mathbf{x}-\mathbf{x}')\,\tau_{ik}(\mathbf{x}')\,d\mathbf{x}' \tag{3.153}$$

h_{ijk} can be derived through integration by parts as

$$h_{ijk}(x) = -\frac{1}{2}\left(G_{jk,i}(\mathbf{x}) + G_{ij,k}(\mathbf{x})\right) \tag{3.154}$$

With the consideration of strain field $\varepsilon_{ij} = 1/2(u_{i,j}+u_{j,i})$ in Eq. (3.153), we get the periodic Lippmann–Schwinger integral equation related to strain field as follows:

$$\varepsilon_{ij}(\mathbf{x}) = \varepsilon_{ij}^0(\mathbf{x}) - \int_\Omega \Gamma_{ijk}(\mathbf{x}-\mathbf{x}')\,\tau_{kl}(\mathbf{x}')\,\mathrm{d}\mathbf{x}' \tag{3.155}$$

where Green tensor operator is

$$\Gamma_{ijkl}(\mathbf{x}) = \frac{1}{2}\left[\frac{\partial h_{ikl}}{\partial \mathbf{x}_j}(\mathbf{x}) + \frac{\partial h_{jkl}}{\partial \mathbf{x}_i}(\mathbf{x})\right] \tag{3.156}$$

When the macroscopic strain field ε_0 is prescribed into the unit cell, without external force, Eq. (3.155) can be written into a terse form, given as follows:

$$\varepsilon(\mathbf{x}) = \varepsilon_0 - G(\mathbf{x}) * \tau(\mathbf{x}) \tag{3.157}$$

where "*" represents the convolution operator. Making use of the convolution theorem, after Fourier transform, Eq. (3.157) attains a form in frequency domain:

$$\hat{\varepsilon}(\xi) = -\hat{\Gamma}(\xi) : \hat{\tau}(\xi), \forall\xi \neq 0,\ \hat{\varepsilon}(0) = \varepsilon_0 \tag{3.158}$$

To calculate the effective elastic tensor, linearly independent unit test strains are prescribed in the unit cell. After obtaining the strain field from Eq. (3.158), the stress is calculated by Eq. (3.137). At last the effective elastic tensor $\mathbf{C}^\mathrm{H}$ can be naturally determined as follows:

$$\mathbf{C}^\mathrm{H} = \int_\Omega \sigma \mathrm{d}\mathbf{x} : \left(\int_\Omega \varepsilon \mathrm{d}\mathbf{x}\right)^{-1} \tag{3.159}$$

In the asymptotic homogenization method, the displacement has a similar expression to Eq. (3.141):

$$\mathbf{u} = \mathbf{u}^0(\mathbf{x},\mathbf{y}) + \gamma\mathbf{u}^1(\mathbf{x},\mathbf{y}) + o\left(\gamma^2\right) \tag{3.160}$$

where $\mathbf{y}=\mathbf{x}/\gamma$ denotes the position relation in macroscopic coordinate and microscopic coordinate with a constant $0<\gamma\ll 1$. The first item in Eq. (3.160) represents the displacement in macroscale, similar to $\mathbf{u}^0$ in Eq. (3.141). The second item $\mathbf{u}^1$, namely the characteristic microscale displacement, is the solution to the following characteristic equation:

$$\int_\Omega C_{ijmn}\varepsilon_{ij}\,\varepsilon_{mn}\left(u_i^{1(kl)}\right) d\Omega = \int_\Omega C_{ijkl}\varepsilon_{ij}\,\varepsilon_{mn}^{0(kl)}\, d\Omega \tag{3.161}$$

where Ω is the domain occupied by the unit cell and the periodic admissible strain ε_{ij} is multiplied on both sides of Eq. (3.161) to obtain the weak form. In two dimensions, there are three linearly independent unit test strains $\varepsilon_{ij}^{0(kl)}$ given as $[1\ 0\ 0]^{\mathrm{T}}$, $[0\ 1\ 0]^{\mathrm{T}}$ and $[0\ 0\ 1]^{\mathrm{T}}$, respectively.

In three dimensions, the number of test strains becomes six. After the characteristic displacements in Eq. (3.161) are solved using FEA [21], the homogenized elasticity tensor is thereafter obtained as follows:

$$C_{ijkl}^{H} = \frac{1}{|\Omega|}\left(\int_{\Omega} C_{ijkl} - C_{ijmn}\frac{\partial u_m^{1(kl)}}{\partial y_n}\right) d\Omega \tag{3.162}$$

where $|\Omega|$ denotes the volume of unit cell. The effective elasticity tensor can be rewritten in the form of strains as follows:

$$C_{ijkl}^{H} = \frac{1}{|\Omega|}\int_{\Omega} C_{mnrs}\left(\varepsilon_{mn}^{0(ij)} - \varepsilon_{mn}^{(ij)}\left(u_m^{1(ij)}\right)\right)\left(\varepsilon_{rs}^{0(kl)} - \varepsilon_{rs}^{(kl)}\left(u_m^{1(kl)}\right)\right) d\Omega \tag{3.163}$$

where u_{m} is the characteristic displacement obtained from the characteristic displacement in terms of the strain–displacement relation.

3.4.2 Implementation of FFT-based Homogenization Method

To obtain the strain in Eq. (3.157), we rearrange it by using Eqs. (3.138) and (3.158) and get

$$(\mathbf{I} + \mathbf{B})\,\varepsilon = \varepsilon_0 \tag{3.164}$$

where **I** stands for a unit matrix and

$$\mathbf{B} = F^{-1}\left[\gamma : F\left(\mathbf{C} - \mathbf{C^0}\right)\right] \tag{3.165}$$

with F and F^{-1} representing Fourier transform and inverse Fourier transform, respectively. In the real domain, the Green tensor operator can be derived as follows:

$$\Gamma_{ijkl} = -\frac{1}{4}\left[\frac{\partial^2 G_{il}}{\partial x_j \partial x_k}(\mathbf{x}) + \frac{\partial^2 G_{ik}}{\partial x_j \partial x_l}(\mathbf{x}) + \frac{\partial^2 G_{jl}}{\partial x_i \partial x_k}(\mathbf{x}) + \frac{\partial^2 G_{jk}}{\partial x_i \partial x_l}(\mathbf{x})\right] \tag{3.166}$$

In the frequency domain, it becomes as follows:

$$\hat{\Gamma}_{ijkl} = \frac{1}{4}\left[\hat{G}_{il}(\xi)\,\xi_j\xi_k + \hat{G}_{ik}(\xi)\,\xi_j\xi_l + \hat{G}_{jl}(\xi)\,\xi_i\xi_k + \hat{G}_{jk}(\xi)\,\xi_i\xi_l\right] \tag{3.167}$$

Rewriting the elastic tensor of the isotropic reference material as follows:

$$C_{ijkl}^{0} = \lambda^0\delta_{ij}\delta_{kl} + \mu^0\left(\delta_{ik}\delta_{jl} + \delta_{il}\delta_{jk}\right) \tag{3.168}$$

where the Lamé parameters of reference material are represented as λ^0 and μ^0, respectively. Substituting Eq. (3.168) into Eq. (3.151), we get the Green's function in frequency domain as

$$\hat{G}_{ij}(\xi) = \frac{1}{\mu^0}\frac{\delta_{ij}}{|\xi|^2} - \frac{\left(\lambda^0 + \mu^0\right)}{\mu^0\left(\lambda^0 + 2\mu^0\right)}\frac{\xi_i\xi_j}{|\xi|^4} \tag{3.169}$$

Therefore, by substituting Eq. (3.169) into Eq. (3.166), the Green tensor operator in frequency domain can be explicitly expressed as follows:

$$\hat{\Gamma}_{ijkl} = \frac{\left(\delta_{ik}\xi_l\xi_j + \delta_{il}\xi_k\xi_j + \delta_{jk}\xi_l\xi_i + \delta_{lj}\xi_i\xi_k\right)}{4\mu^0|\xi|^2} - \frac{\left(\lambda^0 + \mu^0\right)}{\mu^0\left(\lambda^0 + 2\mu^0\right)}\frac{\xi_i\xi_j\xi_k\xi_l}{|\xi|^4} \tag{3.170}$$

When we define $\mathbf{A} = \mathbf{I} + \mathbf{B}$, Eq. (3.164) can be rewritten as

$$\mathbf{A}\varepsilon = \varepsilon_0 \tag{3.171}$$

where the inverse of matrix **A** can be obtained based on the Neumann expansion [69], given as

$$A^{-1} = \frac{1}{I+B} = \sum_{i=0}^{\infty}(-B)^i \tag{3.172}$$

Therefore, an iterative procedure can be determined as

$$\varepsilon^{(i+1)} = (-B)\,\varepsilon^{(i)} + \varepsilon_0 \tag{3.173}$$

Inside the iterative scheme, the optimal rate of convergence is achieved by defining the reference material with specified Lamé parameters as following:

$$\lambda^0 = \frac{\min\{\lambda\} + \max\{\lambda\}}{2}, \mu^0 = \frac{\min\{\mu\} + \max\{\mu\}}{2} \tag{3.174}$$

The natural criteria for the iterative algorithms is based on the difference between $\varepsilon^{(i+1)}$ and $\varepsilon^{(i)}$, which can be expressed by

$$\eta^i = \frac{\|\varepsilon^{(i+1)} - \varepsilon^{(i)}\|^2}{\|\varepsilon^0\|^2} \le \xi \tag{3.175}$$

Although the standard FFT-based algorithm using the Neumann series iterative scheme is very simple, one drawback of the Neumann series expansion is that it suffers slow convergence when the contrast of material properties increases, which is undesirable for porous materials. But this linear algebra problem can be solved by the conjugate gradient method.

3.5 Conclusion

Staggered array of fibers gives higher values of storage modulus as compared to regular array. The higher values probably occur because this type of RVE tends to create long chains of nearly touching fibers with a high degree of overlap. Since, loss factor is inversely proportional to storage moduli, hence, the staggered array of fibers will give lower values of loss factors in comparison to Type 2 and Type 3 RVE's. The longitudinal storage modulus is increasing with l/d, whereas, the other moduli, first decrease rapidly, till $l/d = 100$. After this limit, the moduli become insensitive to variation in l/d. A probable reason for rapid fall of storage moduli (except E'_{22}) till $l/d = 100$ is, increasing the fiber length decreases the number of fiber ends and thus the amount of matrix between the fiber ends. Since the amount of matrix is fixed, it moves toward the space between the fibers and increases the interfiber spacing and thus decreases the values of E'_{22}. Same reasoning can be given for G'_{23} and G'_{12}.

Following conclusions can be made from the above discussion:

(a) Storage moduli
  (i) Staggered array of fibers gives higher values of storage modulus as compared to regular array.
  (ii) All the storage moduli show an increasing trend with increasing fiber volume fraction.
  (iii) Storage moduli, E'_{22}, G'_{23} and G'_{12} first decrease with l/d, till the fiber aspect ratio, $l/d = 100$, and thereafter become constant.
  (iv) Aspect ratio significantly influences the longitudinal storage modulus but has little effect on E'_{22}, G'_{23} and G'_{12}, except at low-aspect ratios.

(v) Values of E'_{22}, G'_{23} and G'_{12} are significantly lower than that obtained for E'_{11}. This is due to the fact that in transverse and shear loading, the properties of the composite are matrix dominated.

(b) Loss factors

(i) Loss factors η_{11}, η_{22}, η_{23}, and η_{12} all decrease with V_f.
(ii) Loss factor η_{11} decreases with the fiber aspect ratio.
(iii) Loss factors η_{22}, η_{23}, and η_{12} all increase with l/d, till $l/d = 100$ and then become constant.
(iv) Fibers having low-aspect ratios give higher values of damping as compared to the fibers having large aspect ratios.
(v) The highest loss factor is obtained at lowest fiber volume fractions. This coincides with the fact that the damping in the composites is matrix-dominated.
(vi) Decrease in damping with increase in fiber volume fraction is more for the fibers having large aspect ratios as compared to the fibers having small aspect ratios.
(vii) A possible reason for improved damping at low-fiber aspect ratios is the presence of shear stress concentration at numerous fiber ends in a short-fiber composite and the resulting stress transfer to the viscoelastic matrix.

References

1 Hashin, Z. (1983). Analysis of composite materials – a survey. *ASME, Journal of Applied Mechanics* 50: 481–505.
2 Hill, R. (1963). Elastic properties of reinforced solids: some theoretical principles. *Journal of Mechanics and Physics of Solids* 11: 357–372.
3 Cox, H.L. (1952). The elasticity and strength of paper and other fibrous materials. *British Journal of Applied Physics* 3: 72–79.
4 Kelly, A. and Tyson, W.R. (1965). Tensile properties of fiber-reinforced metals: copper/tungsten and copper/molybdenum. *Journal of the Mechanics and Physics of Solids* 13 (6): 329–338.
5 Nairn, J. (1997). On the use of shear-lag methods for the analysis of stress transfer in unidirectional composites. *Mechanics of Materials* 26 (2): 63–80.
6 Landis, C.M. and McMeeking R.M. (1999). A shear-lag model for a broken fiber embedded in a composite with a ductile matrix. *Composites Science and Technology* 59 (3): 447–457.
7 Wilczynski, A.P. (1990). A basic theory of reinforcement for unidirectional fibrous composites. *Composites Science and Technology* 38: 327–336.
8 Rosen, B.W. (1964). Tensile failure of fibrous composites. *AIAA Journal*: 1985–1991.
9 Carman, G.P. and Reifsnider, K.L. (1992). Micro-mechanics of short-fiber composites. *Composites Science and Technology* 43: 137–146.
10 Hashin, Z. and Rosen, B.W. (1964). The elastic moduli of fiber-reinforced materials. *ASME, Journal of Applied Mechanics* 31: 223–232.
11 Robinson, I.M. and Robinson, J.M. (1994). The effect of fiber aspect ratio on the stiffness of discontinuous fiber-reinforced composites. *Composites* 25: 499–503.
12 Fukuda, H. and Kawata, K. (1974). On Young's modulus of short fiber composites. *Fiber Science and Technology* 7: 207–222.

13 Eshelby, J.D. (1957). The determination of the elastic field of an ellipsoidal inclusion and related problems. *Proceedings of the Royal Society of London Series A* 241: 376–396.

14 Mura, T. (1982). *Micro-mechanics of Defects in Solids*. The Hague: Martinus Nijho.

15 Russel, W.B. (1973). On the effective moduli of composite materials: effect of fiber length and geometry at dilute concentrations. *Journal of Applied Mathematics and Physics, (JAMP)* 24: 581–600.

16 Chow, T.S. (1977). Elastic moduli of filled polymers: the effect of particle shape. *Journal of Applied Physics* 48: 4072–4075.

17 Steif, P.S. and Hoysan, S.F. (1987). An energy method for calculating the stiffness of aligned short-fiber composites. *Mechanics of Materials* 6: 197–210.

18 Mori, T. and Tanaka, K. (1973). Average stress in matrix and average elastic energy of materials with mis-fitting inclusions. *Acta Materialia* 21: 571–574.

19 Benveniste, Y. (1987). A new approach to the application of Mori–Tanaka's theory in composite materials. *Mechanics of Materials* 6: 147–157.

20 Wakashima, K., Otsuka, M., and Umekawa, S. (1974). Thermal expansion of heterogeneous solids containing aligned ellipsoidal inclusions. *Journal of Composite Materials* 8: 391–404.

21 Taya, M. and Mura, T. (1981). On stiffness and strength of an aligned short-fiber reinforced composite containing fiber-end cracks under uniaxial applied stress. *ASME, Journal of Applied Mechanics* 48: 361–367.

22 Taya, M. and Chou, T.W. (1981). On two kinds of ellipsoidal in-homogeneities in an infinite elastic body: an application to a hybrid composite. *International Journal of Solids and Structures* 17: 553–563.

23 Weng, G.J. (1984). Some elastic properties of reinforced solids, with special reference to isotropic ones containing spherical inclusions. *International Journal of Engineering Science* 22 (7): 845–856.

24 Tandon, G.P. and Weng, G.J. (1984). The effect of aspect ratio of inclusions on the elastic properties of uni-directionally aligned composites. *Polymer Composites* 5: 327–333.

25 Tucker, C.L. and Liang, E. (1999). Stiffness predications for unidirectional short-fiber composites: review and evaluation. *Composites Science and Technology* 59: 655–671.

26 Chow, T.S. (1978). Effect of particle shape at finite concentration on the elastic modulus of filled polymers. *Journal of Polymer Science, Polymer Physics Edition* 16: 959–965.

27 Ferrari, M. (1994). Composite homogenization via the equivalent poly-inclusion approach. *Composites Engineering* 4: 37–45.

28 Advani, S.G. and Tucker, C.L. III, (1987). The use of tensors to describe and predict the fiber orientation in short fiber composites. *Journal of Rheology* 31 (8): 751.

29 Gibson, R.F., Sun, C.T., and Wu, J.K. (1985). Prediction of material damping in randomly oriented short fiber polymer matrix composites. *Journal of Reinforced Plastics and Composites* 4: 262–272.

30 Wu, C.T.D. and Mc Cullough, R.L. (1977). Constitutive relationships for heterogeneous materials. In: *Developments in Composites Materials-1* (ed. G.S. Holister), 118–186. London: Applied Science.

31 Hashin, Z. and Shtrikman, S. (1962). On some variational principles in anisotropic and non-homogeneous elasticity. *Journal of the Mechanics and Physics of Solids* 10: 335–342.

32 Walpole, L.J. (1966). On bounds for the overall elastic moduli for inhomogeneous systems. *Journal of the Mechanics and Physics of Solids* 14: 151–162.

33 Willis, J.R. (1977). Bounds and self-consistent estimates for the overall properties of anisotropic composites. *Journal of Mechanics of Physics of Solids* 25: 185–202.
34 Weng, G.J. (1992). Explicit evaluation of Willis' bounds with ellipsoidal inclusions. *Journal of Engineering Science* 30: 83–92.
35 Eduljee, R.F., McCullough, R.L., and Gillespie, J.W. (1994). The influence of aggregated and dispersed textures on the elastic properties of dis-continuous fiber composites. *Composites Science and Technology* 50: 381–391.
36 Lielens, G., Pirotte, P., Couniot, A. et al. (1997). Prediction of thermo-mechanical properties for compression-moulded composites. *Composites: Part A* 29: 63–70.
37 Hill, R. (1965). A self-consistent mechanics of composite materials. *Journal of Mechanics of Physics of Solids* 13: 213–222.
38 Budiansky, B. (1965). On the elastic moduli of some heterogeneous materials. *Journal of Mechanics of Physics of Solids* 13: 223–227.
39 Laws, N. and McLaughlin, R. (1979). The effect of fiber length on the overall moduli of composite materials. *Journal of Mechanics of Physics of Solids* 27: 1–13.
40 Chou, T.W., Nomura, S., and Taya, M. (1980). A self-consistent approach to the elastic stiffness of short-fiber composites. *Journal of Composite Materials* 14: 178–188.
41 Lin, S.C. and Mura, T. (1973). Elastic fields of inclusions in anisotropic media (II). *Physica Statatus Solidi* 15: 281–285.
42 Chou, T.W., Nomura, S., and Taya, M. (1979). A self-consistent approach to the elastic stiffness of short-fiber composites. In: *Modern Developments in Composite Materials and Structures* (ed. J.R. Vinson), 149–164. New York, NY: ASME.
43 Kerner, E.H. (1956). The elastic and thermo-elastic properties of composite media. *Proceedings of Physics Society B* 69: 808–813.
44 Hermans, J.J. (1967). The elastic properties of fiber reinforced materials when the fibers are aligned. *Proceedings of the Koninklijke Nederlandse Akademie van Wetenschappe B* 65: 1–9.
45 Alexander, H. (1941). *FEM Modeling of Plates and Structures*, 1e. McGraw Hill.
46 Courant, R. (1942). *FEM Modeling of Structures*, 1e. New York: McGraw Hill.
47 Ji, X., Cao, Y., and Feng, X. (2010). Micromechanics prediction of the effective elastic moduli of graphene sheet-reinforced polymer nanocomposites. *Modelling and Simulation in Materials Science and Engineering* 18: 045005.
48 Feng, X., Tian, Z., Liu, Y., and Yu, S. (2004). Effective elastic and plastic properties of interpenetrating multiphase composites. *Applied Composite Materials* 11: 33–55.
49 Lo Cascio, M., Milazzo, A., and Benedetti, I. (2020). Virtual element method for computational homogenization of composite and heterogeneous materials. *Composite Structures* 232: 111523.
50 Bensoussan, A., Lions, J., and Papanicolaou, G. (1978). *Asymptotic Analysis for Periodic Structures*. Amsterdam: North-Holland.
51 Andrianov, I., Bolshakov, V., Danishevs'kyy, V., and Weichert, D. (2007). Asymptotic simulation of imperfect bonding in periodic fibre-reinforced composite materials under axial shear. *International Journal of Mechanical Sciences* 49: 1344–1354.

52 Sridhar, A., Liu, L., Kouznetsova, V., and Geers, M. (2018). Homogenized enriched continuum analysis of acoustic metamaterials with negative stiffness and double negative effects. *Journal of the Mechanics and Physics of Solids* 119: 104–117.

53 Hu, R. and Oskay, C. (2019). Multiscale nonlocal effective medium model for in-plane elastic wave dispersion and attenuation in periodic composites. *Journal of the Mechanics and Physics of Solids* 124: 220–243.

54 Moulinec, H. and Suquet, P. (1994). A fast numerical method for computing the linear and nonlinear mechanical properties of composites. *Comptes Rendus de l'Académie des Sciences* 2: 1417–1423.

55 Moulinec, H. and Suquet, P. (1998). A numerical method for computing the overall response of nonlinear composites with complex microstructure. *Computer Methods in Applied Mechanics and Engineering* 157: 69–94.

56 Moulinec, H., Suquet, P., and Milton, G. (2018). Convergence of iterative methods based on Neumann series for composite materials: theory and practice. *International Journal of Numerical Methods in Engineering* 114: 1103–1130.

57 Michel, J., Moulinec, H., and Suquet, P. (2001). A computational scheme for linear and non-linear composites with arbitrary phase contrast. *International Journal of Numerical Methods in Engineering* 52: 139–158.

58 Zeman, J., Vondřejc, J., Novák, J., and Marek, I. (2010). Accelerating a FFT-based solver for numerical homogenization of periodic media by conjugate gradients. *Journal of Computational Physics* 229: 8065–8071.

59 Gélébart, L. and Mondon-Cancel, R. (2013). Non-linear extension of FFT-based methods accelerated by conjugate gradients to evaluate the mechanical behavior of composite materials. *Computational Materials Science* 77: 430–439.

60 Monchiet, V. and Bonnet, G. (2012). A polarization-based FFT iterative scheme for computing the effective properties of elastic composites with arbitrary contrast. *International Journal of Numerical Methods in Engineering* 89: 1419–1436.

61 Brisard, S. and Dormieux, L. (2010). FFT-based methods for the mechanics of composites: a general variational framework. *Computational Materials Science* 49: 663–671.

62 Eyre, D. and Milton, G. (1999). A fast numerical scheme for computing the response of composites using grid refinement. *European Journal of Physics* 6: 41–47.

63 Bilger, N., Auslender, F., Bornert, M. et al. (2007). Bounds and estimates for the effective yield surface of porous media with a uniform or a non-uniform distribution of voids. *European Journal of Mechanics - A/Solids* 26: 810–836.

64 Idiart, M., Willot, F., Pellegrini, Y., and Castaneda, P. (2009). Infinite-contrast periodic composites with strongly nonlinear behavior: effective-medium theory versus full-field simulations. *International Journal of Solids and Structures* 46: 3365–3382.

65 Vigliotti, A. and Pasini, D. (2012). Linear multiscale analysis and finite element validation of stretching and bending dominated lattice materials. *Mechanics of Materials* 46: 57–68.

66 Willot, F., Gillibert, L., and Jeulin, D. (2013). Microstructure-induced hotspots in the thermal and elastic responses of granular media. *International Journal of Solids and Structures* 50: 1699–1709.

67 Willot, F. and Jeulin, D. (2011). Elastic and electrical behavior of some random multiscale highly contrasted composites. *International Journal of Multiscale Computational Engineering* 9: 305–326.

68 Azzimonti, D., Willot, F., and Jeulin, D. (2013). Optical properties of deposit models for paints: fullfields FFT computations and representative volume element. *Journal of Modern Optics* 60: 519–528.

69 Michel, J., Moulinec, H., and Suquet, P. (1999). Effective properties of composite materials with periodic microstructure: a computational approach. *Computer Methods in Applied Mechanics and Engineering* 172: 109–143.

4

Prediction of Mechanical Properties

4.1 Storage Moduli

In this section, variation of different storage moduli with both the fiber volume fraction as well as aspect ratio of fibers is discussed. Also, the results obtained for storage moduli from different models are compared with each other as well as with the results obtained from finite element modeling (FEM) modeling and the results given by Finegan et al. [1].

4.1.1 Longitudinal Storage Modulus (E'_{11})

4.1.1.1 Variation of E'_{11} with V_f

Figure 4.1 shows the variation of storage modulus (E'_{11}) with fiber volume fraction using different models for $l/d = 19$. Longitudinal storage modulus (E'_{11}) increases with the fiber volume fraction (V_f) for a constant fiber aspect ratio ($l/d = 19$), as shown in Figure 4.1. This is due to the fact that the fibers are aligned in the longitudinal direction. As V_f increases, for a constant aspect ratio ($l/d = 19$), the volume of the fibers is increasing. Since the fibers have high modulus in the longitudinal direction, the composite has improved properties in the said direction. The significance of this increase in the storage modulus is that the amount of energy that can be recovered is increasing with the fiber volume fraction. As shown in Figure 4.1, Halpin–Tsai model gives lower values of E'_{11} as compared to other models. This is probably because of the fact that the Halpin–Tsai model was originally designed for continuous fiber reinforced composites. Halpin and Tsai approximated the value of $\xi = 2l/d$ by comparing the model predictions with available 2D finite element results. Halpin–Tsai equations are known to fit some data very well at low-fiber volume fractions, but they underpredict the stiffness values at high-volume fractions as shown in Figure 4.1. Halpin–Tsai model provide reasonable estimate of the stiffness of short-fiber composites, but the Mori–Tanaka model is more accurate. Cox model treats the fiber as a slender body. The true axial displacements, according to Cox model, lag behind the beam theory predictions. Thus, the results of Halpin–Tsai and Cox are lower than other models as observed in Figure 4.1. Hence, these two models are approximate or crude models.

Eshelby model assumes that the strain is uniform in the fiber and highly nonuniform in the matrix. According to Eshelby, the interaction between the fibers is negligible at small volume fractions as shown in Figure 4.1. Mori–Tanaka model gives better values of E'_{11} at all the fiber volume fractions. Chow model considered Eshelby's inclusion problem and concluded that in a concentrated composite the fiber strain would be the sum of average strain in the matrix and the dilute result given by Eshelby. Eshelby method is found to predict slightly higher

Mechanics of Particle- and Fiber-Reinforced Polymer Nanocomposites: From Nanoscale to Continuum Simulations, First Edition. Sumit Sharma.

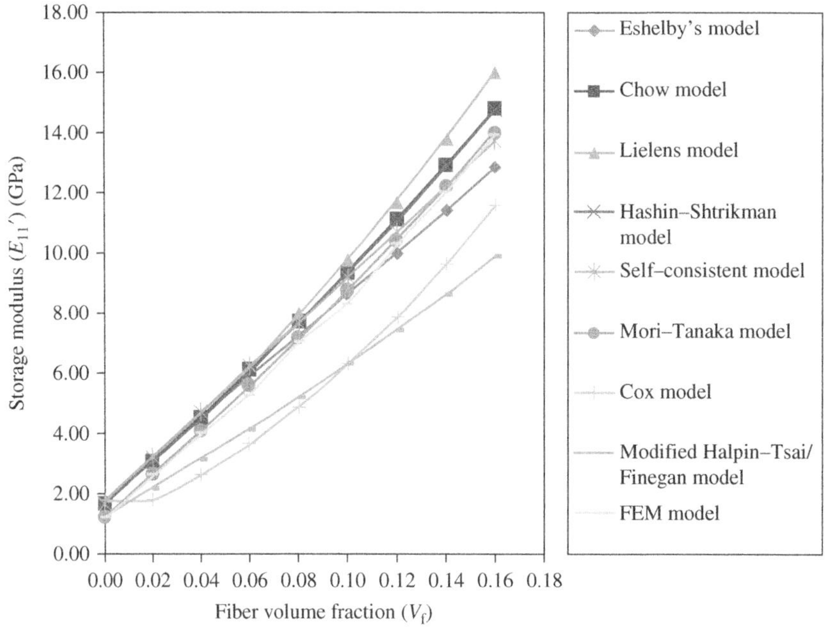

Figure 4.1 Variation of E'_{11} vs. V_f for $l/d = 19$.

values of E'_{11} for high values of V_f. This is because the Eshelby model is strictly valid for inclusion in a far field in which a low-fiber volume fraction is presumed. Another reason for low values of E'_{11} at low values of V_f is that the matrix stiffness strongly influences the composite stiffness at low-fiber volume fractions.

Lielens suggested that at very high-fiber volume fractions, the composite stiffness should be closer to the upper bound or equivalently to the Mori–Tanaka predictions. The Lielens model reproduces the Mori–Tanaka results at low-fiber volume fractions, but gives improved results at high V_f as observed in Figures 4.1. Self-consistent model makes an initial assumption of the embedding matrix properties "C" and "S" to find an improved value of the composite stiffness. Thus, self-consistent model is an approximate model for finding the stiffness and damping of the composites. Because of this reason, the values obtained from this model are not nearer to most of the other models, as observed in Figure 4.1. FEM model gives better corelationship with Mori–Tanaka model for storage modulus, as can be seen from Figure 4.1. This is due to the fact that, here only the actual state of stress is accounted for. Also, periodic arrangement of the fibers is assumed in this work, and in FEM, the effect of inter fiber interaction is taken into account.

Difference between Mori–Tanaka model and FEM values, for storage modulus E'_{11}, at $V_f = 0.02$ is 0.76%. The difference reduces to 0.49% at $V_f = 0.16$. These values are for $l/d = 19$. Similarly, the differences between Mori–Tanaka model and Modified Halpin–Tsai/Finegan model and between FEM model and Modified Halpin–Tsai/Finegan model are shown in Table 4.1. Percentage increase in E'_{11} with increasing V_f for $l/d = 19$, as given by Mori–Tanaka model, Modified Halpin–Tsai/Finegan model, and FEM model, is shown in Table 4.2.

From Table 4.1, it can be observed that difference between Mori–Tanaka model and Modified Halpin–Tsai/Finegan model is less at low-fiber volume fractions than at high values of V_f. Also, FEM results are closer to Mori–Tanaka model. Difference between

Table 4.1 Comparison between E'_{11} values given by different models for increasing V_f at $l/d = 19$.

Fiber volume fraction (V_f)	Difference in E'_{11} values (in %) between		
	Mori–Tanaka and Modified Halpin–Tsai/Finegan model	Mori–Tanaka and FEM model	Modified Halpin–Tsai/Finegan model and FEM model
0.02	17.87	0.76	17.24
0.06	25.80	4.65	22.18
0.16	29.65	0.49	29.30

Table 4.2 Percentage increase in E'_{11} with V_f given by different models for $l/d = 19$.

Range of fiber volume fraction (V_f)	Increase in E'_{11} values (in %)		
	Mori–Tanaka model	Modified Halpin–Tsai/Finegan model	FEM model
0.0–0.06	353.65	236.58	332.52
0.06–0.16	151.43	138.40	162.40

Modified Halpin–Tsai/Finegan model and FEM model is less at low values of fiber volume fractions than at high values of V_f. Table 4.2 shows that Mori–Tanaka model gives the highest percentage increase in E'_{11} at low-fiber volume fractions and the FEM results are nearer to it. Since Modified Halpin–Tsai/Finegan model is semiempirical, the results given by it are lower than both the Mori–Tanaka model and FEM model. At low-fiber volume fractions, all the models give almost same results and E'_{11} increases rapidly (till $V_f = 0.03$). But as the fiber volume fraction increases, the effect of inter fiber interaction makes the storage modulus to increase less rapidly as compared to the increase at lower values of V_f.

4.1.1.2 Variation of E'_{11} with l/d

Figure 4.2 shows the variation of storage modulus (E'_{11}) with fiber aspect ratio using different models for $V_f = 0.03$. It can be observed that E'_{11} increases sharply upto $l/d = 100$, which is same for almost all the models. A possible reason for this is the presence of shear-tress concentration at numerous fiber ends in a short-fiber composite. However, for $l/d > 100$, E'_{11} increases slowly. This is because, after $l/d = 100$, the fibers may be treated as continuous fibers and larger l/d causes higher-fiber axial stress than small aspect ratios.

Difference between Mori–Tanaka model and FEM values, for storage modulus E'_{11} at $V_f = 0.03$ and $l/d = 100$, is 11.16%, whereas at $l/d = 1000$, the difference is 6.2%. Differences between Mori–Tanaka model and Modified Halpin–Tsai/Finegan model and between FEM model and Modified Halpin–Tsai/Finegan model are shown in Table 4.3. Percentage increase in E'_{11} with increasing l/d for $V_f = 0.03$ as given by Mori–Tanaka model, Modified Halpin–Tsai/Finegan model and FEM model is shown in Table 4.4. The FEM model gives lower results for storage modulus. This is due to the fact that only the actual state of stress is accounted for.

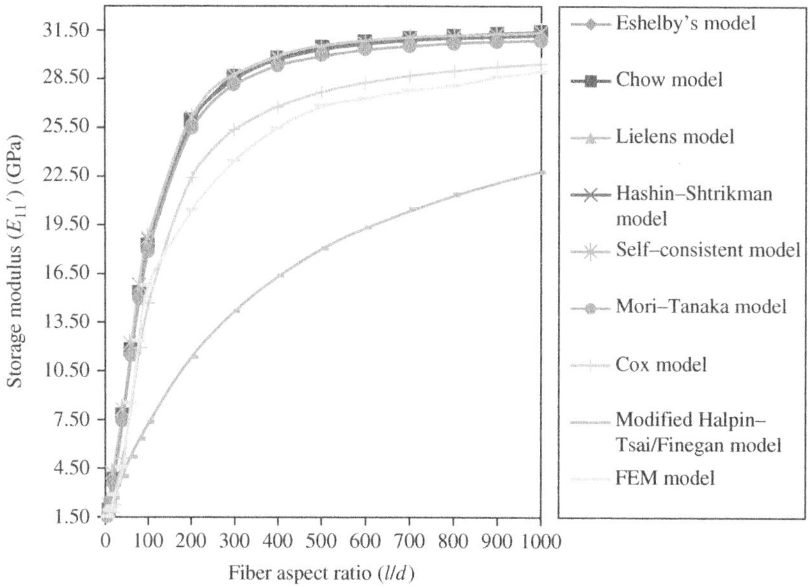

Figure 4.2 Variation of E'_{11} vs. l/d for $V_f = 0.03$.

Table 4.3 Comparison between E'_{11} values given by different models for increasing l/d ratios at $V_f = 0.03$.

	Difference in E'_{11} values (in %) between		
Fiber aspect ratio (l/d)	**Mori–Tanaka and Modified Halpin–Tsai/ Finegan model**	**Mori–Tanaka and FEM model**	**Modified Halpin–Tsai/ Finegan model and FEM model**
5	5.128	2.56	7.89
100	58.81	11.14	53.67
1000	26.19	6.2	21.31

Table 4.4 Percentage increase in E'_{11} with l/d given by different models at $V_f = 0.03$.

	Increase in E'_{11} values (in %)		
Range of fiber aspect ratio (l/d)	**Mori–Tanaka model**	**Modified Halpin–Tsai/ Finegan model**	**FEM model**
5–200	1530.13	591.46	1242.10
200–1000	21.15	100.46	41.66

From Table 4.3, it can be observed that difference between Mori–Tanaka model and Modified Halpin–Tsai/Finegan model is less at low-fiber aspect ratios than at high values of l/d. Also, FEM results are nearer to Mori–Tanaka model. Difference between Modified Halpin–Tsai/Finegan model and FEM model is less at low values of fiber aspect ratios than at high values of l/d. Table 4.4 shows that Mori–Tanaka model gives the highest percentage increase in E'_{11} at low-fiber aspect ratios and the FEM results are nearer to it. Since Modified Halpin–Tsai/Finegan model is semiempirical, the results given by it are lower than both the Mori–Tanaka model and FEM model. At low-fiber aspect ratios, all the models give almost same results and E'_{11} increases rapidly (till $l/d = 100$). But as the fiber aspect ratio increases, larger l/d causes higher fiber axial stress than small aspect ratios and thus, the storage modulus increases less rapidly as compared to the increase at lower values of l/d.

4.1.2 Transverse Storage Modulus (E'_{22})

4.1.2.1 Variation of E'_{22} with V_f

The transverse storage modulus (E'_{22}), is increasing with fiber volume fraction, as shown in Figure 4.3. As V_f increases for a constant aspect ratio ($l/d = 19$), the volume of the fibers is increasing. The fibers have very high modulus as compared to the matrix. Hence, as the volume of fibers increases, the storage modulus increases. But increase in E'_{22} is insignificant as compared to the increase in the values of E'_{11} as can be seen by comparing Figures 4.1 and 4.3. This is due to the fact that reinforcing fibers improve the properties of the composite in the longitudinal direction considerably, whereas the transverse properties are only slightly improved as the force is applied perpendicular to the fibers which are weak in transverse direction.

As shown in Figure 4.3, Halpin–Tsai model gives low values of E'_{22} as compared to other models. This is probably because of the fact that Halpin–Tsai model was originally designed for continuous fiber-reinforced composites. From Figure 4.3, it can be seen that E'_{22} values are starting from different points. This is due to several factors which are explained here. Self-consistent model starts with an assumption of the composite properties and then

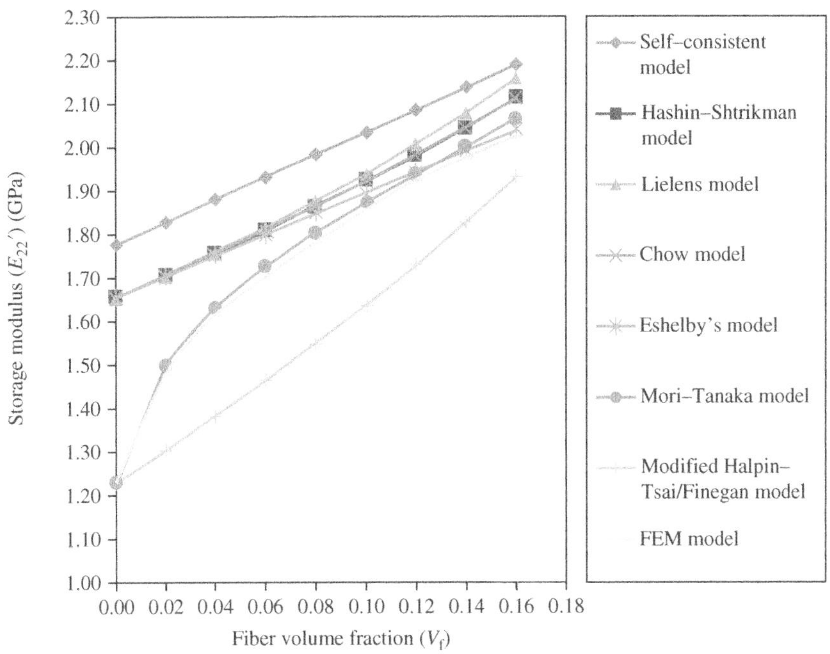

Figure 4.3 Variation of E'_{22} vs. V_f for $l/d = 19$.

improves the stiffness values. Thus, it starts from a different point on the E'_{22} vs. V_f curve. Mori–Tanaka model, Halpin–Tsai/Finegan model, and FEM model values are starting from the same point on the E'_{22} vs. V_f curve. Hashin–Shtrikman model, Chow model, Lielens model, and Eshelby model start from the same point on the E'_{22} vs. V_f curve of Figure 4.3. Lielens model is a reproduction of the Mori–Tanaka model at low-fiber volume fractions and gives improved results at high-volume fractions. Though Mori–Tanaka model and Halpin–Tsai/Finegan model start from the same point, they diverge considerably at high-volume fractions. This is because of the fact that Halpin–Tsai/Finegan model underpredicts the stiffness values at higher-fiber volume fractions as explained earlier.

Difference between Mori–Tanaka model and FEM values, for storage modulus E'_{22}, at $V_f = 0.02$ is 1.26%. The difference reduces to 1.22% at $V_f = 0.16$. These values are for $l/d = 19$. Similarly, the differences between Mori–Tanaka model and Modified Halpin–Tsai/Finegan model and between FEM model and Modified Halpin–Tsai/Finegan model are shown in Table 4.5. Percentage increase in E'_{22} with increasing V_f for $l/d = 19$, as given by Mori–Tanaka model, Modified Halpin–Tsai/Finegan model and FEM model, is shown in Table 4.6. From Table 4.5, it can be observed that Mori–Tanaka model and FEM model give almost same results for transverse loading. The difference between E'_{22} values given by Mori–Tanaka model and Modified Halpin–Tsai/Finegan model at large fiber volume fractions is less as compared to the large difference at low-volume fractions. Table 4.6 shows that Mori–Tanaka model gives highest percentage increase in E'_{22} at low-volume fractions. Also, FEM results are nearer to Mori–Tanaka results. At low-fiber volume fractions, Mori–Tanaka model gives greatest increase in E'_{22}, whereas at high-fiber volume fractions, Modified Halpin–Tsai/Finegan model gives highest increase in E'_{22}.

Table 4.5 Comparison between E'_{22} values given by different models for increasing V_f at $l/d = 19$.

	Difference in E'_{22} values (in %) between		
Fiber volume fraction (V_f)	Mori–Tanaka and Modified Halpin–Tsai/ Finegan model	Mori–Tanaka and FEM model	Modified Halpin–Tsai/ Finegan model and FEM model
0.02	12.88	1.26	11.76
0.06	15.13	1.33	13.97
0.16	6.70	1.22	4.70

Table 4.6 Percentage increase in E'_{22} with V_f given by different models for $l/d = 19$.

	Increase in E'_{22} values (in %)		
Range of fiber volume fraction (V_f)	Mori–Tanaka model	Modified Halpin–Tsai/ Finegan model	FEM model
0.0–0.06	40.24	19.02	38.41
0.06–0.16	19.65	31.76	18.90

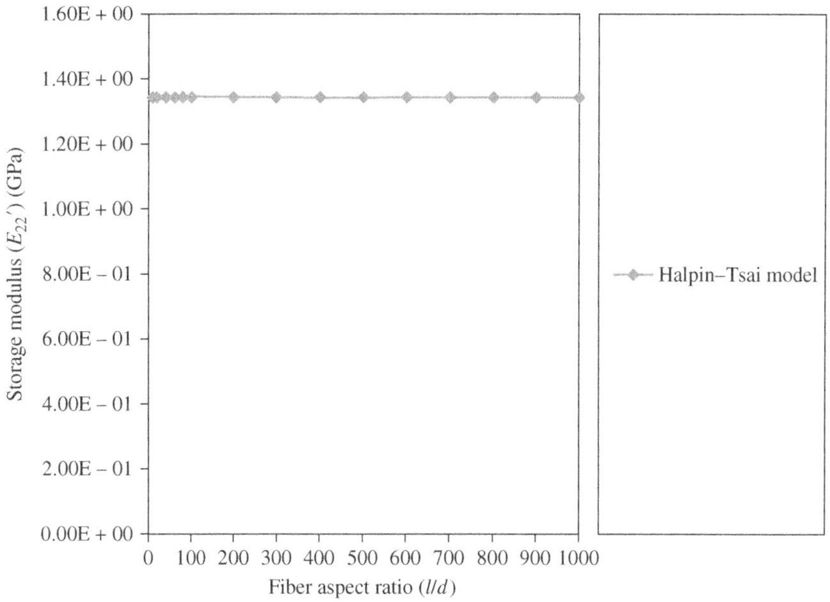

Figure 4.4 Variation of E'_{22} vs. l/d for $V_f = 0.03$ using Halpin–Tsai model.

4.1.2.2 Variation of E'_{22} with l/d

Halpin–Tsai/Finegan model shows no dependence of E'_{22} on l/d ratio as shown in Figure 4.4. This is because according to this model, E'_{22} does not depend on fiber aspect ratio (Eq. (3.121)), whereas Figure 4.5 shows that E'_{22} values first decrease at low-fiber aspect ratios (<100) and then become approximately constant at high-aspect ratios. This difference is due to the fact that other models consider the effect of interfiber interaction, whereas Halpin–Tsai model does not consider this effect. Another reason for the fall in E'_{22} with l/d (till $l/d = 100$) is that at low-fiber aspect ratios, number of fibers in the composite will be more for same volume fraction. More the number of discontinuous fibers, more is the number of fiber-ends available for stress concentration. Thus at low values of l/d, stress concentration at numerous fiber ends make the values of E'_{22} to fall. As the fiber aspect ratio increases, the modulus E'_{22} becomes a constant as shown in Figure 4.5.

Difference between Mori–Tanaka model and FEM values, for storage modulus E'_{22} for $V_f = 0.03$ and at $l/d = 100$, is 0.136% and at $l/d = 1000$ the difference is 0.098%. Differences between Mori–Tanaka model and Modified Halpin–Tsai/Finegan model and between FEM model and Modified Halpin–Tsai/Finegan model are shown in Table 4.7. Percentage decrease in E'_{22} with increasing l/d for $V_f = 0.03$ as given by Mori–Tanaka model, Modified Halpin–Tsai/Finegan model and FEM model is shown in Table 4.8. FEM model gives lower results for storage modulus E'_{22}. This is due to the fact that only actual state of stress is considered in FEM modeling.

From Table 4.7, it can be observed that difference between Mori–Tanaka model and Modified Halpin–Tsai/Finegan model almost remains the same for low-fiber aspect ratios as well as for high values of l/d. Similarly, the difference between Modified Halpin–Tsai/Finegan model and FEM model also remains the same for low as well as high values of fiber aspect ratios. Also, FEM results are nearer to Mori–Tanaka model. Table 4.8

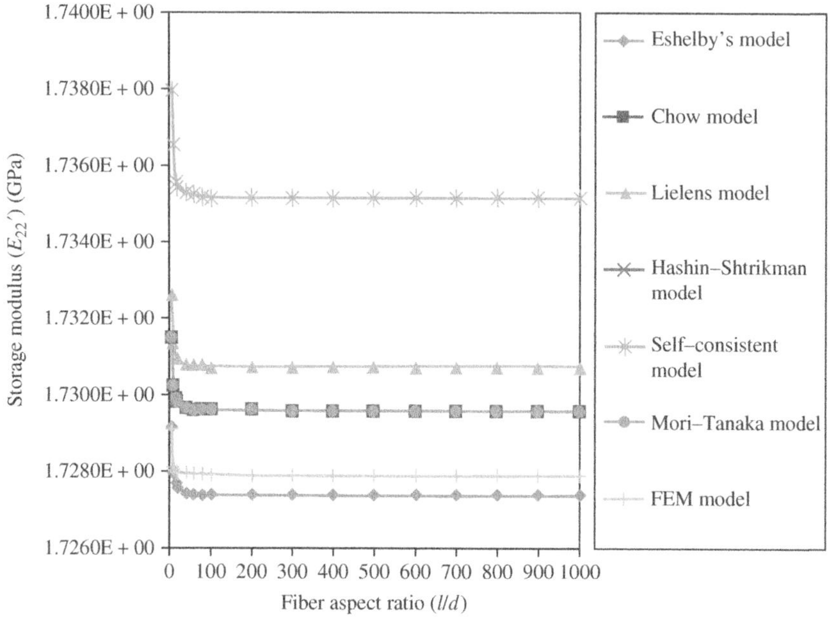

Figure 4.5 Variation of E'_{22} vs. l/d for $V_f = 0.03$.

Table 4.7 Comparison between E'_{22} values given by different models for increasing l/d ratios at $V_f = 0.03$.

	Difference in E'_{22} values (in %) between		
Fiber aspect ratio (l/d)	**Mori–Tanaka and Modified Halpin–Tsai/Finegan model**	**Mori–Tanaka and FEM model**	**Modified Halpin–Tsai/Finegan model and FEM model**
5	22.39	0.136	22.29
100	22.31	0.098	22.23
1000	22.31	0.098	22.23

Table 4.8 Percentage decrease in E'_{22} with l/d given by different models at $V_f = 0.03$.

	Decrease in E'_{22} values (in %)		
Range of fiber aspect ratio (l/d)	**Mori–Tanaka model**	**Modified Halpin–Tsai/Finegan model**	**FEM model**
5–200	0.108	Constant	0.0695
200–1000	0.000 65	Constant	0.0013

shows that Mori–Tanaka model gives the highest percentage decrease in E'_{22} at low-fiber aspect ratios, and the FEM results are nearer to it. Since Modified Halpin–Tsai/Finegan model is semiempirical, the results given by it are lower than both the Mori–Tanaka model and FEM model.

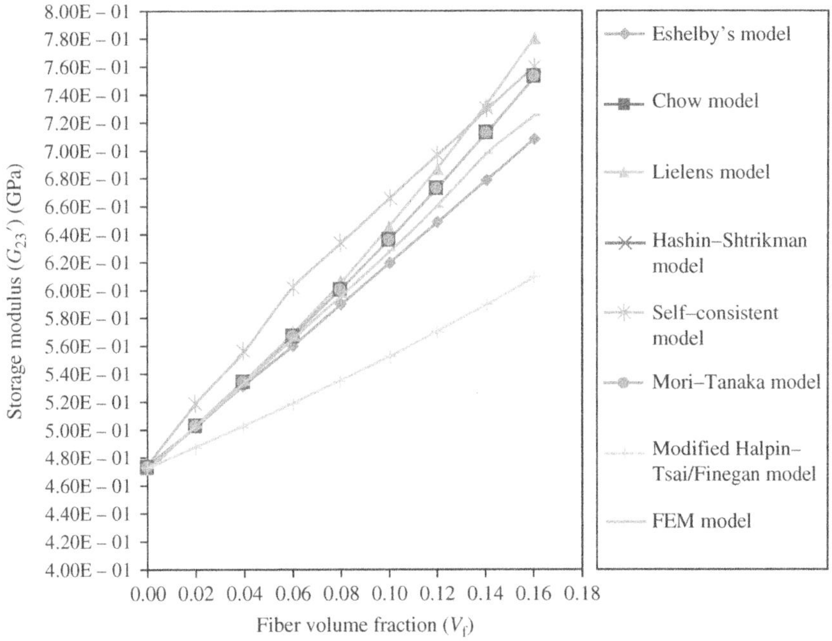

Figure 4.6 Variation of G'_{23} vs. V_f for $l/d = 19$.

4.1.3 Transverse Shear Storage Modulus (G'_{23})

4.1.3.1 Variation of G'_{23} with V_f

Shear storage modulus (G'_{23}), increases with fiber volume fraction (V_f), for a constant fiber aspect ratio ($l/d = 19$), as shown in Figure 4.6, but the increase is very small as compared to longitudinal storage modulus. This is due to the fact that the properties of the composite in shear are matrix-dominated. Here again, results of Halpin–Tsai/Finegan model are lower than other models. This is due to the fact explained earlier, i.e. Halpin–Tsai model is a semiempirical model.

As V_f increases, for a constant aspect ratio ($l/d = 19$), the volume of the fibers is increasing. Fibers have very high modulus as compared to the matrix. Hence, as the volume of fibers increases, the storage modulus increases. But increase in G'_{23} is insignificant as compared to the increase in the values of E'_{11} as can be seen by comparing Figures 4.1 and 4.6. This is due to the fact that reinforcing fibers improve the properties of the composite in the longitudinal direction considerably, whereas the transverse shear properties are only slightly improved.

Difference between Mori–Tanaka model and FEM values, for storage modulus G'_{23}, at $V_f = 0.02$ is 0.0755%. The difference increases to 3.704% at $V_f = 0.16$. These values are for $l/d = 19$. Similarly, the differences between Mori–Tanaka model and Modified Halpin–Tsai/Finegan model and between FEM model and Modified Halpin–Tsai/Finegan model are shown in Table 4.9. Percentage increase in G'_{23} with increasing V_f for $l/d = 19$, as given by Mori–Tanaka model, Modified Halpin–Tsai/Finegan model and FEM model, is shown in Table 4.10.

From Table 4.9, it can be observed that Mori–Tanaka model and FEM model give almost same results for transverse shear loading. Difference between G'_{23} values given by Mori–Tanaka model and Modified Halpin–Tsai/Finegan model at higher-fiber volume

Table 4.9 Comparison between G'_{23} values given by different models for increasing V_f at $l/d = 19$.

	Difference in G'_{23} values (in %) between		
Fiber volume fraction (V_f)	**Mori–Tanaka and Modified Halpin–Tsai/ Finegan model**	**Mori–Tanaka and FEM model**	**Modified Halpin–Tsai/ Finegan model and FEM model**
0.02	3.05	0.076	2.98
0.06	8.64	0.53	8.11
0.16	19.04	3.70	16.21

Table 4.10 Percentage increase in G'_{23} with V_f given by different models for $l/d = 19$.

	Increase in G'_{23} values (in %)		
Range of fiber volume fraction (V_f)	**Mori–Tanaka model**	**Modified Halpin–Tsai/ Finegan model**	**FEM model**
0.0–0.06	19.85	9.68	19.36
0.06–0.16	32.84	17.49	28.44

fractions is more as compared to the one at low-volume fractions. Table 4.10 shows that Mori–Tanaka model gives highest percentage increase in G'_{23} at both, low- as well as high-fiber volume fractions. Also, FEM results are nearer to Mori–Tanaka results. It also shows that the percentage increase in G'_{23} given by Modified Halpin–Tsai/Finegan model is less as compared to other models. Thus, it can be concluded that Modified Halpin–Tsai/Finegan model underpredicts the values of G'_{23}.

4.1.3.2 Variation of G'_{23} with l/d

Storage modulus G'_{23} first decreases with l/d (till $l/d = 100$) for a constant fiber volume fraction ($V_f = 0.03$) and then becomes constant as shown in Figure 4.7. This is probably due to the fact that at low-fiber aspect ratios number of fibers in the composite will be more for the same volume fraction. More the number of discontinuous fibers, more is the number of fiber-ends available for stress concentration. Thus at low values of l/d, the stress concentration at numerous fiber ends leads to fall in the values of G'_{23}. As the fiber aspect ratio increases, modulus G'_{23} becomes constant as shown in Figure 4.7. Modified Halpin–Tsai/Finegan model gives no variation of G'_{23} with l/d, as shown in Figure 4.8. This is because according to this model, shear modulus G'_{23} does not depend on fiber aspect ratio (Eq. (3.123)).

Difference between Mori–Tanaka model and FEM values, for storage modulus G'_{23} at $V_f = 0.03$ and $l/d = 100$, is 0.159%, whereas at $l/d = 1000$, the difference is 0.188%. Differences between Mori–Tanaka model and Modified Halpin–Tsai/Finegan model and between FEM model and Modified Halpin–Tsai/Finegan model are shown in Table 4.11. Percentage decrease in G'_{23} with increasing l/d for $V_f = 0.03$ as given by Mori–Tanaka model, Modified Halpin–Tsai/Finegan model and FEM model is shown in Table 4.12. FEM model gives

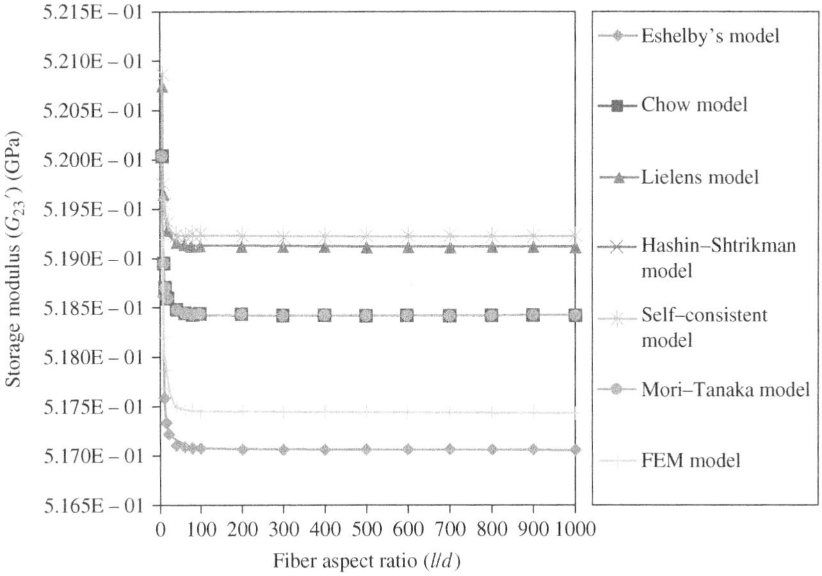

Figure 4.7 Variation of G'_{23} vs. l/d for $V_f = 0.03$.

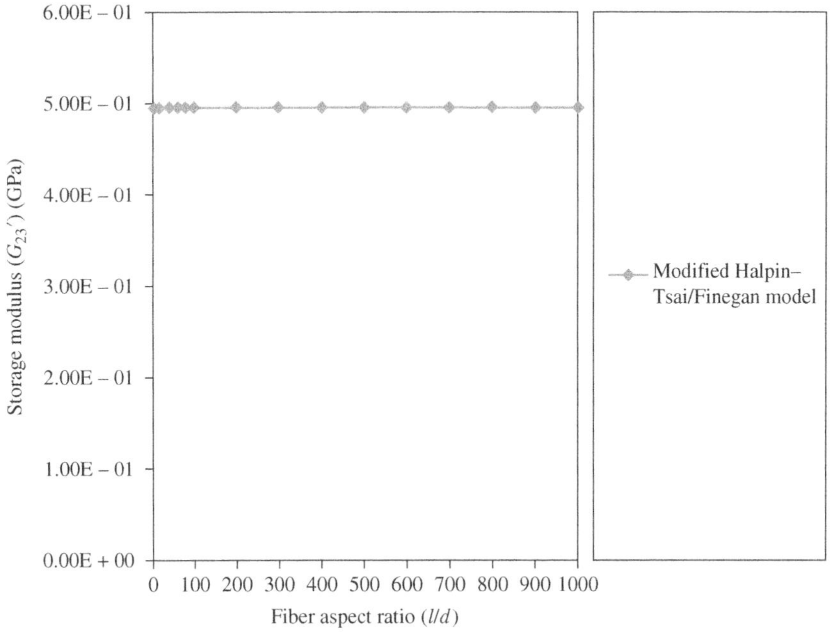

Figure 4.8 Variation of G'_{23} vs. l/d for $V_f = 0.03$ (Modified Halpi-Tsai Finegan model).

lower results for storage modulus. This is due to the fact that only the actual state of stress is considered in FEM modeling.

From Table 4.11, it can be observed that difference between Mori–Tanaka model and Modified Halpin–Tsai/Finegan model almost remains the same for low-fiber aspect ratios as well as for high values of l/d. Similarly, the difference between Modified Halpin–Tsai/Finegan

Table 4.11 Comparison between G'_{23} values given by different models for increasing l/d ratios at $V_f = 0.03$.

	Difference in G'_{23} values (in %) between		
Fiber aspect ratio (l/d)	Mori–Tanaka and Modified Halpin–Tsai/ Finegan model	Mori–Tanaka and FEM model	Modified Halpin–Tsai/ Finegan model and FEM model
5	4.76	0.159	4.60
100	4.46	0.187	4.28
1000	4.46	0.187	4.28

Table 4.12 Percentage decrease in G'_{23} with l/d given by different models at $V_f = 0.03$.

	Decrease in G'_{23} values (in %)		
Range of fiber aspect ratio (l/d)	Mori–Tanaka model	Modified Halpin–Tsai/ Finegan model	FEM model
5–200	0.308	Constant	0.307
200–1000	0.0019	Constant	0.0011

model and FEM model also remains the same for low as well as high values of fiber aspect ratios. Also, FEM results are nearer to Mori–Tanaka model. Table 4.12 shows that Mori–Tanaka model gives the highest percentage decrease in G'_{23} at low-fiber aspect ratios and the FEM results are nearer to it. Since Modified Halpin–Tsai/Finegan model is semiempirical, the results predicted by it are lower than both Mori–Tanaka model and FEM model.

4.1.4 Longitudinal Shear Storage Modulus (G'_{12})

4.1.4.1 Variation of G'_{12} with V_f

As compared to G'_{23}, the values of G'_{12} are higher as can be seen by comparing Figures 4.6 and 4.9. As shown in Figure 4.9, shear storage modulus (G'_{12}), increases with the fiber volume fraction (V_f) for a constant fiber aspect ratio ($l/d = 19$), but the increase is very small as compared to longitudinal storage modulus. This is again due to the fact that properties of composite in shear are matrix-dominated. As shown in Figure 4.10, results of Halpin–Tsai/Finegan are lower than other models. This is because it is a semiempirical model where one parameter is obtained by comparing experimental results with available 2-D finite element results.

Difference between Mori–Tanaka model and FEM values, for storage modulus G'_{12} at $V_f = 0.02$ is 0.078%. Difference in results increases to 4.14% at $V_f = 0.16$. These values are for $l/d = 19$. Similarly, the differences between Mori–Tanaka model and Modified Halpin–Tsai/Finegan model and between FEM model and Modified Halpin–Tsai/Finegan model are shown in Table 4.13. Percentage increase in G'_{23} with increasing V_f for $l/d = 19$, as given by Mori–Tanaka model, Modified Halpin–Tsai/Finegan model, and FEM model, is shown in Table 4.14.

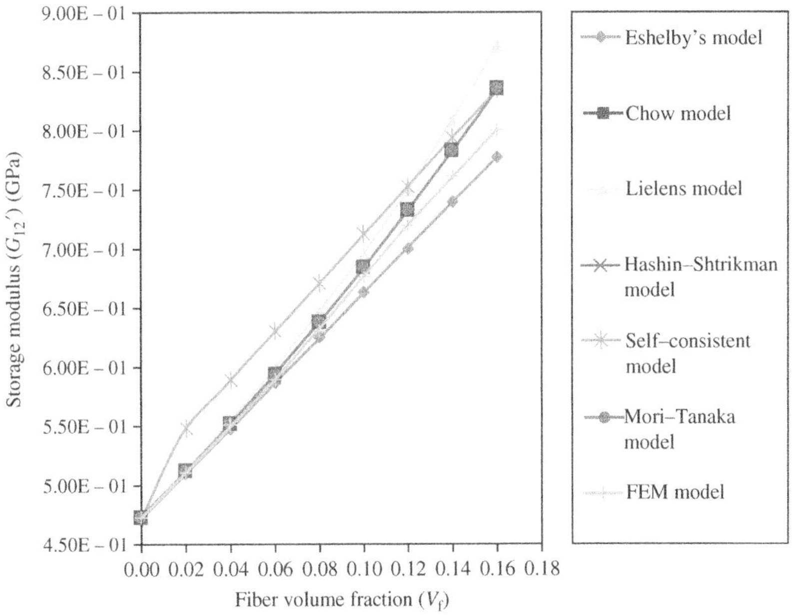

Figure 4.9 Variation of G'_{12} vs. V_f for $l/d = 19$.

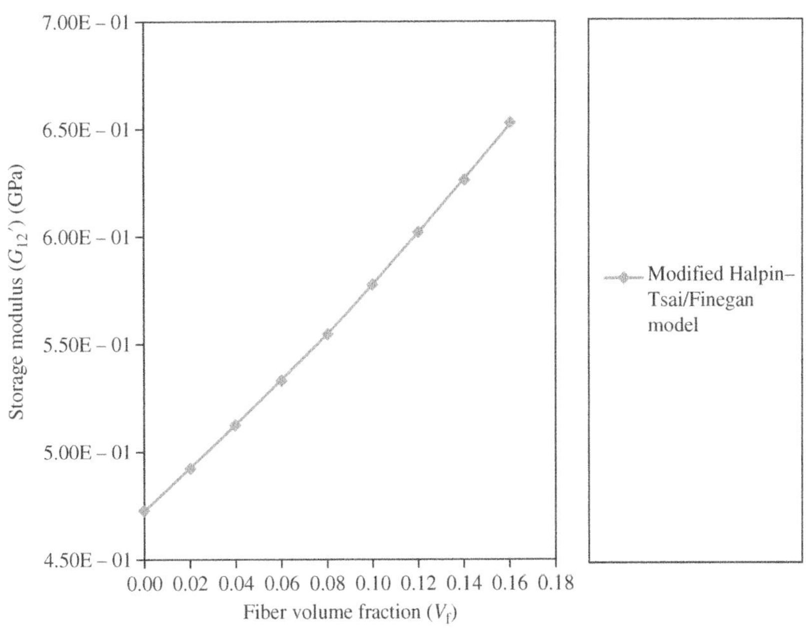

Figure 4.10 Variation of G'_{12} vs. V_f for $l/d = 19$ (Modified Halpi-Tsai Finegan model).

Table 4.13 Comparison between G'_{12} values given by different models for increasing V_f at $l/d = 19$.

Fiber volume fraction (V_f)	Difference in G'_{12} values (in %) between: Mori–Tanaka and Modified Halpin–Tsai/Finegan model	Mori–Tanaka and FEM model	Modified Halpin–Tsai/Finegan model and FEM model
0.02	3.81	0.078	3.72
0.06	10.27	0.66	9.67
0.16	21.81	4.14	18.44

Table 4.14 Percentage increase in G'_{12} with V_f given by different models for $l/d = 19$.

Range of fiber volume fraction (V_f)	Increase in G'_{12} values (in %): Mori–Tanaka model	Modified Halpin–Tsai/Finegan model	FEM model
0.0–0.06	25.65	12.73	24.80
0.06–0.16	40.46	22.40	35.55

From Table 4.13, it can be observed that Mori–Tanaka model and FEM model give almost same results for longitudinal shear loading. Difference between G'_{12} values given by Mori–Tanaka model and Modified Halpin–Tsai/Finegan model at large fiber volume fractions, is more as compared to the small difference at low volume fractions. Table 4.14 shows that Mori–Tanaka model gives highest percentage increase in G'_{12} at both, low- as well as high-volume fractions. Also, FEM results are nearer to Mori–Tanaka results. It also shows that the percentage increase in G'_{12} given by Modified Halpin–Tsai/Finegan model is less as compared to other models. Thus, it can be concluded that Modified Halpin–Tsai/Finegan model underpredicts the values of G'_{12}.

4.1.4.2 Variation of G'_{12} with l/d

Storage modulus (G'_{12}) first decreases sharply with fiber aspect ratio (l/d) for fiber volume fraction, $V_f = 0.03$ and then becomes constant as shown in Figure 4.11. This is probably due to the fact that at low-fiber aspect ratios, number of fibers in the composite will be more. More the number of discontinuous fibers, more is the number of fiber-ends available for stress concentration. Thus at low values of l/d, the stress concentration at numerous fiber ends makes the values of G'_{12} to fall. As the fiber aspect ratio increases, the modulus G'_{12} becomes constant as shown in Figure 4.11. Modified Halpin–Tsai/Finegan model gives no variation of G'_{12} with l/d, as shown in Figure 4.12. This is because according to this model, shear modulus G'_{12} does not depend on fiber aspect ratio (Eq. (3.122)).

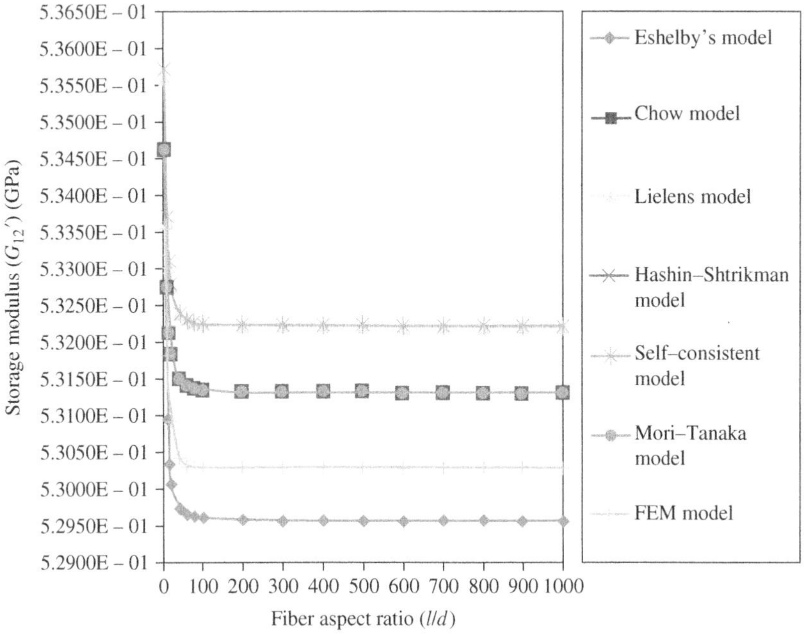

Figure 4.11 Variation of G'_{12} vs. l/d for $V_f = 0.03$.

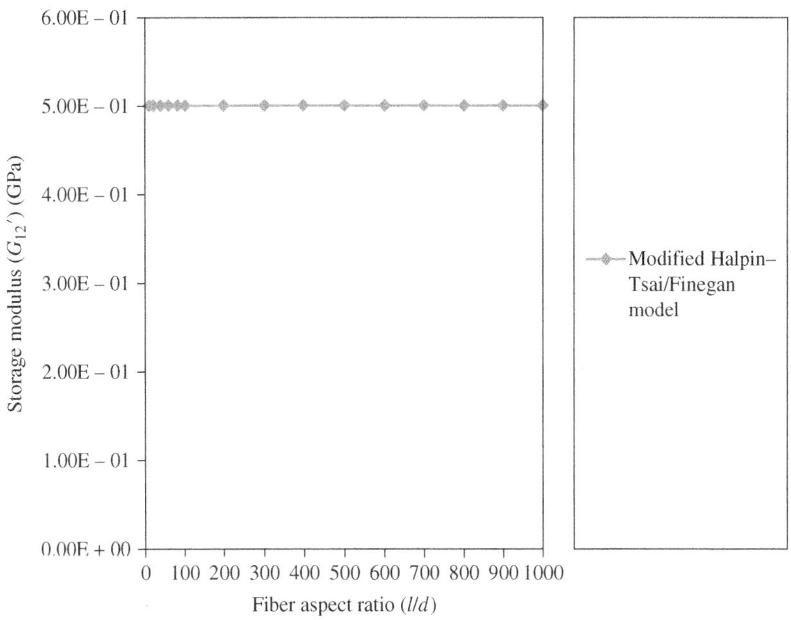

Figure 4.12 Variation of G'_{12} vs. l/d for $V_f = 0.03$ (Modified Halpi-Tsai Finegan model).

Table 4.15 Comparison between G'_{12} values given by different models for increasing l/d ratios at $V_f = 0.03$.

	Difference in G'_{12} values (in %) between		
Fiber aspect ratio (l/d)	Mori–Tanaka and Modified Halpin–Tsai/Finegan model	Mori–Tanaka and FEM model	Modified Halpin–Tsai/Finegan model and FEM model
5	4.76	0.159	4.60
100	4.46	0.187	4.28
1000	4.46	0.187	4.28

Table 4.16 Percentage decrease in G'_{12} with l/d given by different models at $V_f = 0.03$.

	Decrease in G'_{12} values (in %)		
Range of fiber aspect ratio (l/d)	Mori–Tanaka model	Modified Halpin–Tsai/Finegan model	FEM model
5–200	0.308	Constant	0.307
200–1000	0.0019	Constant	0.0011

Difference between Mori–Tanaka model and FEM values, for storage modulus G'_{12} for $V_f = 0.03$ and $l/d = 100$ is 0.196%, whereas at $l/d = 1000$, the difference is 0.192%. Differences between Mori–Tanaka model and Modified Halpin–Tsai/Finegan model and between FEM model and Modified Halpin–Tsai/Finegan model are shown in Table 4.15. Percentage decrease in G'_{12} with increasing l/d for $V_f = 0.03$ as given by Mori–Tanaka model, Modified Halpin–Tsai/Finegan model and FEM model is shown in Table 4.16. FEM model gives lower results for storage modulus. This is due to the fact that only the actual state of stress is considered in FEM modeling.

From Table 4.15, it can be observed that difference between Mori–Tanaka model and Modified Halpin–Tsai/Finegan model almost remains the same for low as well as for high values of l/d. Similarly, the difference between Modified Halpin–Tsai/Finegan model and FEM model also remains the same for low as well as high values of fiber aspect ratios. Also, FEM results are nearer to Mori–Tanaka model. Table 4.16 shows that Mori–Tanaka model gives the highest percentage decrease in G'_{12} at low-fiber aspect ratios, and the FEM results are nearer to it. Since Modified Halpin–Tsai/Finegan model is semiempirical, the results given by it are lower than both Mori–Tanaka model and FEM model.

4.2 Loss Factors

In this section, variation of different loss factors with both the fiber volume fraction as well as aspect ratio of fibers is discussed. Also, the results obtained for loss factor from different models are compared with each other as well as with the results obtained from FEM modeling and the results given by Finegan et al. [1].

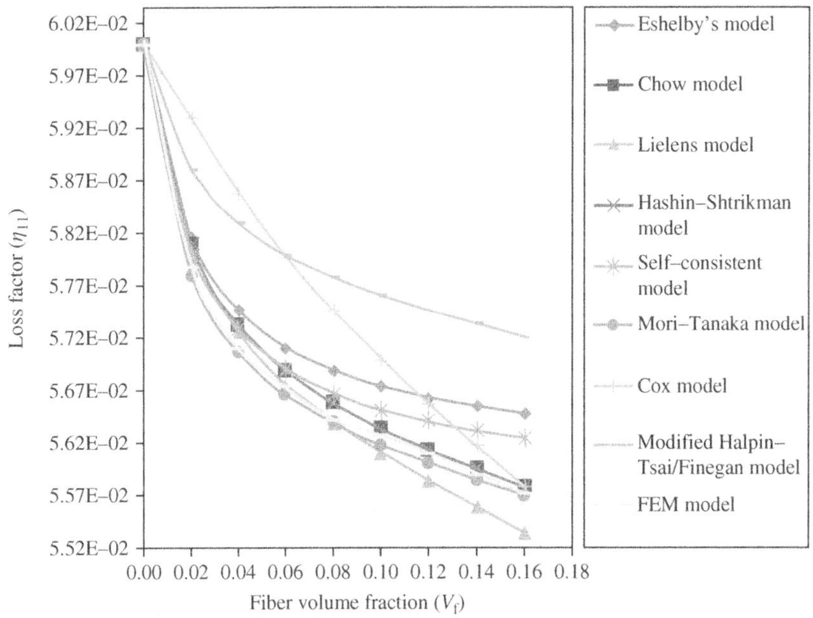

Figure 4.13 Variation of η_{11} vs. V_f for $l/d = 19$.

4.2.1 Longitudinal Loss Factor (η_{11})

4.2.1.1 Variation of η_{11} with V_f

From Figure 4.13, it can be observed that loss factor η_{11} decreases with V_f for a constant aspect ratio of $l/d = 19$. This is due to the fact that the relation between η_{11} and E'_{11} is

$$\eta_{11} \propto \frac{1}{E'_{11}} \tag{4.1}$$

Thus as E'_{11} increases, η_{11} goes on decreasing.

Difference between Mori–Tanaka model and FEM values for loss factor η_{11}, at $V_f = 0.02$ is 0.14%. The difference increases to 0.359% at $V_f = 0.16$. These values are at $l/d = 19$. Similarly, the differences between Mori–Tanaka model and Modified Halpin–Tsai/Finegan model and between FEM model and Modified Halpin–Tsai/Finegan model are shown in Table 4.17. Percentage decrease in η_{11} with increasing V_f at $l/d = 19$, as given by Mori–Tanaka model, Modified Halpin–Tsai/Finegan model and FEM model, is shown in Table 4.18.

Table 4.17 Comparison between η_{11} values given by different models for increasing V_f at $l/d = 19$.

	Difference in η_{11} values (in %) between		
Fiber volume fraction (V_f)	**Mori–Tanaka and Modified Halpin–Tsai/ Finegan model**	**Mori–Tanaka and FEM model**	**Modified Halpin–Tsai/ Finegan model and FEM model**
0.02	1.76	0.14	1.62
0.06	2.34	0.14	2.20
0.16	2.71	0.36	2.67

Table 4.18 Percentage decrease in η_{11} with V_f given by different models for $l/d = 19$.

Range of fiber volume fraction (V_f)	Decrease in η_{11} values (in %)		
	Mori–Tanaka model	Modified Halpin–Tsai/ Finegan model	FEM model
0.0–0.06	5.56	3.35	5.43
0.06–0.16	1.70	1.34	1.59

From Table 4.17, it can be observed that difference between Mori–Tanaka model and Modified Halpin–Tsai/Finegan model is less at low-fiber volume fractions than at high values of V_f. Also, FEM results are closer to Mori–Tanaka model. Difference between Modified Halpin–Tsai/Finegan model and FEM model is less at low values of fiber volume fractions than at high values of V_f. Table 4.18 shows that Mori–Tanaka model gives the highest percentage decrease in η_{11} at low-fiber volume fractions, and the FEM results are nearer to it. Since Modified Halpin–Tsai/Finegan model is semiempirical, the results given by it are lower than both the Mori–Tanaka model and FEM model. At low-fiber volume fractions all the models give almost same results and η_{11} decreases rapidly (till $V_f = 0.03$). But as the fiber volume fraction increases, the effect of interfiber interaction makes the loss factor to decrease slowly as compared to the decrease at lower values of V_f.

4.2.1.2 Variation of η_{11} with l/d

Figure 4.14 shows the variation of loss factor (η_{11}) with fiber aspect ratio using different models for $V_f = 0.03$. Because of the inverse relation between η_{11} and E'_{11}, loss factor η_{11}

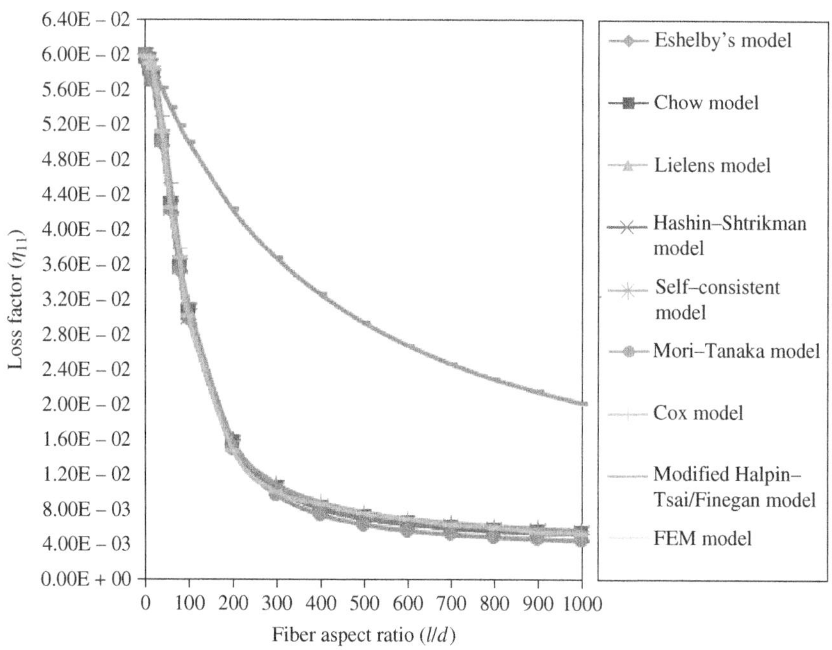

Figure 4.14 Variation of η_{11} vs. l/d for $V_f = 0.03$.

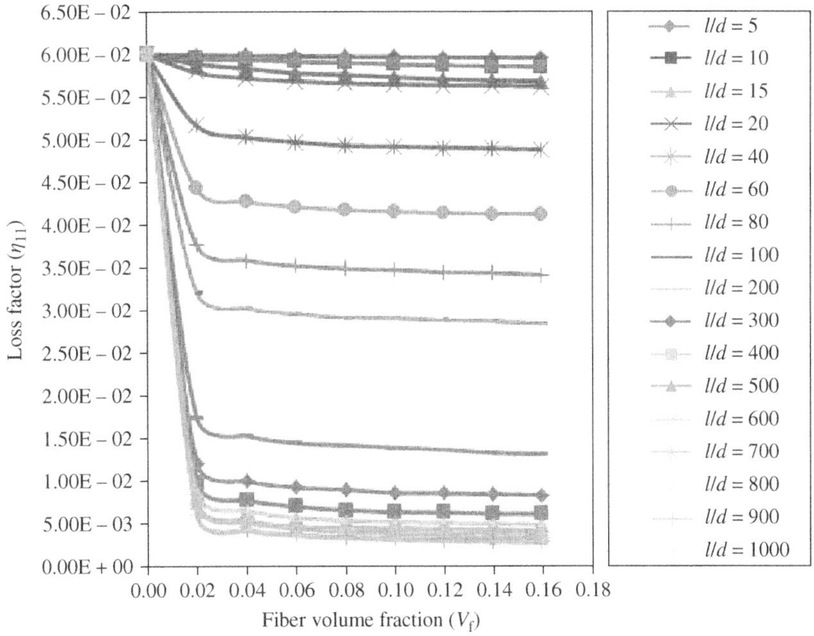

Figure 4.15 Mori–Tanaka model: variation of η_{11} vs. V_f for different l/d ratios.

decreases with increasing l/d ratios as can be observed from Figure 4.14. It can be observed that η_{11} decreases sharply upto $l/d = 100$, which is same for almost all the models. However, for $l/d > 100$, η_{11} decreases slowly. This is because, after $l/d = 100$, the fibers may be treated as continuous fibers and larger l/d causes higher-fiber axial stress than small aspect ratios. Figure 4.15 shows the variation of η_{11} with V_f for different l/d ratios using Mori–Tanaka model. It shows that fibers having low-aspect ratios give better values of damping as compared to the fibers having large aspect ratios. A possible reason for improved damping at low-fiber aspect ratios is the presence of shear stress concentration at numerous fiber ends in a short-fiber composite and the resulting stress transfer to the viscoelastic matrix. Also, the decrease in damping with increase in fiber volume fraction is more for the fibers having large aspect ratios as compared to the fibers having small aspect ratios as can be observed from Figure 4.15.

Difference between Mori–Tanaka model and FEM values, for loss factor η_{11} at $V_f = 0.03$ and $l/d = 100$ is 1.04%, whereas at $l/d = 1000$ the difference is 18.52%. Differences between Mori–Tanaka model and Modified Halpin–Tsai/Finegan model and between FEM model and Modified Halpin–Tsai/Finegan model are shown in Table 4.19. Percentage decrease in η_{11} with increasing l/d for $V_f = 0.03$ as given by Mori–Tanaka model, Modified Halpin–Tsai/Finegan model and FEM model is shown in Table 4.20.

From Table 4.19, it can be observed that difference between Mori–Tanaka model and Modified Halpin–Tsai/Finegan model is less at lower-fiber aspect ratio than at higher values of l/d. Also, FEM results are nearer to Mori–Tanaka model. Difference between Modified Halpin–Tsai/Finegan model and FEM model is also less at low values of fiber aspect ratios than at high values of l/d. Table 4.20 shows that Mori–Tanaka model gives the highest percentage decrease in η_{11} at low-fiber aspect ratios and the FEM results are nearer to it. Since Modified Halpin–Tsai/Finegan model is semiempirical, the results given by it are lower than both the Mori–Tanaka model and FEM model. At low-fiber aspect ratios all the models give

Table 4.19 Comparison between η_{11} values given by different models for increasing l/d ratios at $V_f = 0.03$.

	Difference in η_{11} values (in %) between		
Fiber aspect ratio (l/d)	Mori–Tanaka and Modified Halpin–Tsai/ Finegan model	Mori–Tanaka and FEM model	Modified Halpin–Tsai/ Finegan model and FEM model
5	0.12	0.06	0.17
100	67.76	1.04	66.04
1000	351	18.52	280.77

Table 4.20 Percentage decrease in η_{11} with l/d given by different models at $V_f = 0.03$.

	Increase in η_{11} values (in %)		
Range of fiber aspect ratio (l/d)	Mori–Tanaka model	Modified Halpin–Tsai/ Finegan model	FEM model
5–200	75.0	29.18	75.0
200–1000	70.07	52.29	64.27

almost same results and η_{11} decreases rapidly (till $l/d = 100$). But as the fiber aspect ratio increases, larger l/d causes higher-fiber axial stress than small aspect ratios and thus, the loss factor decreases less rapidly as compared to the decrease at lower values of l/d.

4.2.2 Transverse Loss Factor (η_{22})

4.2.2.1 Variation of η_{22} with V_f

The transverse loss factor (η_{22}) is decreasing with fiber volume fraction, as shown in Figure 4.16. As V_f increases for an aspect ratio $l/d = 19$, the volume of the fibers is increasing. The fibers have very high modulus as compared to the matrix. Hence, as the volume of fibers increases, the loss factor decreases. But decrease in η_{22} is insignificant as compared to the decrease in the values of η_{11} as can be seen by comparing Figures 4.13 and 4.16. This is due to the fact that reinforcing fibers improve the properties of the composite in the longitudinal direction considerably, whereas the transverse properties are only slightly improved as the force is applied perpendicular to the fibers which are weak in transverse direction. As shown in Figure 4.16, Halpin–Tsai model gives low values of η_{22} as compared to other models. This is probably because of the fact that Halpin–Tsai model was originally designed for continuous fiber-reinforced composites. The loss factor η_{22} is decreasing with V_f for a constant aspect ratio of $l/d = 19$. This is due to the fact that relation between η_{22} and E'_{22} is

$$\eta_{22} \propto \frac{1}{E'_{22}} \tag{4.2}$$

Difference between Mori–Tanaka model and FEM values for loss factor η_{22}, at $V_f = 0.02$ is 0.0017%. The difference becomes 0.0020% at $V_f = 0.16$. These values are at $l/d = 19$. Similarly,

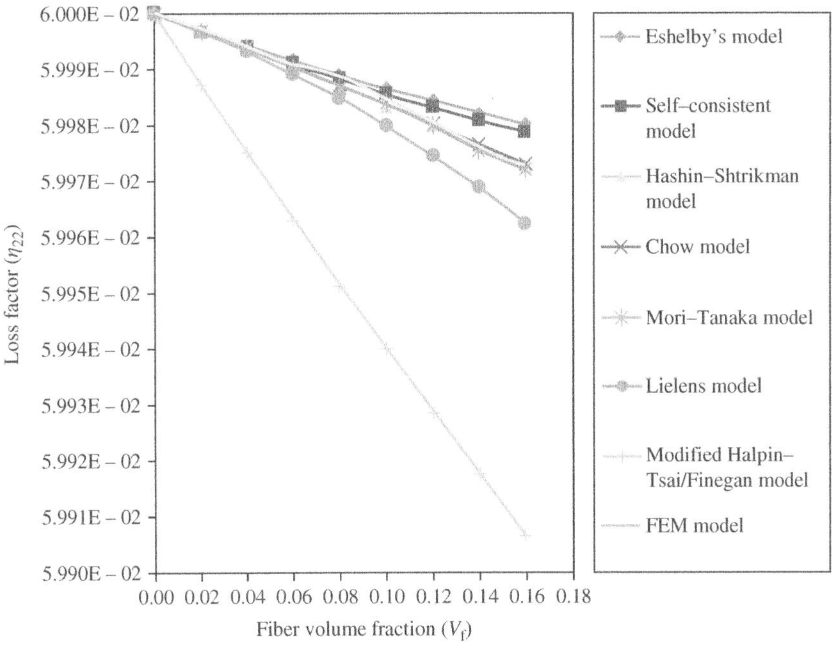

Figure 4.16 Variation of η_{22} vs. V_f for $l/d = 19$.

Table 4.21 Comparison between η_{22} values given by different models for increasing V_f at $l/d = 19$.

	Difference in η_{22} values (in %) between		
Fiber volume fraction (V_f)	**Mori–Tanaka and Modified Halpin–Tsai/ Finegan model**	**Mori–Tanaka and FEM model**	**Modified Halpin–Tsai/ Finegan model and FEM model**
0.02	0.015	0.0017	0.0175
0.06	0.045	0.0017	0.0465
0.16	0.110	0.0020	0.1107

the differences between Mori–Tanaka model and Modified Halpin–Tsai/Finegan model and between FEM model and Modified Halpin–Tsai/Finegan model are shown in Table 4.21. Percentage decrease in η_{22} with increasing V_f at $l/d = 19$, as given by Mori–Tanaka model, Modified Halpin–Tsai/Finegan model and FEM model, is shown in Table 4.22.

From Table 4.21, it can be observed that Mori–Tanaka model and FEM model give almost same results for transverse loading. Difference between η_{22} values given by Mori–Tanaka model and Modified Halpin–Tsai/Finegan model at higher-fiber volume fractions is more as compared to the one at low-fiber volume fractions. Table 4.22 shows that Modified Halpin–Tsai/Finegan model gives highest percentage decrease in η_{22} at low-volume fractions. Also, FEM results are nearer to Mori–Tanaka results.

4.2.2.2 Variation of η_{22} with l/d

Halpin–Tsai/Finegan model shows no dependence of η_{22} on l/d ratio as shown in Figure 4.17. This is because according to this model, η_{22} does not depend on fiber aspect ratio (Eq. (3.124)),

Table 4.22 Percentage decrease in η_{22} with V_f given by different models for $l/d = 19$.

Range of fiber volume fraction (V_f)	Decrease in η_{22} values (in %) Mori–Tanaka model	Modified Halpin–Tsai/ Finegan model	FEM model
0.0–0.06	0.0157	0.0614	0.0149
0.06–0.16	0.0310	0.0942	0.0298

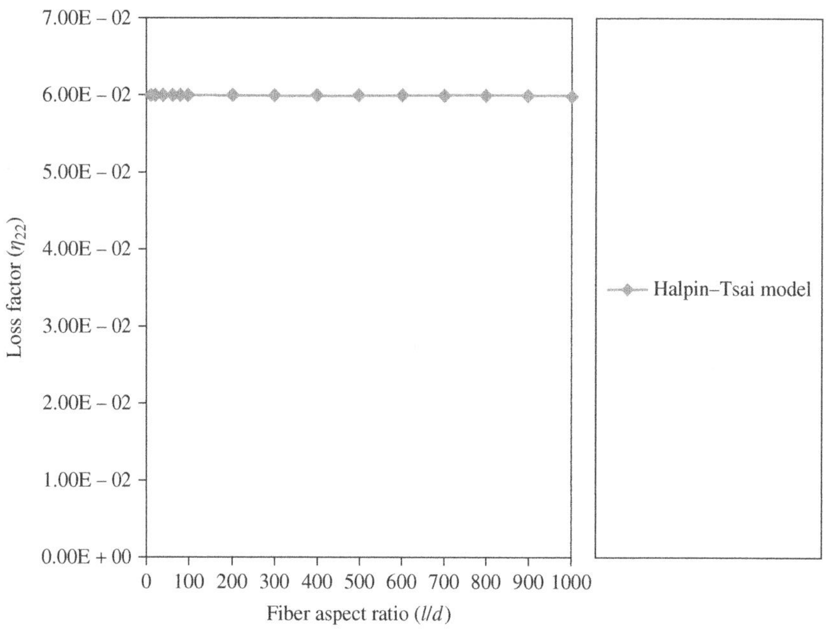

Figure 4.17 Variation of η_{22} vs. l/d for $V_f = 0.03$ using Halpin–Tsai model.

whereas Figure 4.18 shows that η_{22} values first increase at low-fiber aspect ratios (<100) and then become approximately constant at high aspect ratios. This difference is due to the fact that other models consider the effect of interfiber interaction, whereas Halpin–Tsai model does not consider this effect. Another reason for the rise in η_{22} with l/d (till $l/d = 100$) is that at low-fiber aspect ratios, number of fibers in the composite will be more for same volume fraction. More the number of discontinuous fibers, more will be the number of fiber-ends available for stress concentration. Thus at low values of l/d, stress concentration at numerous fiber ends make the values of η_{22} to rise. As the fiber aspect ratio increases, the modulus η_{22} becomes constant as shown in Figure 4.18. Figure 4.19 shows the variation of η_{22} with V_f for different l/d ratios using Mori–Tanaka model. It shows that l/d has little effect on η_{22} with varying fiber volume fraction. At all l/d ratios, η_{22} remains almost the same.

Difference between Mori–Tanaka model and FEM values for loss factor η_{22} at $V_f = 0.03$ and $l/d = 100$ is 0.000 287%, whereas at $l/d = 1000$ the difference is 0.000 281%. Differences between Mori–Tanaka model and Modified Halpin–Tsai/Finegan model and between FEM

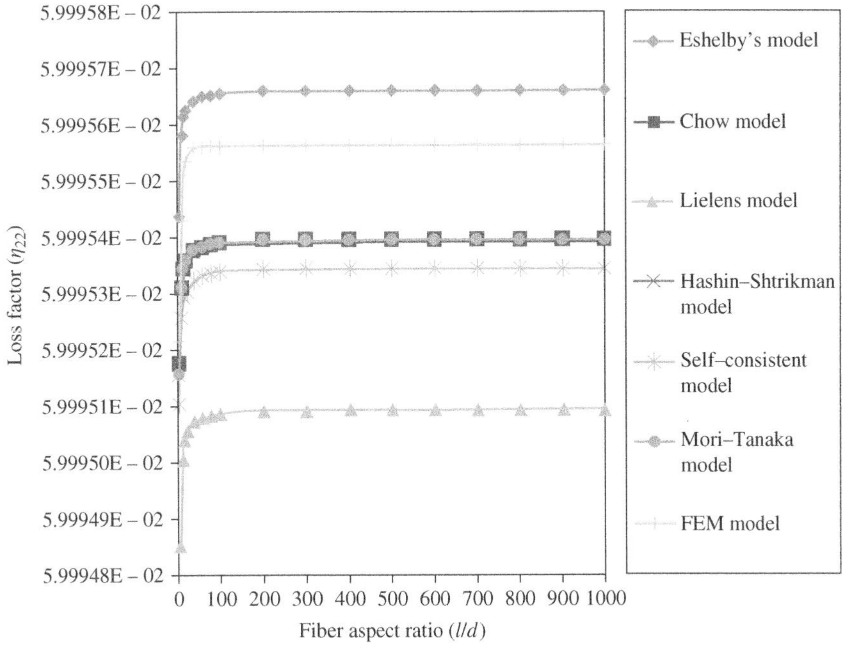

Figure 4.18 Variation of η_{22} vs. l/d for $V_f = 0.03$.

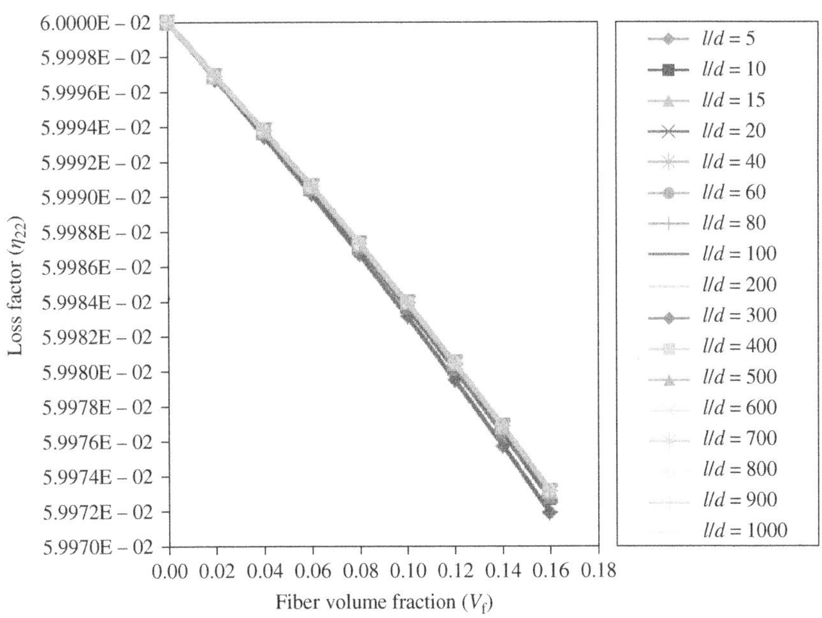

Figure 4.19 Variation of η_{22} vs. V_f for different l/d using Mori–Tanaka model.

Table 4.23 Comparison between η_{22} values given by different models for increasing l/d ratios at $V_f = 0.03$.

Fiber aspect ratio (l/d)	Difference in η_{22} values (in %) between: Mori–Tanaka and Modified Halpin–Tsai/Finegan model	Mori–Tanaka and FEM model	Modified Halpin–Tsai/Finegan model and FEM model
5	0.023	0.000 096	0.023
100	0.024	0.000 28	0.024
1000	0.024	0.000 28	0.024

Table 4.24 Percentage increase in η_{22} with l/d given by different models at $V_f = 0.03$.

Range of fiber aspect ratio (l/d)	Increase in η_{22} values (in %): Mori–Tanaka model	Modified Halpin–Tsai/Finegan model	FEM model
5–200	0.0038	Constant	0.000 58
200–1000	0.000 01	Constant	0.000 005

model and Modified Halpin–Tsai/Finegan model are shown in Table 4.23. Percentage increase in η_{22} with increasing l/d at $V_f = 0.03$ as given by Mori–Tanaka model, Modified Halpin–Tsai/Finegan model, and FEM model is shown in Table 4.24. FEM model gives higher results for loss factor η_{22}. This is due to the fact that there is an inverse relation between E'_{22} and η_{22} (Eq. (4.2)).

From Table 4.23, it can be observed that difference between Mori–Tanaka model and Modified Halpin–Tsai/Finegan model almost remains the same for low-fiber aspect ratios as well as for high values of l/d. Similarly, the difference between Modified Halpin–Tsai/Finegan model and FEM model also remains the same for low as well as high values of fiber aspect ratios. Also, FEM results are nearer to Mori–Tanaka model. Table 4.24 shows that Mori–Tanaka model gives the highest percentage increase in η_{22} at low-fiber aspect ratios, and FEM results are nearer to it. It also shows that for Modified Halpin–Tsai/Finegan model, η_{22} does not vary with l/d.

4.2.3 Transverse Shear Loss Factor (η_{23})

4.2.3.1 Variation of η_{23} with V_f

Shear loss factor (η_{23}), decreases with fiber volume fraction (V_f), for a constant fiber aspect ratio ($l/d = 19$), as shown in Figure 4.20, but the decrease is very small as compared to longitudinal loss factor. This is due to the fact that the properties of the composite in shear are matrix-dominated. Figure 4.20 shows that the shear loss factor (η_{23}) decreases with V_f. This is due to the relation given by Eq. (4.3):

$$\eta_{23} \propto \frac{1}{G'_{23}} \tag{4.3}$$

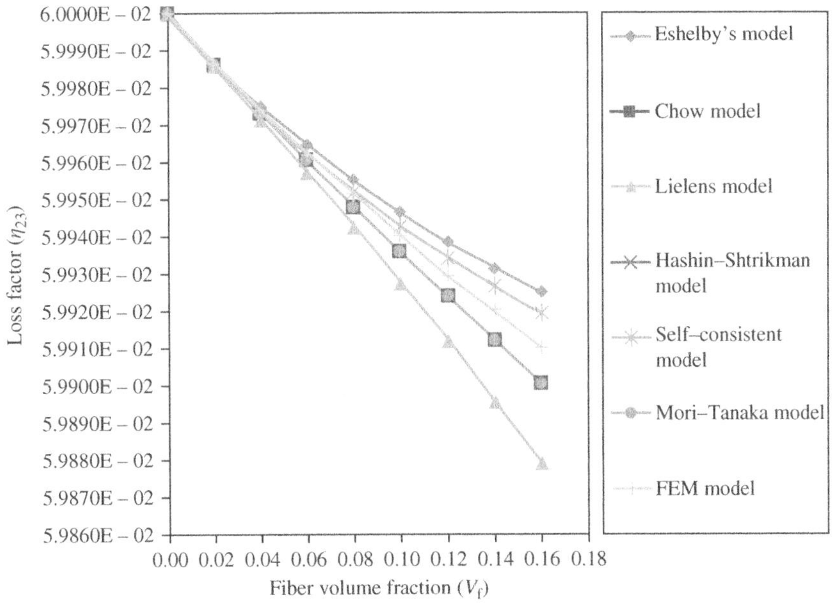

Figure 4.20 Variation of η_{23} vs. V_f for $l/d = 19$.

As V_f increases for a constant aspect ratio ($l/d = 19$), the volume of the fibers is increasing. Fibers have very high modulus as compared to the matrix. Hence, as the volume of fibers increases, the loss factor decreases. But decrease in η_{23} is insignificant as compared to the decrease in the values of η_{11} as can be seen by comparing Figures 4.13 and 4.20. This is due to the fact that reinforcing fibers improve the properties of the composite in the longitudinal direction considerably, whereas the transverse shear properties are only slightly improved. Figure 4.21 shows the variation of η_{23} with V_f for Modified Halpin–Tsai/Finegan model. Since the values of G'_{23} for this model are lower than other models, η_{23} values for this model will be higher, as observed by comparing Figures 4.20 and 4.21.

Difference between Mori–Tanaka model and FEM values for loss factor η_{23} at $V_f = 0.02$ is 0.0005%. The difference increases to 0.016% at $V_f = 0.16$. These values are at $l/d = 19$. Similarly, the differences between Mori–Tanaka model and Modified Halpin–Tsai/Finegan model and between FEM model and Modified Halpin–Tsai/Finegan model are shown in Table 4.25. Percentage decrease in η_{23} with increasing V_f for $l/d = 19$, as given by Mori–Tanaka model, Modified Halpin–Tsai/Finegan model, and FEM model is shown in Table 4.26.

From Table 4.25, it can be observed that Mori–Tanaka model and FEM model give almost same results for transverse shear loading. Difference between η_{23} values given by Mori–Tanaka model and Modified Halpin–Tsai/Finegan model at large-fiber volume fractions is more as compared to the small difference at low-volume fractions. Table 4.26 shows that Mori–Tanaka model gives highest percentage decrease in η_{23} at both, low- as well as high-volume fractions. Also, FEM results are nearer to Mori–Tanaka results. It also shows that the percentage decrease in η_{23} given by Modified Halpin–Tsai/Finegan model is less as compared to other models. Thus, it can be concluded that Modified Halpin–Tsai/Finegan model underpredicts the values of η_{23}. Here also, results of Modified Halpin–Tsai/Finegan model are lower than other models.

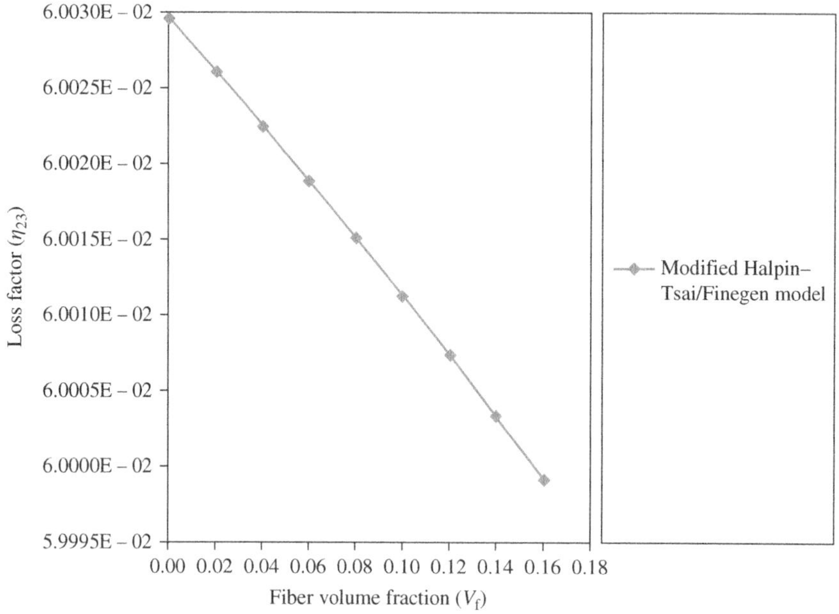

Figure 4.21 Variation of η_{23} vs. V_f for $l/d = 19$ using Halpin–Tsai model.

Table 4.25 Comparison between η_{23} values given by different models for increasing V_f at $l/d = 19$.

	Difference in η_{23} values (in %) between		
Fiber volume fraction (V_f)	**Mori–Tanaka and Modified Halpin–Tsai/ Finegan model**	**Mori–Tanaka and FEM model**	**Modified Halpin–Tsai/ Finegan model and FEM model**
0.02	0.0664	0.0005	0.0658
0.06	0.0976	0.0029	0.0947
0.16	0.1650	0.0162	0.1488

Table 4.26 Percentage decrease in η_{23} with V_f given by different models for $l/d = 19$.

	Decrease in η_{23} values (in %)		
Range of fiber volume fraction (V_f)	**Mori–Tanaka model**	**Modified Halpin–Tsai/ Finegan model**	**FEM model**
0.0–0.06	0.066	0.012	0.063
0.06–0.16	0.100	0.032	0.086

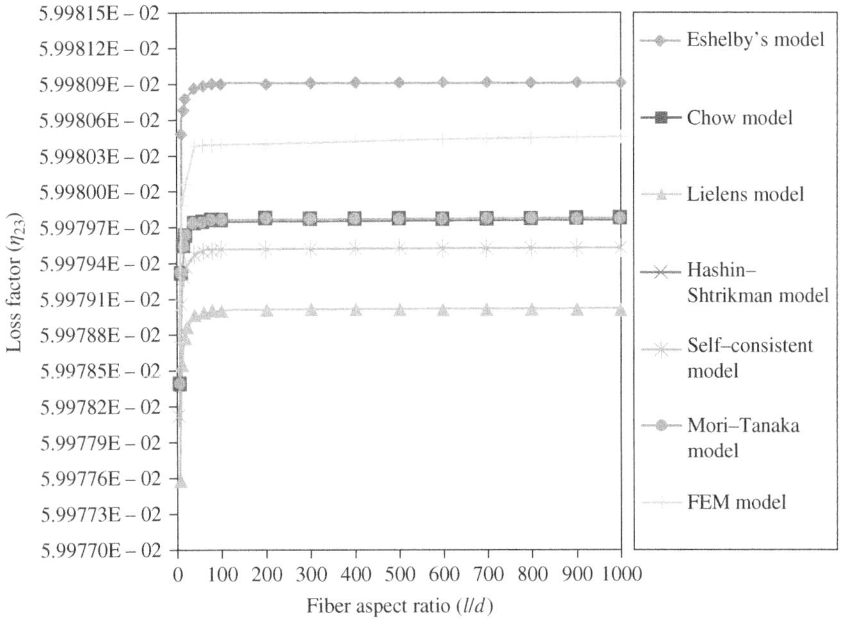

Figure 4.22 Variation of shear loss factor (η_{23}) vs. l/d for $V_f = 0.03$.

4.2.3.2 Variation of η_{23} with l/d

Loss factor η_{23} first increases with l/d (till $l/d = 100$) for a constant fiber volume fraction ($V_f = 0.03$) and then becomes constant as shown in Figure 4.22. This is probably due to the fact that at low-fiber aspect ratios number of fibers in the composite will be more for the same volume fraction. More the number of discontinuous fibers, more is the number of fiber-ends available for stress concentration. Thus at low values of l/d, the stress concentration at numerous fiber ends make the values of η_{23} to rise. As the fiber aspect ratio increases, η_{23} becomes a constant as shown in Figure 4.22. Modified Halpin–Tsai/Finegan model gives no variation of η_{23} with l/d, as shown in Figure 4.23. This is because according to this model, η_{23} does not depend on fiber aspect ratio (Eq. (3.125)). Figure 4.24 shows the variation of η_{23} with V_f for different l/d ratios using Mori–Tanaka model. It shows that l/d has little effect on η_{23} with varying fiber volume fraction. At all l/d ratios, η_{23} remains almost the same.

Difference between Mori–Tanaka model and FEM values for loss factor η_{23} at $V_f = 0.03$ and $l/d = 100$ is 0.0011%, whereas at $l/d = 1000$ difference remains the same. Differences between Mori–Tanaka model and Modified Halpin–Tsai/Finegan model and between FEM model and Modified Halpin–Tsai/Finegan model are shown in Table 4.27. Percentage increase in η_{23} with increasing l/d at $V_f = 0.03$ as given by Mori–Tanaka model, Modified Halpin–Tsai/Finegan model and FEM model is shown in Table 4.28. FEM model gives higher results for loss factor. This is due to the fact that storage modulus G'_{23} values given by FEM are lower and G'_{23} is inversely proportional to η_{23} (Eq. (4.3)).

From Table 4.27, it can be observed that difference between Mori–Tanaka model and Modified Halpin–Tsai/Finegan model almost remains the same for low-fiber aspect ratios as well as for high values of l/d. Similarly, the difference between Modified Halpin–Tsai/Finegan model and FEM model also remains the same for low as well as high values of fiber aspect ratios. Also, FEM results are nearer to Mori–Tanaka model. Table 4.28 shows

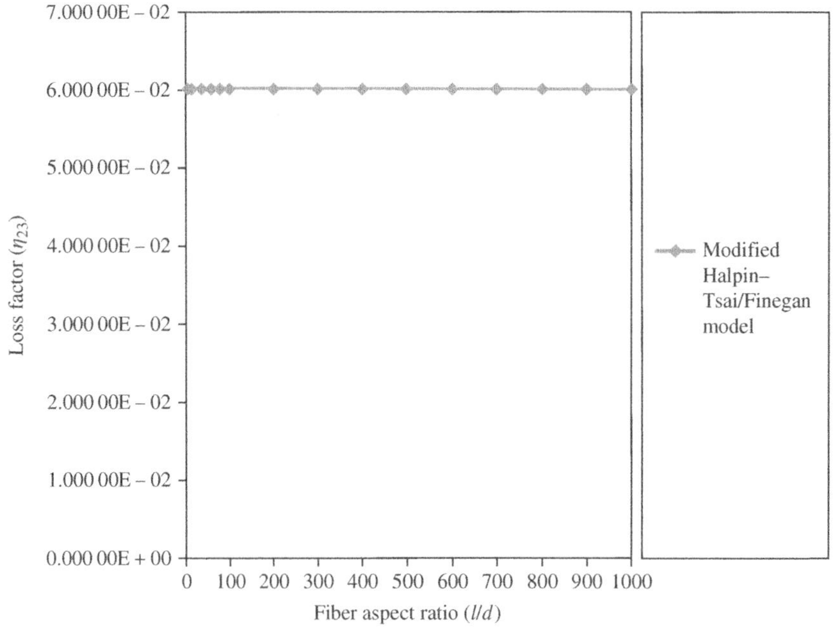

Figure 4.23 Variation of shear loss factor (η_{23}) vs. l/d for $V_f = 0.03$ (Modified Halpi-Tsai Finegan model).

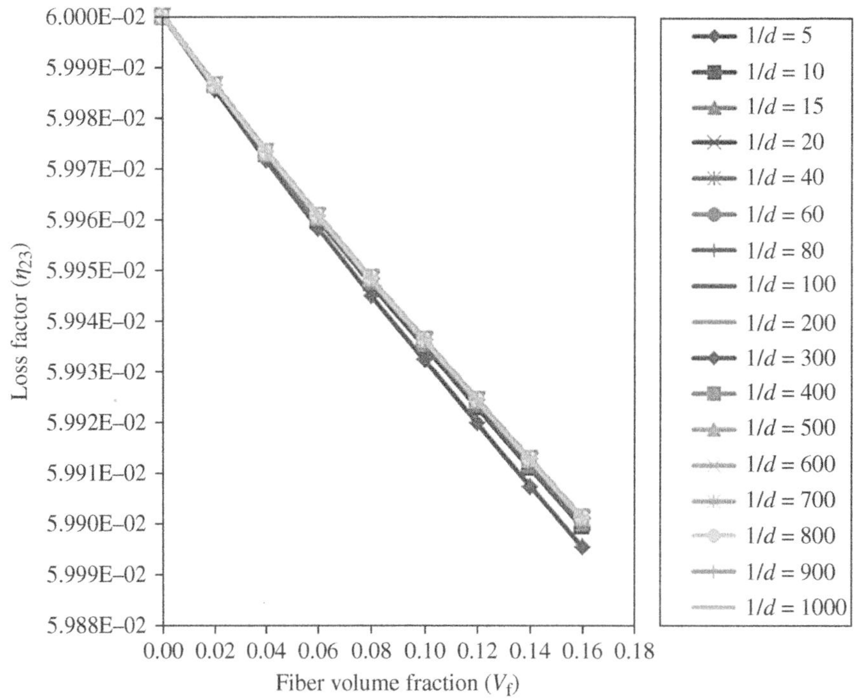

Figure 4.24 Variation of η_{23} vs. V_f for different l/d using Mori–Tanaka model.

Table 4.27 Comparison between η_{23} values given by different models for increasing l/d ratios at $V_f = 0.03$.

	Difference in η_{23} values (in %) between		
Fiber aspect ratio (l/d)	Mori–Tanaka and Modified Halpin–Tsai/ Finegan model	Mori–Tanaka and FEM model	Modified Halpin–Tsai/ Finegan model and FEM model
5	0.0765	0.0006	0.0759
100	0.0742	0.0011	0.0732
1000	0.0742	0.0011	0.0732

Table 4.28 Percentage increase in η_{23} with l/d given by different models at $V_f = 0.03$.

	Increase in η_{23} values (in %)		
Range of fiber aspect ratio (l/d)	Mori–Tanaka model	Modified Halpin–Tsai/ Finegan model	FEM model
5–200	0.0023	Constant	0.0023
200–1000	0.000 015	Constant	0.000 01

that Mori–Tanaka model gives the highest percentage increase in η_{23} at low-fiber aspect ratios and the FEM results are nearer to it. Since Modified Halpin–Tsai/Finegan model is semiempirical, the results given by it are lower than both Mori–Tanaka model and FEM model.

4.2.4 Longitudinal Shear Loss Factor (η_{12})

4.2.4.1 Variation of η_{12} with V_f

As compared to η_{23}, values of η_{12} are lower as can be seen by comparing Figures 4.20 and 4.25. As shown in Figure 4.25, shear loss factor (η_{12}) decreases with fiber volume fraction (V_f) for a constant fiber aspect ratio ($l/d = 19$), but the decrease is very small as compared to $\eta_{11.}$ This is again due to the fact that properties of composite in shear are matrix-dominated. Figure 4.26 shows that the shear loss factor (η_{12}) given by Modified Halpin–Tsai/Finegan model decreases with V_f. This is due to the relation given by Eq. (4.4):

$$\eta_{12} \propto \frac{1}{G'_{12}} \tag{4.4}$$

As shown in Figure 4.26, results of Modified Halpin–Tsai/Finegan model are higher than other models. This is because G'_{12} is inversely proportional to η_{12} (Eq. (4.4)).

Difference between Mori–Tanaka model and FEM values for loss factor η_{12} at $V_f = 0.02$ is 0.000 76%. This difference increases to 0.040% at $V_f = 0.16$. These values are for $l/d = 19$. Similarly, the differences between Mori–Tanaka model and Modified Halpin–Tsai/Finegan model and between FEM model and Modified Halpin–Tsai/Finegan model are shown in Table 4.29. Percentage decrease in η_{12} with increasing V_f for $l/d = 19$, as given by Mori–Tanaka model, Modified Halpin–Tsai/Finegan model and FEM model, is shown in Table 4.30.

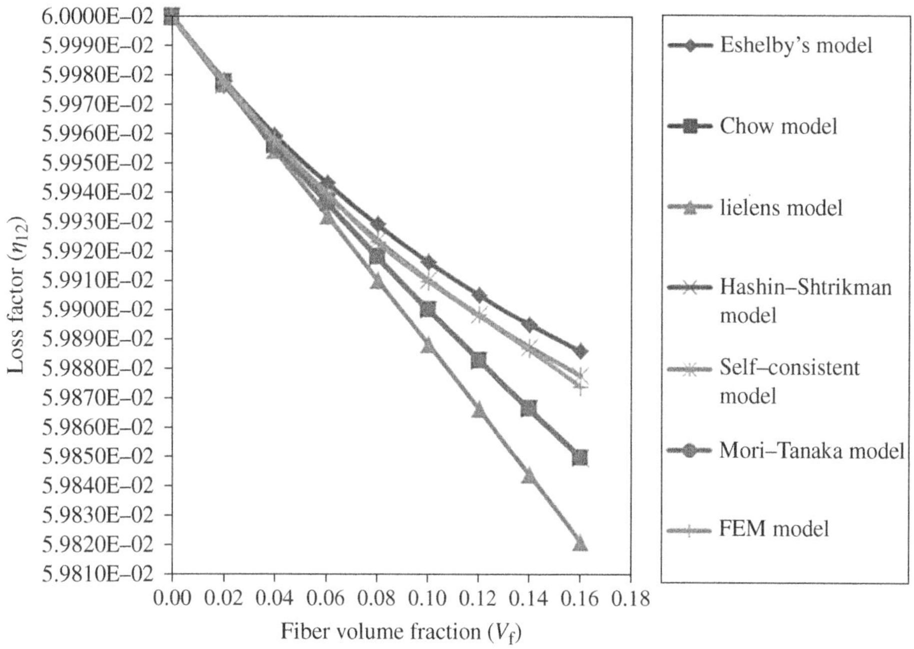

Figure 4.25 Variation of η_{12} vs. V_f for $l/d = 19$.

Table 4.29 Comparison between η_{12} values given by different models for increasing V_f at $l/d = 19$.

	Difference in η_{12} values (in %) between		
Fiber volume fraction (V_f)	Mori–Tanaka and Modified Halpin–Tsai/ Finegan model	Mori–Tanaka and FEM model	Modified Halpin–Tsai/ Finegan model and FEM model
0.02	0.0772	0.000 76	0.074 69
0.06	0.1258	0.005 17	0.120 62
0.16	0.2209	0.0401	0.180 80

From Table 4.29, it can be observed that Mori–Tanaka model and FEM model give almost same results for η_{12}. Difference between η_{12} values given by Mori–Tanaka model and Modified Halpin–Tsai/Finegan model at large-fiber volume fractions is more as compared to the small difference at low-volume fractions. Table 4.30 shows that Mori–Tanaka model gives highest percentage decrease in η_{12} at both, low- as well as high-volume fractions. Also, FEM results are nearer to Mori–Tanaka results. It also shows that the percentage decrease in η_{12} given by Modified Halpin–Tsai/Finegan model is less as compared to other models. Thus, it can be concluded that Modified Halpin–Tsai/Finegan model underpredicts the values of η_{12}.

4.2.4.2 Variation of η_{12} with l/d

Loss factor (η_{12}) first increases (till $l/d = 100$) with fiber aspect ratio (l/d) for fiber volume fraction ($V_f = 0.03$) and then becomes constant as shown in Figure 4.27. This is probably due

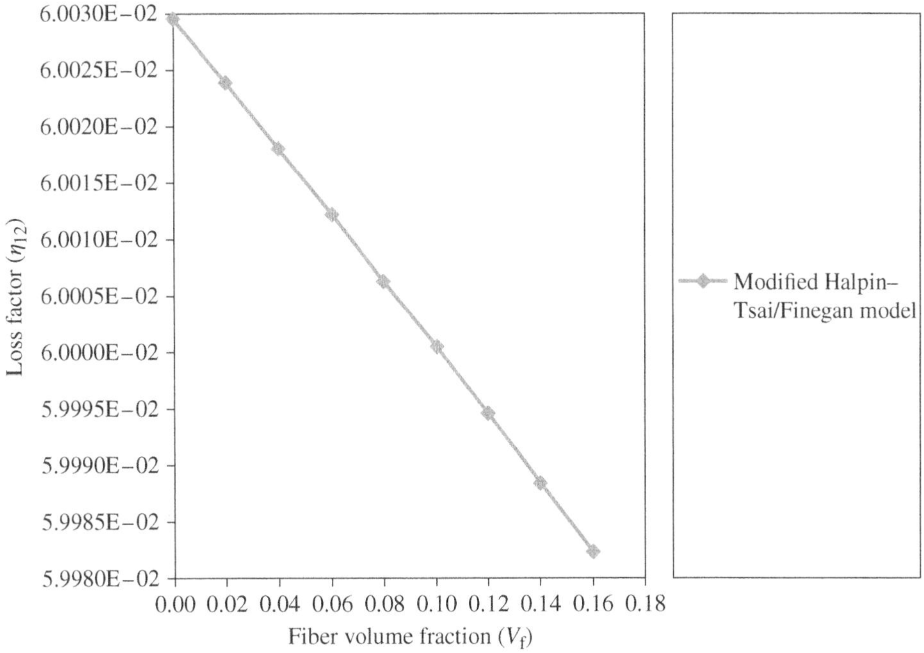

Figure 4.26 Variation of η_{12} vs. V_f for $l/d = 19$ (Modified Halpi-Tsai Finegan model).

Table 4.30 Percentage decrease in η_{12} with V_f given by different models for $l/d = 19$.

	Decrease in η_{12} values (in %)		
Range of fiber volume fraction (V_f)	**Mori–Tanaka model**	**Modified Halpin–Tsai/ Finegan model**	**FEM model**
0.0–0.06	0.105	0.019	0.100
0.06–0.16	0.144	0.050	0.110

to the fact that at low-fiber aspect ratios, number of fibers in the composite will be more. More the number of discontinuous fibers, more is the number of fiber-ends available for stress concentration. Thus at low values of l/d, the stress concentration at numerous fiber ends make the values of η_{12} to rise. As the fiber aspect ratio increases, η_{12} becomes a constant as shown in Figure 4.27. Modified Halpin–Tsai/Finegan model gives no variation of η_{12} with l/d, as shown in Figure 4.28. This is because according to this model, η_{12} does not depend on fiber aspect ratio (Eq. (3.124)). Figure 4.29 shows the variation of η_{12} with V_f for different l/d ratios using Mori–Tanaka model. It shows that l/d has little effect on η_{12} with varying fiber volume fraction. At all l/d ratios, η_{12} remains almost the same.

Difference between Mori–Tanaka model and FEM values, for loss factor η_{12} for $V_f = 0.03$ and $l/d = 100$ is 0.001%, whereas at $l/d = 1000$ the difference is almost the same. Differences between Mori–Tanaka model and Modified Halpin–Tsai/Finegan model and between FEM model and Modified Halpin–Tsai/Finegan model are shown in Table 4.31. The percentage

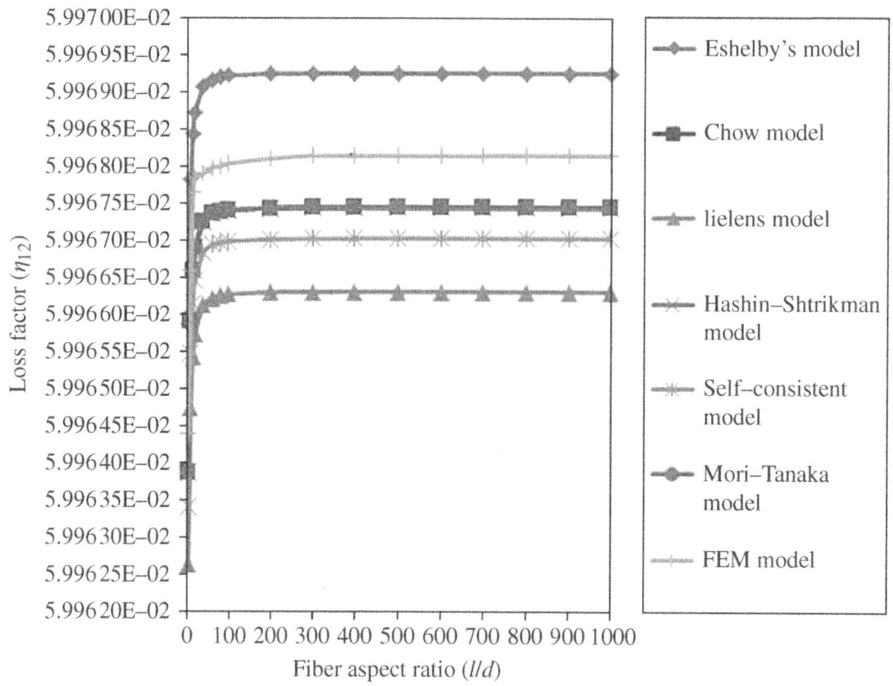

Figure 4.27 Variation of η_{12} vs. l/d for $V_f = 0.03$.

Table 4.31 Comparison between η_{12} values given by different models for increasing l/d ratios at $V_f = 0.03$.

	Difference in η_{12} values (in %) between		
Fiber aspect ratio (l/d)	**Mori–Tanaka and Modified Halpin–Tsai/ Finegan model**	**Mori–Tanaka and FEM model**	**Modified Halpin–Tsai/ Finegan model and FEM model**
5	0.095	0.0009	0.094
100	0.089	0.0010	0.088
1000	0.089	0.0012	0.088

increase in η_{12} with increasing l/d at $V_f = 0.03$ given by the Mori–Tanaka model, Modified Halpin–Tsai/Finegan model and FEM model is shown in Table 4.32. FEM model gives higher results for η_{12}. This is due to the fact that only the actual state of stress is considered in FEM modeling and G'_{12} values obtained by FEM models are lower and therefore η_{12} values are higher.

From Table 4.31, it can be observed that difference between Mori–Tanaka model and Modified Halpin–Tsai/Finegan model almost remains the same for low as well as for high values of l/d. Similarly, the difference between Modified Halpin–Tsai/Finegan model and FEM model also remains the same for low as well as high values of fiber aspect ratios. Also, FEM results are nearer to Mori–Tanaka model. Table 4.32 shows that Mori–Tanaka model gives the highest percentage decrease in η_{12} at low-fiber aspect ratios and the FEM results are nearer to

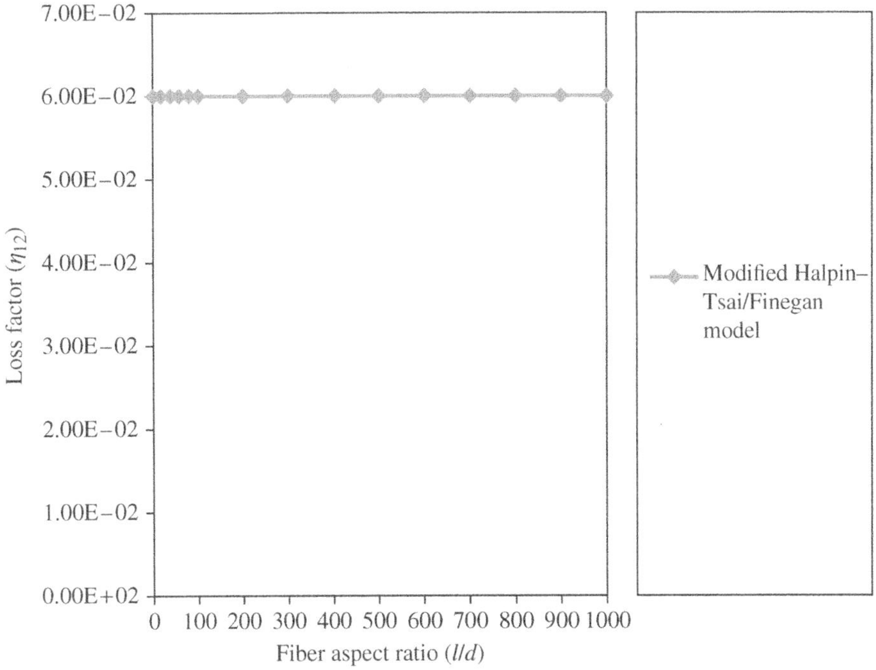

Figure 4.28 Variation of η_{12} vs. l/d for $V_f = 0.03$ (Modified Halpi-Tsai Finegan model).

Table 4.32 Percentage increase in η_{12} with l/d given by different models at $V_f = 0.03$.

	Increase in η_{12} values (in %)		
Range of fiber aspect ratio (l/d)	**Mori–Tanaka model**	**Modified Halpin–Tsai/ Finegan model**	**FEM model**
5–200	0.006	Constant	0.006
200–1000	0.0006	Constant	0.0002

it. Since Modified Halpin–Tsai/Finegan model is semiempirical, the results given by it are lower than both Mori–Tanaka model and FEM model.

4.3 Conclusions

Storage moduli and loss factors of vapor grown carbon fiber (VGCF)/pp nanocomposites have been predicted for different fiber volume fractions and fiber aspect ratios using different micromechanical models such as Eshelby's model, Mori–Tanaka model, Cox shear lag model, Modified Halpin–Tsai or Finegan model, Hashin–Shtrikman model, Chow model, Self-consistent model, and Lielens model. Results predicted by above-listed models have been compared with those predicted by FEM/strain energy approach and also with the results predicted by Finegan et al. [1]. Predictions have been made for longitudinal storage modulus

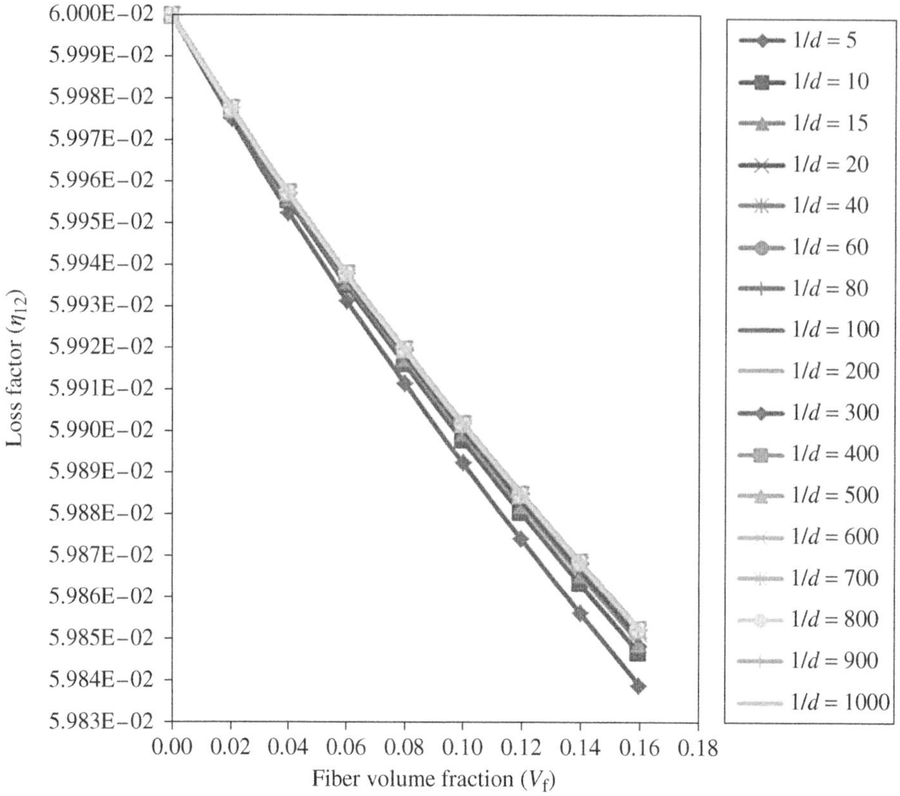

Figure 4.29 Variation of η_{12} vs. V_f for different l/d using Mori–Tanaka model.

(E'_{11}), transverse storage modulus (E'_{22}), transverse shear storage modulus (G'_{23}), and longitudinal shear storage modulus (G'_{12}). Also, predictions have been made for longitudinal loss factor (η_{11}), transverse loss factor (η_{22}), transverse shear loss factor (η_{23}), and longitudinal shear loss factor (η_{12}). On the basis of above discussion, it can be concluded that most of the models give a good job in approximating modulus and damping properties of short fiber-reinforced composites, and the results of FEM and Finegan et al. [1] are in good agreement with the presented work. Following conclusions can be drawn from the above discussion.

(a) Storage moduli
   (i) At low-fiber volume fractions, the matrix stiffness strongly influences the composite stiffness and hence low values of storage modulus at low-fiber volume fractions.
   (ii) Halpin–Tsai model gives satisfactory results for short fibers but falls below for longer fibers. Halpin–Tsai model contain no dependence on aspect ratio for E'_{22}, G'_{23}, and G'_{12}. Halpin–Tsai equations underpredict the values of G'_{23} and G'_{12}.
   (iii) Lielens model reproduces Mori–Tanaka results at low-fiber volume fractions, but gives improved results at high-volume fractions. Self-consistent model usually gives higher results as it starts with making an assumption of the composite properties. Cox model gives good results for longer fibers, but low for short fibers. Eshelby's model is found to predict slightly higher values of E'_{11} for higher V_f because it is

strictly valid for inclusion in a far field, in which a low-fiber volume fraction is presumed.

(iv) Aspect ratio significantly influences the longitudinal storage modulus, but has little effect on E'_{22}, G'_{23}, and G'_{12}.

(v) Mori–Tanaka and Lielens models are the most accurate in predicting E'_{11} and E'_{22}, although, most of the models give a good prediction of these two storage moduli.

(vi) FEM results are lower than the predicted values for storage modulus.

(vii) All the storage moduli increase with the increase in fiber volume fraction.

(viii) Storage moduli E'_{22}, G'_{23}, and G'_{12}, first decrease, with increase in l/d, till $l/d = 100$, for a constant fiber volume fraction ($V_f = 0.03$), and then becomes constant.

(ix) As compared to G'_{23}, the values of G'_{12} are higher.

(x) Values of E'_{22}, G'_{23}, and G'_{12} are significantly lower than that obtained for E'_{11}.

(xi) Consideration of variable stress (in the FEM method) and the additional effect of fiber-to-fiber interaction (in Mori–Tanaka method) render results from these theories rather more accurate.

(b) Loss factors

(i) Highest loss factor is obtained at lowest fiber volume fractions. This coincides with the fact that the damping in the composites is matrix-dominated.

(ii) Decrease in damping with increase in fiber volume fraction is more for the fibers having large aspect ratios as compared to the fibers having small aspect ratios.

(iii) FEM model results for η_{23} and η_{12} lie between the Eshelby model and the Mori–Tanaka model.

(iv) FEM model-based damping results are for single representative volume element (RVE). Thus, the loss factors are higher for all loading conditions due to the fact that both actual state of stress as well as effect of shear stress concentration are considered.

(v) There is better correlation of the loss factor predictions made by Eshelby's method, and the Hashin–Shtrikman model. The only exception is the case of the longitudinal loss factor (η_{11}), where deviation is observed in the prediction based on Eshelby's method. The other methods give better results.

(vi) Analytically, Mori–Tanaka model may be considered as the best choice for estimating the properties of aligned short-fiber composites.

Reference

1 Finegan, I.C., Tibbetts, G.G., and Gibson, R.F. (2003). Modeling and characterization of damping in carbon nanofiber/polypropylene composites. *Composite Science and Technology* 63: 1629–1635.

5

Experimental Work

Currently many types of reinforcing fibers are used in composite materials. Glass fibers are the most common reinforcing fibers for polymer matrix composites due to their low cost and high strength, and are used in many high-volume applications, particularly the automotive industry. The disadvantage of glass fibers are that they have a relatively low modulus and poor abrasion resistance, which decreases its potential strength, and poor adhesion to polymer matrix resins, especially in the presence of moisture. Other composites made from carbon, boron, or other high-modulus fibers are used primarily for aerospace and military applications where their higher costs can be justified by improved performance. Carbon fibers are currently the most widely used advanced fibers and are generally manufactured by pyrolysis of a poly-acrylonitrile (PAN) or a pitch precursor. Vapor grown carbon fibers (VGCF) are highly graphitic with superior mechanical properties and have good thermal and electrical conductivity. The bulk properties of composites are not only functions of the fiber and polymeric matrix materials, but also the state of fiber orientation. Random orientation is not always undesirable, but usually composites are designed to maximize the performance along a given direction. In this chapter, the principle of Dynamic Mechanical Analyser (DMA) and its working is explained. Specimen of VGCF/polypropylene (PP) composite is tested on DMA. The storage modulus of the nanocomposite in fiber direction and damping loss factor are then evaluated and the results are compared with FEM and Mori–Tanaka model.

5.1 Materials

Fiber used here is VGCF, and the matrix material is PP. Table 5.1 shows the physical and mechanical properties of VGCF. Table 5.2 represents the various properties of PP. Figure 5.1 shows the VGCF/PP nanocomposite and neat PP samples. A single screw laboratory extruder is used to fabricate nanocomposite samples containing two different weight fractions such as 1 wt% VGCF and 0.5 wt% VGCF. After hot pressing the samples, the specimens are converted into rectangular sectioned thin strips of size: length $L = 45$ mm, width = 4.45 mm, and thickness = 0.7075 mm. These specimens conform to ASTM test method D 882. A DMA manufactured by Tritec 2000 DMA is used to analyze the dynamic characteristics such as damping and stiffness of VGCF/PP nanocomposite specimens. DMA is an analytical instrument used to measure the modulus (stiffness) and damping (energy dissipation) of the samples.

Mechanics of Particle- and Fiber-Reinforced Polymer Nanocomposites: From Nanoscale to Continuum Simulations, First Edition. Sumit Sharma.

Table 5.1 Physical and mechanical properties of VGCF.

Property	Pyrograph III® PR-24-PS
Ultimate strength, GPa	7.0
Tensile modulus, GPa	600
Diameter, μm	0.060–0.2
Length, μm	30–100
Density, g/cm^3	1.95
Electrical resistivity, μΩ-cm	55
Loss factor (η)	0.0015 (Finegan et al. [1])

Source: Modified from Zhou et al. [2].

Table 5.2 Properties of polypropylene (PP).

Property	Polypropylene
Storage modulus, GPa	1.23 (Finegan et al. [1])
Loss factor, η	0.06 (Finegan et al. [1])
Density, g/cm^3	0.900 (Alam et al. [3])
Poisson's ratio, ν^m	0.30

5.2 Principles of DMA – Forced Nonresonance Technique

Forced nonresonance technique is one of the simpler dynamic mechanical methods to understand. In most commercially available instruments, a force is applied to a sample and the amplitude and phase of the resultant displacement are measured. All of these instruments employ a linear actuator where the force applied is calculated from knowledge of the input signal to the electromagnet coils in the driver. An alternative to the above is where a force transducer is used to measure the applied load, with the sample between this transducer and the magnetic driver. These are the two types of arrangements that are found with the forced nonresonance technique. In each case, the sample is driven at a frequency below that of the test arrangement. Typical frequency range of such instruments is 0.001–1000 Hz. Any measurements below 0.01 Hz take too long for most analytical experiments, especially if data are required as a function of temperature and resonance often occurs at frequencies >100 Hz, depending upon the sample stiffness.

5.2.1 Terms and Definitions

In a dynamic mechanical test, it is the sample stiffness and loss modulus that are being measured. The sample stiffness will depend upon its modulus of elasticity and its geometry or shape. The modulus measured depends upon the choice of geometry, Young's modulus (E^*) for tension, compression, and bending, and Shear modulus (G^*) for torsion. The modulus is defined as the stress per unit area divided by the strain resulting from the applied force. Therefore, it is a measure of the material's resistance to deformation, the higher the modulus, the

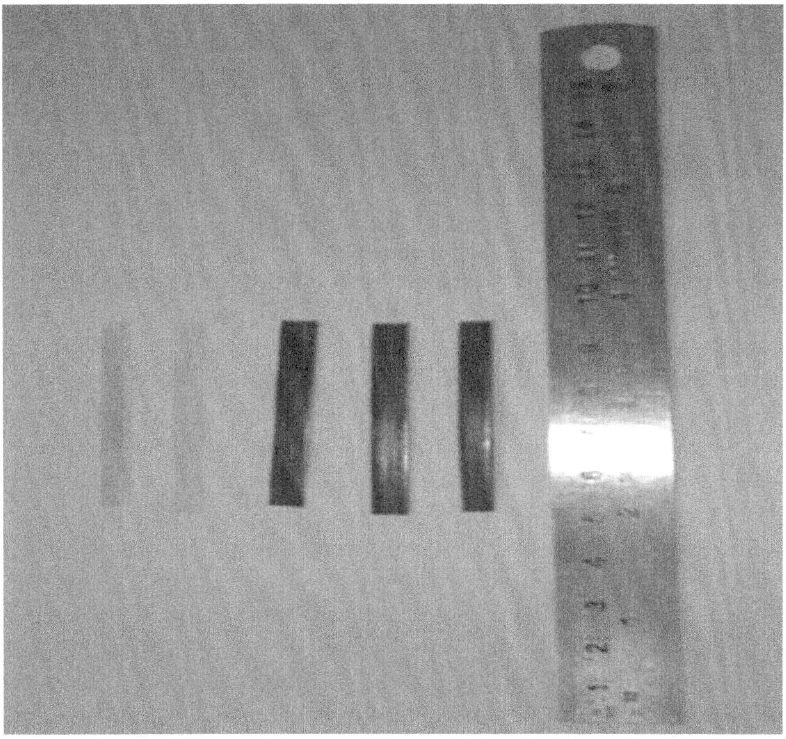

Figure 5.1 VGCF/pp nanocomposite and neat polypropylene samples. Source: Dr. Sumit Sharma.

more rigid the material is. For materials that exhibit time-invariant deformation, for example metals and ceramics at room temperature, any measurement of strain will lead to a constant value of modulus. Dynamic mechanical testers apply a periodic stress or strain to a sample and measure the resulting strain or stress response. Due to the time-dependent properties of polymers the resultant response is out-of-phase with the applied stimulus. The Complex Modulus M* is defined as the instantaneous ratio of the stress/strain. In order to understand the deformational mechanisms occurring in the material, this is resolved into an in-phase and out-of-phase response. This is equivalent to a complex number, where M' is the in-phase or elastic response this being the recoverable or stored energy. M'' is the imaginary or viscous response, this being proportional to the irrecoverable or dissipated energy. Thus, for a completely elastic material $M^* = M'$, while for a totally viscous material $M^* = M''$* δ is the measured phase lag between the applied stimulus and the response. Tan δ is given by the ratio M''/M' and is proportional to the ratio of energy dissipated/energy stored. This is called the loss tangent or damping factor, which is one of the key parameters in dynamic mechanical testing, since it is seen to increase during transitions between different deformational mechanisms.

5.2.2 Choice of Sample Geometry

Often the choice of geometry is dictated by the sample being investigated. For example thin films can only be measured accurately in tension. Fortunately, all good dynamic mechanical testers perform well in tension and should deal with the necessary pretension forces fully automatically, including those associated with large modulus changes that may occur at the glass transition. Pretension is necessary in order to maintain the sample under a net tension to prevent buckling that would otherwise occur. Tension should be the first choice for any sample less than 1 mm thick. Samples with thicker than 1 mm will probably be too stiff for the instrument in tension and bending mode would be preferable in this case. Materials that creep excessively, such as polyethylene, may be difficult to test in tension, due to creep under the pretension force. Bending mode is probably the most accommodating geometry, in that common-sized bars ($50 \times 10 \times 2$ mm) of material are readily tested. Such sizes are within the ranges of most commercial dynamic mechanical testers. Clamped modes yield better results over the whole temperature range, but suffer from clamping effects, whilst simply supported modes (three-point bending) yield the most accurate moduli. Torsion is a good choice of geometry, but since this has a low inherent stiffness, it necessitates reasonably large samples. Also few dynamic mechanical testers have a torsional capability. Simple shear is an excellent means of measuring low-modulus materials, such as rubbers, gels, and pastes. Glassy materials will be too stiff for most dynamic mechanical testers in this mode. Compression is the worst choice for any sample. It is the mode with the most geometrical errors (assumption of perfect lubrication at surface), but is often the only way to measure irregularly shaped samples. Under these circumstances an accurate modulus cannot be obtained, but transition information should not be compromised. Again, due to instrument range, it is only suitable for rubbers, gels, and pastes. The geometry chosen is dictated by the nature and size of sample being analyzed. The stiffness range of the instrument dictates if a particular sample can be successfully run using the chosen geometry. Selection of the geometry automatically enters the geometry constant onto the experiment method spreadsheet, along with the sample dimensions. The maximum and minimum modulus measurable for this sample is displayed on the experiment method spreadsheet alongside the geometry constant and peak strain. These values are calculated on a minimum and maximum stiffness limit of $10^{2.4}$ and 10^8 N/m, respectively. If the modulus values are known to fall outside of this range, then a different size or geometry mode must be chosen, in order to obtain satisfactory data. If a sufficiently low minimum modulus cannot be achieved, for example, make the sample thicker or shorter. Alternatively, if bending geometry is being used, consider using tension mode. The Modulus (Pa) is calculated using:

$$E = \frac{S}{k} = \frac{\mathrm{N/m}}{\mathrm{m}} = \mathrm{N/m^2} \tag{5.1}$$

where S is the measured stiffness in N/m and k is the geometry constant in meter (m). For a typical polymer sample, the modulus ranges from about 10^6 Pa for the rubbery state to $10^{9.5}$ Pa for the glassy state. Therefore, the optimum geometry constant should lie between $10^{-3.5}$ to 10^{-2} m. The geometry constant (k) is calculated using the following basic equations:

(i) *Tension and compression*: $k = A/l$, where A is the cross-sectional area and l is the free length.
(ii) *Single cantilever*: $k = w(t/l)^3$, where w is the width, t is the thickness, and l is the free length again.
(iii) *Dual cantilever*: $k = 2w(t/l)^3$
(iv) *Three-point bending*: $k = w/2(t/l)^3$
(v) *Simple shear*: $k = 2A/t$

5.2.3 Geometry Choice Guidelines

Various choices for mounting of the specimen on DMA are discussed below:

(i) *Single cantilever mode*: It is excellent for general characterization of most polymeric bar samples above and below their T_g. Note that reclamping at the lowest temperature is recommended if the sample has been cooled from ambient.
(ii) *Dual cantilever mode*: It is useful for low-stiffness samples, such as thin films, using a small free length. The smallest free length is 5 mm, but this may be reduced to 1 or 2 mm by using the extension clamps. Note that tension mode is often better for such samples.
(iii) *Three-point bending mode*: It is used for accurate modulus work on stiff bar samples, e.g. composites or thermoplastics, not subjected to temperatures above their T_g.
(iv) *Shear mode*: It is useful, if examining, low-stiffness materials such as elastomers, pressure sensitive adhesives, bitumen and tar, or studying the cure of materials such as epoxy resins. Note that this mode is not recommended for glassy materials, since these materials will be too stiff in this configuration. Displacement levels of around 10 μm generally yield good data.
(v) *Tension mode*: It is very useful for the analysis of thin films and fibers or can be used for bar samples with no static force if expansion information is required. It is useful to keep the dynamic displacement to a small value in tension, since this requires a lower level of static force preload, which in turn minimizes creep, suggest 10 μm.
(vi) *Compression mode*: It is least used, but may find application with polymer foams, gels, and natural materials such as bread dough, meat, and confectionery. As in tension mode, small displacements (about 10 μm) usually result in better data.

Set up of dual cantilever mode used for testing the specimen is shown in Figures 5.2 and 5.3. Dual cantilever geometry requires the specimen to be true and to be loaded with the clamps perpendicular to the long axis of the sample. In addition, care must be taken to clamp the specimen evenly with similar forces and to not introduce a twisting or distortion in clamping. Moduli from dual cantilever fixtures tend to run 10–30% different than the same material measured in three-point bending at best. This is due to shearing strain induced by clamping the specimen in place at the ends and center, which makes the sample more difficult to deform.

5.3 Experimental Procedure for Dual Cantilever Mode

Following steps are followed to test the samples on DMA:

(i) *Starting the program*: The software can be started from the Start menu program; alternatively a shortcut can be created on the desktop. As the program starts, the calibration values are read from the instrument and checked against the calibration file. If differences are found, the user is prompted to choose between the instruments stored value and the calibration file. The user is then asked if the values should be updated, in order to synchronize the values. If no differences are found, the user will be asked to choose an oven.
(ii) *Choose oven*: Up to three oven configurations can be used. Separate PID calibration values, temperature probe type, and temperature range limits can be stored for each

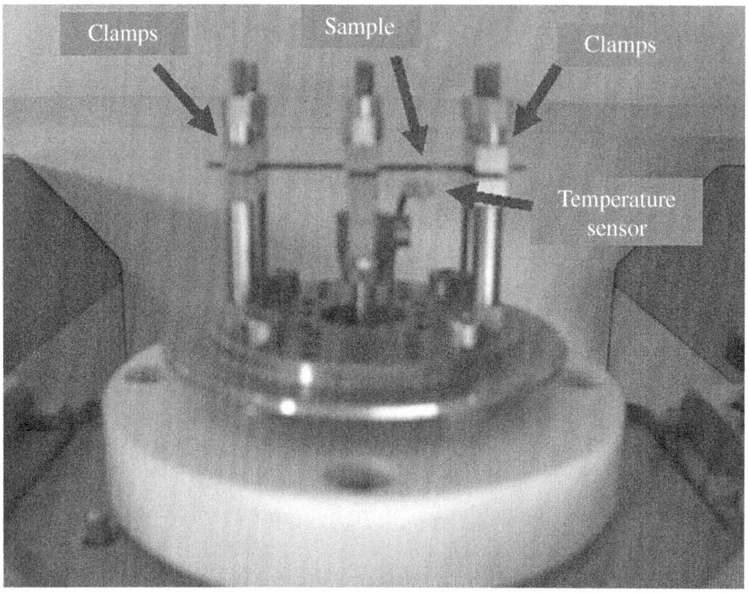

Figure 5.2 Set up for dual cantilever bending mode (Tritec DMA 2000). Source: Dr. Sumit Sharma.

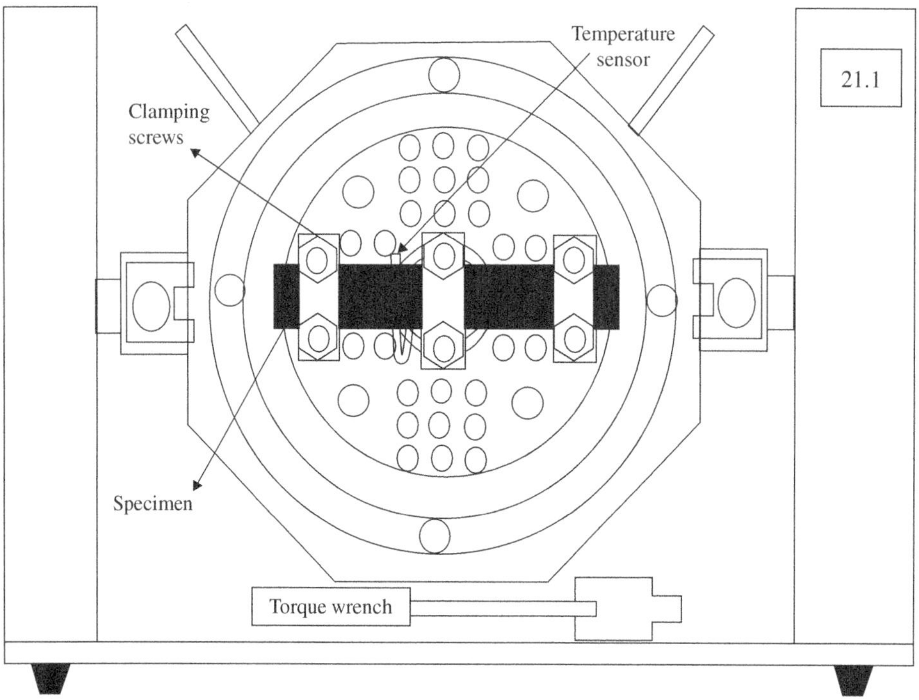

Figure 5.3 Set up of dual cantilever mode on DMA.

configuration. For dual cantilever bending mode, standard oven selection is used which is selected default in the instrument setup menu.

(iii) *Main menu*: The Triton DMA main menu appears. The main menu screen displays the current temperature and frequency, and a graphical representation of the force and displacement. If available the display also shows the experiment filename, and the geometry constant.

(iv) *Display force and displacement values*: The above function is available from the Tools pull-down menu. This option adds the numeric values of the dynamic force (*N*), dynamic displacement amplitude (mm), and the static displacement (mm).

(v) The next step is to properly measure the width and thickness of the sample. After measuring the above dimensions, the rectangular specimen is set into the DMA dual cantilever mode geometry. In the DMA measurements, by clamping the specimen in place at the ends and center, an extra-shearing strain is introduced, which makes the sample more difficult to deform. With the help of torque wrench, 1 N-m torque is applied at side clamps screws, but the center drive shaft clamps are finger tight. Make sure the sample is properly tight ended between the clamps.

(vi) *Experiment wizard*: The experiment wizard button is used to create a new experiment. The experiment conditions window will appear. Fill the experiment title and choose the experiment type. For this study, time scan type is used for 30 minutes.

(vii) *Time/temperature scans*: The default conditions are set on this form to yield reasonable results for most materials. They may be changed if required. They are frequency, strain or displacement amplitude, temperature range, etc. For this experiment, the default values are followed.

(viii) *Geometry*: The experiment wizard defaults the driveshaft control mode for all geometries. This is suitable for most samples, but it may be changed to other modes if desired. Select the deformation modes as required for the experiment. It may be noted that in this project dual cantilever bending mode is selected. Rectangular cross-section is chosen for dual cantilever mode because circular type specimens are not tested on this mode. Properly fill the sample size such as length, width, and thickness. Note that the sample length, l is always taken as the distance between the fixed clamp and the driveshaft clamp. Therefore, samples in single and dual cantilever mode will both have the same length for a given clamp spacing.

(ix) Now save the experiment for same name as used in experiment title in Step (vi). Now start the experiment by choosing option Go from main menu. When the experiment starts, the Temperature Controller Stabilization box will sometimes appear. This is indicating that the system is conditioning itself for the latest experiment. It may take several minutes on occasions to establish sensible start conditions prior to starting the experiment. BE PATIENT. The progress of stabilization is shown with a bar, which displays "Ready" message when complete and the run will automatically start. Now the excel sheet window will appear with graph plotting windows. The graph between desired properties is clearly shown in windows.

5.4 Theoretical Formulations/Modeling

Following mathematical formulas are used to determine the fiber volume fraction and stiffness of the composite in fiber direction:

(a) *Determination of fiber volume fraction*: The composite volume fraction is a function of the weight of VGCF and PP and also the densities of both materials:

$$V_f = \frac{w_f \rho_m}{w_f \rho_m + w_m \rho_f} \tag{5.2}$$

where

V_f = Volume fraction of fiber;
$w_f(g)$ = Weight of VGCF;
$w_m(g)$ = Weight of matrix;
$\rho_f(g/cm^3)$ = density of VGCF;
$\rho_m(g/cm^3)$ = density of matrix.

(b) *Determination of stiffness of the composite in fiber direction*: Stiffness of the composites for aligned fibers is predicted by using Mori–Tanaka model as given by Eqs. (3.21) and (3.57).

(c) *Finite element modeling*: RVEs for finite element modeling are selected corresponding to the volume fractions of the specimens, i.e. $V_f = 0.231\%$ and 0.464%. FE modeling of square RVE by using the properties of PR-24-PS VGCF and PP nanocomposite is analyzed and longitudinal storage modulus and longitudinal loss factor is predicted. Properties are taken from Tables 5.1 and 5.2. Hexahedron elements are used for meshing the RVE. Boundary conditions are same as used in Chapter 3 (Table 3.3). Number of elements in the fiber are 1107 and total number of elements in the RVE are 2565 and the number of nodes are 3040. The strain energy pattern is shown in Figures 5.4 and 5.5 for different fiber volume fractions. Strain energy is taken from the numerically integrated method for system analysis (NISA) output file. Output of strain energy for both fiber volume fractions is given in Table 5.3. Results show that the strain energy is varying from one volume fraction to another. Its magnitude is changing within the fiber.

5.5 Results and Discussion

Figures 5.6 and 5.7 show the comparison of experimental results with Mori–Tanaka model and FEM/NISA model for elastic modulus E'_{11} and loss factor η_{11}. Figure 5.6 shows the variation of the predicted and measured elastic modulus E'_{11} of VGCF/PP composites for different fiber volume fractions. As expected, longitudinal storage modulus E'_{11} increases linearly with increase in fiber volume fraction for both measured and predicted results. Table 5.4 shows the percentage increase in measured values of E'_{11} in comparison to predictions made by FEM and Mori–Tanaka models for the fiber volume fractions $V_f = 0.231\%$

Table 5.3 Magnitude of strain energy for different V_f.

V_f (%)	W (TSE)	W_f (SEF)	W_m (SEM)
0.231	$7.436\,823 \times 10^{-7}$	4.33×10^{-7}	3.11×10^{-7}
0.464	$5.966\,568 \times 10^{-7}$	4.28×10^{-7}	1.64×10^{-7}

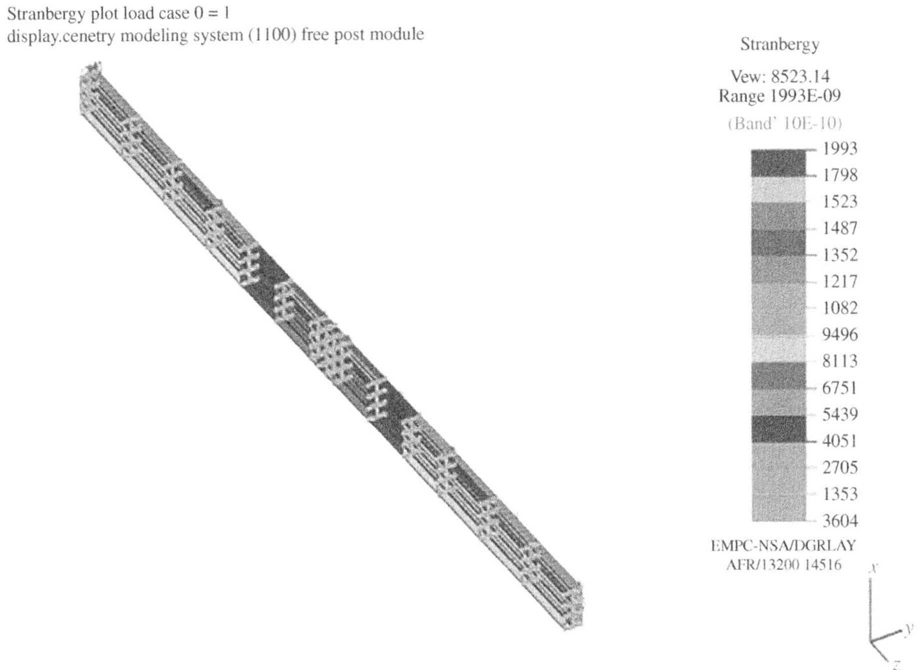

Figure 5.4 Strain energy in RVE when $V_f = 0.231\%$ and $l/d = 500$.

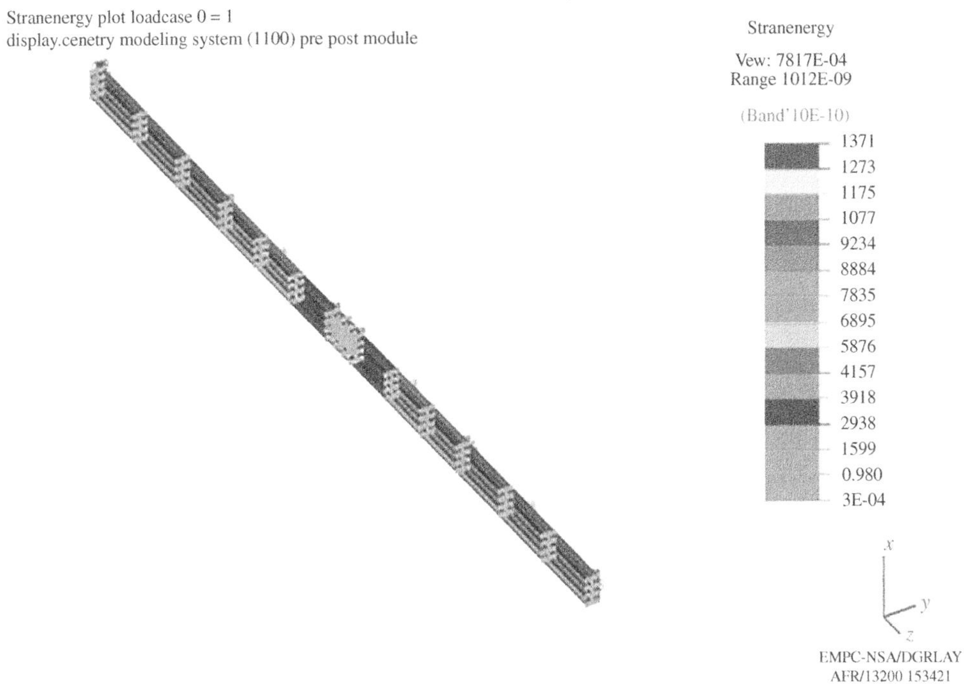

Figure 5.5 Strain energy in RVE when $V_f = 0.464\%$ and $l/d = 500$.

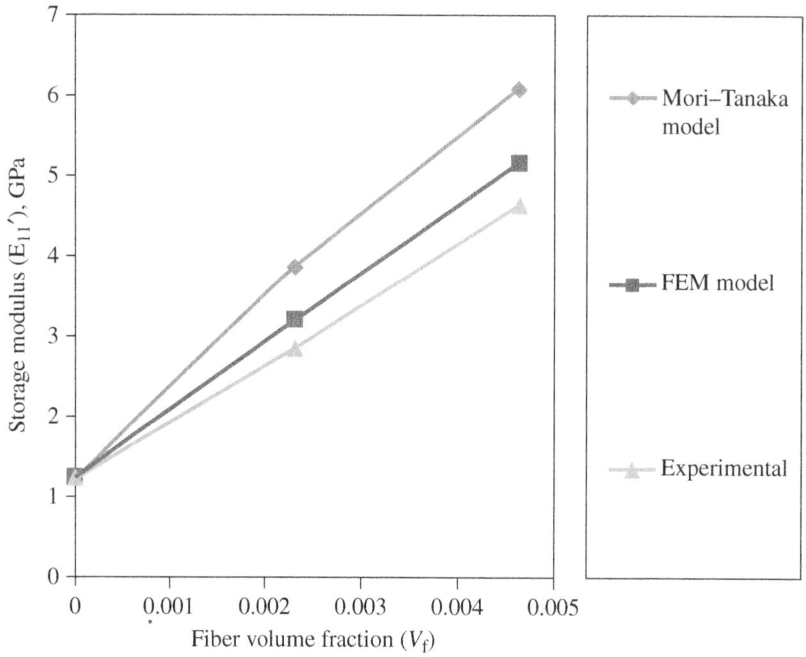

Figure 5.6 Variation of E'_{11} vs. V_f for $l/d = 500$.

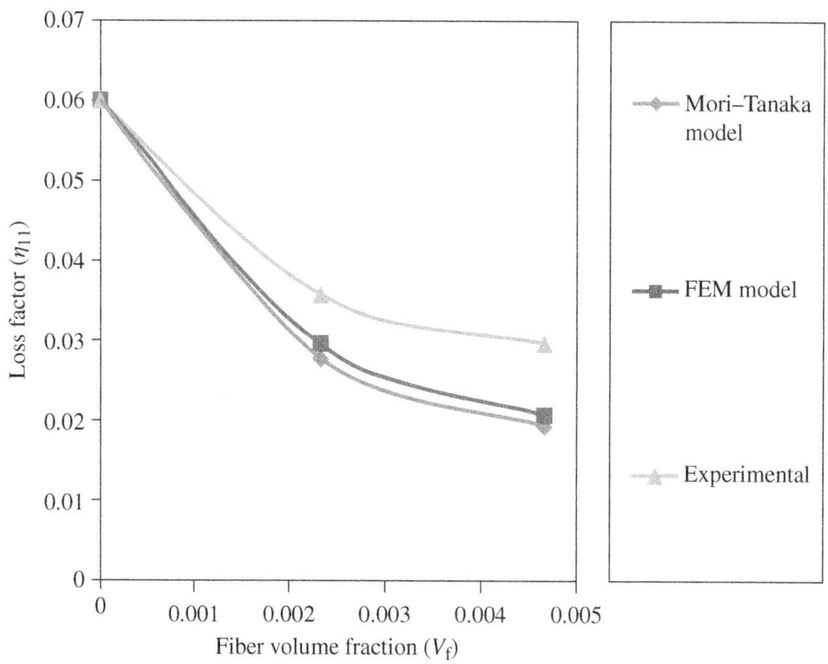

Figure 5.7 Variation of η_{11} vs. V_f for $l/d = 500$.

Table 5.4 Percentage increase in E'_{11} with V_f given by different models for $l/d = 500$.

Range of fiber volume fraction (V_f)	Increase in E'_{11} values (in %)		
	Mori–Tanaka model	FEM model	Experimental
0.0–0.002 31	213.94	161.38	132.19
0.002 31–0.004 64	57.63	56.96	57.58

Table 5.5 Comparison between E'_{11} values given by different models for increasing V_f at $l/d = 500$.

Fiber volume fraction (V_f)	Difference in E'_{11} values (in %) between		
	Mori–Tanaka and FEM	Mori–Tanaka and experimental	FEM model and experimental
0.002 31	16.84	26.16	11.07
0.004 64	14.97	23.68	10.25

and 0.464%. Table 5.5 shows comparison of E'_{11} values between experimental and theoretical models namely, Mori–Tanaka and FEM models. Further, it can be observed that FEM results correlate better with those experimentally obtained.

Table 5.4 shows that longitudinal storage modulus increases 132.19% from neat PP after addition of 0.231% volume fraction of fiber, whereas it increases 57.58% at fiber volume fraction 0.464%. Table 5.5 shows that for both $V_f = 0.231\%$ and 0.464%, there is difference between predicted and experimental values. This difference between the results is due to following reasons:

(i) In theoretical models (Mori–Tanaka model), no shear effect is introduced, whereas actual state of stress is considered in both FEM/strain energy method as well as experimental measurements.
(ii) In DMA measurements by clamping the specimen in place at the ends and center, we introduce an extra shearing strain, which makes the sample more difficult to deform.
(iii) It is also seen that the sample geometry has an important effect on stress and strain values, because small dimensional changes have large consequences. For each test that uses the DMA, a sample geometry is proposed. Each geometry has a different set of equations for calculating stress and strain from force and deformation which can be seen from the Eq. (5.1). The calculation of the storage modulus for each type of tests used in the DMA testing is using a geometry factor that involves squared and cubed terms of the width and thickness of the samples. Small errors in the sample dimensions can generate large errors in the properties.

Figure 5.7 shows variation of predicted and measured loss factor η_{11} of VGCF/PP composites for different fiber volume fractions. Trend of the loss factor is opposite to the trend of elastic modulus of the composite, i.e. as the fiber volume fraction increases the damping decreases for both experimental and predicted models. For both fiber volume fractions, i.e. $V_f = 0.231\%$ and $V_f = 0.464\%$, there is some difference between FEM/strain energy method

Table 5.6 Percentage decrease in η_{11} with V_f given by different models for $l/d = 500$.

Range of fiber volume fraction (V_f)	Decrease in η_{11} values (in %)		
	Mori–Tanaka model	FEM model	Experimental
0.0–0.002 31	53.34	51.67	40.54
0.002 31–0.004 64	32.14	30.43	17.18

Table 5.7 Comparison between η_{11} values given by different models for increasing V_f at $l/d = 500$.

Fiber volume fraction (V_f)	Difference in η_{11} values (in %) between		
	Mori–Tanaka and FEM	Mori–Tanaka and experimental	FEM and experimental
0.002 31	5.39	27.17	20.66
0.004 64	6.68	53.23	43.63

and experimental results. This difference is due to the fact that the calculation of the storage modulus involves the geometry factor; small errors in the sample dimensions can generate large errors in the dynamic properties.

Table 5.6 shows the percentage decrease in measured values of η_{11} in comparison to predictions made by FEM and Mori–Tanaka models for the fiber volume fractions $V_f = 0.231\%$ and 0.464%. Table 5.7 shows the comparison for η_{11} values between experimental and theoretical models namely, Mori–Tanaka and FEM models. It can be observed that FEM results correlate better with those experimentally obtained. Table 5.6 shows that longitudinal loss factor η_{11} decreases 40.54% from neat PP after addition of 0.231% volume fraction of fiber where as 17.18% at volume fraction 0.464%. Table 5.7 shows that for both $V_f = 0.231\%$ and 0.464%, there is difference between predicted and experimental loss factor values.

5.6 Conclusions

Using DMA, longitudinal storage modulus (E'_{11}) and longitudinal loss factor (η_{11}) have been predicted for VGCF/PP nanocomposites for two different fiber volume fractions at $l/d = 500$. Results predicted by DMA are then compared with Mori–Tanaka model and also with FEM models. Following conclusions can be drawn from the experimental work:

(i) Experimentally measured longitudinal storage modulus E'_{11} shows good agreement with theoretical models such as FEM and Mori–Tanaka models for VGCF/PP nanocomposites.
(ii) Experimentally measured longitudinal loss factor η_{11} shows decreasing trend as the fiber volume fraction increases and show only 20.66% difference with FEM at $V_f = 0.464\%$, whereas for the same volume fraction, the experimental loss factor shows 27.17% difference with Mori–Tanaka model.

References

1 Finegan, I.C., Tibbetts, G.G., and Gibson, R.F. (2003). Modeling and characterization of damping in carbon nanofiber/polypropylene composites. *Composite Science and Technology* 63: 1629–1635.
2 Zhou, Y., Shaik, J., and Hasan, M.M. (2007). Thermal and tensile properties of aligned carbon nano-fiber reinforced polypropylene. *Materials Letters* 61: 1134–1136.
3 Alam, M.K., Kuriger, R.J., Anderson, D.P., and Jacobsen, R.L. (2002). Processing and characterization of aligned vapor grown carbon fiber reinforced polypropylene. *Applied Science and Manufacturing* 33: 53–62.

6

Molecular Dynamics Simulation

In the modern nanotechnology age, microscopic analysis methods are necessary in order to generate new functional materials and investigate physical phenomena on a molecular level. In these methods, the constituent species of a system, such as molecules and fine particles, are considered. Macroscopic and microscopic quantities of interest are derived from analyzing the behavior of these species. These approaches, called "molecular simulation methods," are represented by the Monte Carlo (MC) and molecular dynamics (MD) methods. MC methods exhibit a powerful ability to analyze thermodynamic equilibrium, but they are unsuitable for investigating dynamic phenomena. MD methods are useful for thermodynamic equilibrium but are more advantageous for investigating the dynamic properties of a system in a nonequilibrium situation. Besides the above stated methods, there exist other methods also which can be used to investigate the physical phenomenon on a molecular level. Some of these are as follows:

(i) Brownian dynamics (BD) methods, which can simulate the Brownian motion of dispersed particles.
(ii) Dissipative particle dynamics (DPD) and lattice Boltzmann methods in which a liquid system is regarded as composed of virtual fluid particles.

Simulation methods using the concept of virtual fluid particles are generally used for pure liquid systems, but they are useful for simulating particle dispersions.

6.1 Molecular Dynamics

The concept of the MD method is straightforward and logical. The motion of molecules is generally governed by Newton's equations of motion in classical theory. In MD simulations, particle motion is simulated on a computer according to the equations of motion. If one molecule moves solely on a classical mechanics level, a computer is unnecessary because mathematical calculation with pencil and paper is sufficient to solve the motion of the molecule. However, since molecules in a real system are numerous and interact with each other, such mathematical analysis is impracticable. In this situation, therefore, computer simulations become a powerful tool for a microscopic analysis.

In MD simulation, there are different ways through which the position of a particle at a subsequent time can be evaluated. These are listed as below:

(i) *Verlet method*: If the velocity terms are not required for determining the molecular position at the next time step, then this scheme is called the "Verlet method."

Mechanics of Particle- and Fiber-Reinforced Polymer Nanocomposites: From Nanoscale to Continuum Simulations, First Edition. Sumit Sharma.

(ii) *Velocity verlet method*: In some cases, a scheme using the positions and velocities simultaneously may be more desirable in order to keep the system temperature constant. If both the current position and velocity of the particle are required to evaluate the position of the particle at the next time step, then this scheme is known as velocity verlet method.

(iii) *Leap frog method*: This name arises from the evaluation of the positions and forces, and then the velocities, by using time steps in a leapfrog manner. This method is also a significantly superior scheme in regard to stability and accuracy, comparable to the velocity Verlet method.

The MD method is applicable to both equilibrium and nonequilibrium physical phenomena, which makes it a powerful computational tool that can be used to simulate many physical phenomena. The main procedure for conducting the MD simulation using the velocity Verlet method is shown in the following steps:

(i) Specify the initial position and velocity of all molecules.
(ii) Calculate the forces acting on molecules.
(iii) Evaluate the positions of all molecules at the next time step.
(iv) Evaluate the velocities of all molecules at the next time step.
(v) Repeat the procedures from step 2.

In the above procedure, the positions and velocities will be evaluated at every time interval "h" in the MD simulation.

6.2 Monte Carlo Simulation

In the MD method, the motion of molecules is simulated according to the equations of motion and therefore it is applicable to both thermodynamic equilibrium and nonequilibrium phenomena. In contrast, the MC method generates a series of microscopic states under a certain stochastic law, irrespective of the equations of motion of particles. Since the MC method does not use the equations of motion, it cannot include the concept of explicit time, and thus is only a simulation technique for phenomena in thermodynamic equilibrium. Hence, it is unsuitable for the MC method to deal with the dynamic properties of a system, which are dependent on time. In this method, a series of microscopic states is generated using a probability density function (ρ). The main procedure for the MC simulation of a nonspherical particle system is as follows:

(i) Specify the initial position and direction of all particles.
(ii) Regard this state as microscopic state i, and calculate the interaction energy U_i.
(iii) Choose an arbitrary particle in order or randomly and call this particle "particle α."
(iv) Make particle α move translationally using random numbers and calculate the interaction energy U_j for this new configuration.
(v) Adopt this new microscopic state for the case of $U_j \leq U_i$ and go to step 7.
(vi) Calculate ρ_j/ρ_i for the case of $U_j > U_i$ and take a random number R_1 from a uniform random number sequence distributed from zero to unity. If $R_1 \leq \rho_j/\rho_i$, adopt this microscopic state j and go to step 7. If $R_1 > \rho_j/\rho_i$, reject this microscopic state, regard previous state i as new microscopic state j, and go to step 7.

(vii) Change the direction of particle α using random numbers and calculate the interaction energy U_k for this new state.
(viii) If $U_k \leq U_j$, adopt this new microscopic state and repeat from step 2.
(ix) If $U_k > U_j$, calculate ρ_k/ρ_j and take a random number R_2 from the uniform random number sequence. If $R_2 \leq \rho_k/\rho_j$ adopts this new microscopic state k and repeat from step 2. If $R_2 > \rho_k/\rho_j$, rejects this new state, and regard previous state j as new microscopic state k, then repeat from step 2.

6.3 Brownian Dynamics

A dispersion or suspension composed of fine particles dispersed in a base liquid is a difficult case to be treated by simulations in terms of the MD method, because the characteristic time of the motion of the solvent molecules is considerably different from that of the dispersed particles. Simply speaking, if we observe such dispersion based on the characteristic time of the solvent molecules, we can see only the active motion of solvent molecules around the quiescent dispersed particles. Clearly, the MD method is quite unrealistic as a simulation technique for particle dispersions. One approach to overcome this difficulty is to not focus on the motion of each solvent molecule, but regard the solvent molecules as a continuum medium and consider the motion of dispersed particles in such a medium. In this approach, the influence of the solvent molecules is included into the equations of motion of the particles as random forces. The BD method simulates the random motion of dispersed particles that is induced by the solvent molecules; thus, such particles are called "Brownian particles." The main procedure for conducting the BD simulation is shown below:

(i) Specify the initial position of all particles.
(ii) Calculate the forces acting on each particle.
(iii) Generate the random displacements using uniform random numbers.
(iv) Calculate all the particle positions at the next time step.
(v) Return to step 2 and repeat.

6.4 Dissipative Particle Dynamics

It is not realistic to use the MD method to simulate the motion of solvent molecules and dispersed particles simultaneously, since the characteristic time of solvent molecules is much shorter than that of dispersed particles. Hence, in the BD method, the motion of solvent molecules is not treated, but a fluid is regarded as a continuum medium. The influence of the molecular motion is combined into the equations of motion of dispersed particles as stochastic random forces. Are there any simulation methods to simulate the motion of both the solvent molecules and the dispersed particles? As far as we treat the motion of real solvent molecules, the development of such simulation methods may be impractical. However, if groups or clusters of solvent molecules are regarded as virtual fluid particles, such that the characteristic time of the motion of such fluid particles is not so different from that of dispersed particles, then it is possible to simulate the motion of the dispersed and the fluid particles simultaneously. These virtual fluid particles are expected to exchange their momentum, exhibit a random motion similar to Brownian particles, and interact with each other by particle–particle potentials. We call these virtual fluid particles “dissipative particles,” and the simulation technique of treating the motion of dissipative particles instead of the solvent molecules is called the “dissipative particle dynamics (DPD) method.” The main procedure for conducting the DPD simulation is quite similar to BD simulations.

6.5 Lattice Boltzmann Method

In the lattice Boltzmann method, a fluid is assumed to be composed of virtual fluid particles, and such fluid particles move and collide with other fluid particles in a simulation region. A simulation area is regarded as a lattice system, and fluid particles move from site to site; that is, they do not move freely in a region. The most significant difference of this method in relation to the MD method is that the lattice Boltzmann method treats the particle distribution function of velocities rather than the positions and the velocities of the fluid particles.

6.6 Basic Concepts

Following are some of the common terms used in MD simulation:

6.6.1 Force Field

In the context of molecular modeling, force field refers to the form and parameters of mathematical functions used to describe the potential energy of a system of particles (typically molecules and atoms). Force field functions and parameter sets are derived from both experimental work and high-level quantum mechanical calculations. "All-atom" force fields provide parameters for every type of atom in a system, including hydrogen, while "united-atom" force fields treat the hydrogen and carbon atoms in methyl and methylene groups as a single interaction center. "Coarse-grained" force fields, which are frequently used in long-time simulations of proteins, provide even more crude representations for increased computational efficiency. The usage of the term "force field" in chemistry and computational biology differs from the standard usage in physics. In chemistry, it is a system of potential energy functions rather than the gradient of a scalar potential, as defined in physics.

The basic functional form of a force field encapsulates both bonded terms relating to atoms that are linked by covalent bonds, and nonbonded (also called "noncovalent") terms describing the long-range electrostatic and van der Waals forces. The specific decomposition of the terms depend on the force field, but a general form for the total energy in an additive force field can be written as

$$E_{\text{total}} = E_{\text{bonded}} + E_{\text{nonbonded}} \tag{6.1}$$

where the components of the covalent and noncovalent contributions are given by the following summations:

$$E_{\text{bonded}} = E_{\text{bond}} + E_{\text{angle}} + E_{\text{dihedral}} \tag{6.2}$$

$$E_{\text{nonbonded}} = E_{\text{electrostatic}} + E_{\text{vanderWaals}} \tag{6.3}$$

The bond and angle terms are usually modeled as harmonic oscillators in force fields that do not allow bond breaking. A more realistic description of a covalent bond at higher stretching is provided by the more expensive Morse potential. The functional form for the rest of the bonded terms is highly variable. Proper dihedral potentials are usually included. Additionally, "improper torsional" terms may be added to enforce the planarity of aromatic rings and other conjugated systems, and "cross-terms" that describe coupling of different internal variables, such as angles and bond lengths. Some force fields also include explicit terms for hydrogen bonds. The nonbonded terms are most computationally intensive because they include many more interactions per atom. A popular choice is to limit interactions to pair wise energies. The van der Waals term is usually computed with a Lennard–Jones potential and the electrostatic term with Coulomb's law, although both can be buffered or scaled by a constant factor to account for electronic polarizability and produce better agreement with experimental observations. In addition to the functional form of the potentials, a force field defines a set of parameters for each type of atom. For example, a force field would include distinct parameters for an oxygen atom in a carbonyl functional group and in a hydroxyl group. The typical parameter set includes values for atomic mass, van der Waals radius, and partial charge for individual atoms, and equilibrium values of bond lengths, bond angles, and dihedral angles for pairs, triplets, and quadruplets of bonded atoms, and values corresponding to the effective spring constant for each potential.

Parameter sets and functional forms are defined by force field developers to be self-consistent. Because the functional forms of the potential terms vary extensively between even closely related force fields (or successive versions of the same force field), the parameters from one force field should never be used in conjunction with the potential from another. Some of the popular force fields have been discussed below:

(i) *AMBER*: AMBER (an acronym for Assisted Model Building with Energy Refinement) is a family of force fields for MD simulation of biomolecules originally developed by the late Peter Kollman's [1] group at the University of California, San Francisco. AMBER is also the name for the MD software package that simulates these force fields. It is maintained by an active collaboration between David Case at Rutgers University, Tom Cheatham at the University of Utah, Tom Darden at NIEHS, Ken Merz at Florida, Carlos Simmerling at Stony Brook University, Ray Luo at UC Irvine, and Junmei Wang at Encysive Pharmaceuticals.

The term "AMBER force field" generally refers to the functional form used by the family of AMBER force fields. This form includes a number of parameters; each member of the family of AMBER force fields provides values for these parameters and has its own name. The functional form of the AMBER force field is shown below:

$$
\begin{aligned}
V(r^N) = {} & \sum_{\text{bonds}} \frac{1}{2} k_b (l - l_0)^2 + \sum_{\text{angles}} \frac{1}{2} k_a (\theta - \theta_0)^2 \\
& + \sum_{\text{torsions}} \frac{1}{2} V_n (1 + \cos(n\omega - \gamma)) \\
& + \sum_{j=1}^{N-1} \sum_{i=j+1}^{N} \left(\varepsilon_{i,j} \left[\left(\frac{r_{0ij}}{r_{ij}} \right)^{12} - 2 \left(\frac{r_{0ij}}{r_{ij}} \right)^{6} \right] + \frac{q_i q_j}{4\pi\varepsilon_0 r_{ij}} \right)
\end{aligned}
\tag{6.4}
$$

where

k_a and k_b = force constants determined experimentally;
l = bond length after stretching;
l_0 = equilibrium bond length;
θ = bond angle after bending;
θ_0 = equilibrium bond angle;
V_n = energy term;
n = periodicity parameter;
ω = dihedral bond torsion angle;
γ = phase parameter;
$\varepsilon_{i,j}$ = well depth;
r_{0ij} = distance at which the interaction energy between the two atoms is zero;
r_{ij} = separation between the molecules;
q_i, q_j = atomic charges on the atoms/molecules;
ε_0 = permittivity of free space;

Note that despite the term force field, this equation defines the potential energy of the system; the force is the derivative of this potential with respect to position.

The meanings of right hand side terms are

(a) First term (summing over bonds) represents the energy between covalently bonded atoms. This harmonic (ideal spring) force is a good approximation near the equilibrium bond length but becomes increasingly poor as atoms separate.
(b) Second term (summing over angles) represents the energy due to the geometry of electron orbitals involved in covalent bonding.
(c) Third term (summing over torsions) represents the energy for twisting a bond due to bond order (e.g. double bonds) and neighboring bonds or lone pairs of electrons. Note that a single bond may have more than one of these terms, such that the total torsional energy is expressed as a Fourier series.
(d) Fourth term (double summation over i and j) represents the nonbonded energy between all atom pairs, which can be decomposed into van der Waals (first term of summation) and electrostatic (second term of summation) energies.

The form of the van der Waals energy is calculated using the equilibrium distance (r_{0ij}) and well depth (ε). The factor of 2 ensures that the equilibrium distance is r_{0ij}. The energy is sometimes reformulated in terms of σ, where $r_{0ij} = 2^{1/6}(\sigma)$, as used, e.g. in the implementation of the soft core potentials. The form of the electrostatic energy used here assumes that the charges due to the protons and electrons in an atom can be represented by a single point charge (or in the case of parameter sets that employ lone pairs, a small number of point charges.)

To use the AMBER force field, it is necessary to have values for the parameters of the force field (e.g. force constants, equilibrium bond lengths and angles, and charges). A fairly large number of these parameter sets exist and are described in detail in the AMBER software user manual. Each parameter set has a name and provides parameters for certain types of molecules.

(a) Peptide, protein, and nucleic acid parameters are provided by parameter sets with names beginning with "ff" and containing a two digit year number, for instance "ff99."

(b) GAFF (Generalized AMBER force field) provides parameters for small organic molecules to facilitate simulations of drugs and small molecule ligands in conjunction with biomolecules.
(c) The GLYCAM force fields have been developed by Rob Woods for simulating carbohydrates.

(ii) *COMPASS*: Condensed-phase Optimized Molecular Potentials for Atomistic Simulation Studies [2] is a member of the consistent family of force fields (CFF91, PCFF, CFF, and COMPASS), which are closely related second-generation force fields. They were parameterized against a wide range of experimental observables for organic compounds containing H, C, N, O, S, P, halogen atoms and ions, alkali metal cations, and several biochemically important divalent metal cations. PCFF is based on CFF91, extended so as to have a broad coverage of organic polymers and (inorganic) metals. COMPASS is the first force field that has been parameterized and validated using condensed phase properties in addition to various and empirical data for molecules in isolation. Consequently, this force field enables accurate and simultaneous prediction of structural, conformational, vibrational, and thermophysical properties for a broad range of molecules in isolation and in condensed phases.

The COMPASS force field consists of terms for bonds (b), angles (θ), dihedrals (φ), out-of-plane angles (χ) as well as cross-terms, and two nonbonded functions, a Coulombic function for electrostatic interactions and a 9-6 Lennard–Jones potential for van der Waals interactions:

$$E_{\text{total}} = E_b + E_\theta + E_\varphi + E_\chi + E_{b,b\prime} + E_{b,\theta} + E_{b,\varphi} + E_{\theta,\varphi} + E_{\theta,\theta\prime} + E_{\theta,\theta\prime,\varphi} + E_q + E_{\text{vdW}} \tag{6.5}$$

Where,

$$E_b = \sum_b [k_2(b - b_0)^2 + k_3(b - b_0)^3 + k_4(b - b_0)^4] \tag{6.6}$$

$$E_\theta = \sum_\theta [k_2(\theta - \theta_0)^2 + k_3(\theta - \theta_0)^3 + k_4(\theta - \theta_0)^4] \tag{6.7}$$

$$E_\emptyset = \sum_\emptyset [k_1(1 - \cos\emptyset) + k_2(1 - \cos 2\emptyset) + k_3(1 - \cos 3\emptyset)] \tag{6.8}$$

$$E_\chi = \sum_\chi k_2\chi^2 \tag{6.9}$$

$$E_{b,b'} = \sum k(b - b_0)(b' - b'_0) \tag{6.10}$$

$$E_{b,\theta} = \sum_{b,\theta} k(b - b_0)(\theta - \theta_0) \tag{6.11}$$

$$E_{b,\emptyset} = \sum_{b,\emptyset} (b - b_0)[k_1 \cos\emptyset + k_2 \cos 2\emptyset + k_3 \cos 3\emptyset] \tag{6.12}$$

$$E_{\theta,\emptyset} = \sum_{\theta,\emptyset} (\theta - \theta_0)[k_1 \cos\emptyset + k_2 \cos 2\emptyset + k_3 \cos 3\emptyset] \tag{6.13}$$

$$E_{\theta,\theta'} = \sum_{\theta,\theta'} k(\theta - \theta_0)(\theta' - \theta'_0) \tag{6.14}$$

$$E_{\theta,\theta',\varphi} = \sum_{\theta,\theta',\varphi} k(\theta - \theta_0)(\theta' - \theta'_0)\cos\varphi \tag{6.15}$$

$$E_q = \sum_{ij} \frac{q_i q_j}{r_{ij}} \tag{6.16}$$

$$E_{\text{vdW}} = \sum_{ij} \varepsilon_{ij} \left[2\left(\frac{r_{ij}^0}{r_{ij}}\right)^9 - 3\left(\frac{r_{ij}^0}{r_{ij}}\right)^6 \right] \tag{6.17}$$

where

$k, k_1, k_2, k_3,$ and k_4	=	force constants determined experimentally;
b, θ	=	bond length and bond angle after stretching and bending, respectively;
b_0, θ_0	=	equilibrium bond length and equilibrium bond angle, respectively;
φ	=	bond torsion angle;
χ	=	out of plane inversion angle;
$E_{b,b'}, E_{\theta,\theta'}, E_{b,\theta}, E_{b,\varphi}, E_{\theta,\varphi}, E_{\theta,\theta',\varphi}$	=	cross terms representing the energy due to interaction between bond stretch-bond stretch, bond bend–bond bend, bond stretch–bond bend, bond stretch–bond torsion, bond bend–bond torsion, and bond bend–bond bend–bond torsion, respectively;
$\varepsilon_{i,j}$	=	well depth;
r_{ij}^0	=	distance at which the interaction energy between the two atoms is zero;
r_{ij}	=	separation between the atoms/molecules;
q_i, q_j	=	atomic charges on the atoms/molecules;
ε_0	=	permittivity of free space.

(iii) CHARMM force field: CHARMM (CHemistry at HARvard Macromolecular Mechanics), is a highly flexible molecular mechanics and dynamics program originally developed in the laboratory of Dr. Martin Karplus at Harvard University [3]. It was parameterized on the basis of ab initio energies and geometries of small organic models. The CHARMM force fields include the CHARMM (Momany & Rone) force field developed at Accelrys and the academic force fields, primarily developed by Prof. Alex MacKerell and coworkers (charmm19, charmm22, charmm27). All versions of the CHARMM force fields are available from Accelrys.

CHARMM uses a flexible and comprehensive empirical energy function that is a summation of many individual energy terms. The energy function is based on separable internal coordinate terms and pair-wise nonbond interaction terms. The total energy is expressed by the Eq. (6.18). The electrostatic term can be scaled to mimic solvent effects. The van der Waals combination rules and functional form are derived from rare-gas potentials:

$$E_{\text{pot}} = \sum k_b(r - r_0)^2 + \sum k_\theta(\theta - \theta_0)^2 + \sum |k_\emptyset| - k_\emptyset \cos(n\emptyset) + \sum k_\chi(\chi - \chi_0)^2 + \sum \frac{q_i q_j}{4\pi\varepsilon_0 r_{ij}} + \sum \left(\frac{A_{ij}}{r_{ij}^{12}}\right) - \left(\frac{B_{ij}}{r_{ij}^{6}}\right) sw(r_{ij}^2, r_{\text{on}}^2, r_{\text{off}}^2) + E_{\text{constraint}} + E_{\text{user}} \tag{6.18}$$

where

k_b, k_θ, k_φ, and k_χ = force constants for bond stretching, bond bending, bond torsion, and out of plane inversion respectively;
r, θ, and χ = bond length, bond angle, and inversion angle, respectively;
r_0, θ_0, and χ_0 = equilibrium bond length, equilibrium bond angle, and equilibrium inversion angle, respectively;
r_{ij} = separation between the atoms/molecules;
n = periodicity parameter;
q_i, q_j = atomic charges on the atoms/molecules;
ε_0 = permittivity of free space;
A_{ij} and B_{ij} = distances over which the interatomic interaction is zero;
$E_{\text{constraint}}$ and E_{user} = energies arising due to constraint (if applied) and certain user-defined terms (to customize the force field), respectively.

The r^{-12} term is the repulsive term, describing Pauli repulsion at short ranges due to overlapping electron orbitals and the r^{-6} term is the attractive long-range term. Hydrogen bond energy is *not* included as a default energy term. The current CHARMM parameter set has been derived in such a way that hydrogen bond effects are described by the combination of electrostatic and van der Waals forces.

(iv) *Consistent valence force field (CVFF)*: The consistent-valence force field, the original force field, provided with the Discover program is a generalized valence force field. Parameters are provided for amino acids, water, and a variety of other functional groups. CVFF [4] also has the ability to use automatic parameters (Automatic assignment of values for missing parameters) when no explicit parameters are present. These are noted in the output file from the calculation. CVFF was fit to small organic (amides, carboxylic acids, etc.) crystals and gas phase structures. It handles peptides, proteins, and a wide range of organic systems. As the default force field in Discover, it has been used extensively for many years. It is primarily intended for studies of structures and binding energies, although it predicts vibrational frequencies and conformational energies reasonably well.

The out-of-plane energy for the CVFF force field is calculated as an improper torsion. An improper torsion views three connected atoms and a central atom as if it were torsion. There are three possible improper torsions that can be generated for a particular out-of-plane, based on permutations of the connected atoms. For CVFF, only one of these improper torsions is used. The analytic form of the energy expression used in CVFF is shown in Eq. (6.19).

Terms 1–4 are commonly referred to as the diagonal terms of the valence force field and represent the energy of deformation of bond lengths, bond angles, torsion angles, and out-of-plane interactions, respectively. A Morse potential (Term 1) is used for the bond-stretching term. Discover program also supports a simple harmonic potential for this term:

$$
\begin{aligned}
E_{\text{pot}} = {} & \sum D_b \left(1 - e^{-a(b-b_0)}\right) + \sum_{\theta} H_\theta(\theta - \theta_0)^2 + \sum_{\emptyset} H_\emptyset(1 + s\cos(n\emptyset)_0) \\
& + \sum H_\chi \chi^2 + \sum_b \sum_{b'} F_{bb'}(b - b_0)(b' - b'_0) + \sum_\theta \sum_{\theta'} F_{\theta\theta'}(\theta - \theta_0)(\theta' - \theta'_0) \\
& + \sum_b \sum_{\theta'} F_{b\theta}(b - b_0)(\theta - \theta_0) + \sum_{\emptyset} F_{\emptyset\theta\theta'} \cos(\theta - \theta_0)(\theta' - \theta'_0) + \sum_\chi \sum_{\chi'} F_{\chi\chi'} \chi\chi
\end{aligned}
$$

$$+\sum \varepsilon\left[\left(\frac{r^*}{r}\right)^{12}-2\left(\frac{r^*}{r}\right)^{6}\right]+\sum \frac{q_i q_j}{\varepsilon r_{ij}} \tag{6.19}$$

where

D_b	=	well depth or bond dissociation energy;
α	=	parameter that controls the width of the potential well (the smaller the "α" the larger the well);
b and b_0	=	separation between atoms and equilibrium bond distance, respectively;
θ and θ_0	=	bond angle and equilibrium bond angle, respectively;
χ	=	inversion angle;
n	=	periodicity parameter;
q_i, q_j	=	atomic charges on the atoms/molecules;
ε	=	well depth (in term 10) and permittivity of free space (in term 11);
H_θ, H_φ, and H_χ	=	constants associated with bond bending, torsion and out of plane inversion, respectively;
$F_{b,b'}$, $F_{\theta,\theta'}$, $F_{b,\theta}$, $F_{\varphi,\theta,\theta'}$, $F_{\chi,\chi'}$	=	cross terms representing the energy due to interaction between bond stretch–bond stretch, bond bend–bond bend, bond stretch–bond bend, bond torsion–bond bend–bond bend, out of plane-out of plane inversion, respectively;
r^*	=	separation at which interatomic potential is zero;
r	=	separation between the atoms.

The Morse form is computationally more expensive than the harmonic form. Since the number of bond interactions is usually negligible relative to the number of nonbond interactions, the additional cost of using the more accurate Morse potential is insignificant, so this is the default option. When the model being simulated is high in energy (caused, for example, by overlapping atoms or a high-target temperature), a Morse-style function might allow bonded atoms to drift unrealistically far apart. This is not desirable unless we are intending to study bond breakage. Terms 5–9 are off-diagonal (or cross) terms and represent couplings between deformations of internal coordinates. For example, Term 5 describes the coupling between stretching of adjacent bonds. These terms are required to accurately reproduce experimental vibrational frequencies and, therefore, the dynamic properties of molecules. In some cases, research has also shown them to be important in accounting for structural deformations. However, cross terms can become unstable when the structure is far from a minimum. Terms 10–11 describe the nonbond interactions. Term 10 represents the van der Waals interactions with a Lennard–Jones function. Term 11 is the Coulombic representation of electrostatic interactions. The dielectric constant "ε" can be made distance dependent (i.e. a function of r_{ij}). In the CVFF force field, hydrogen bonds are a natural consequence of the standard van der Waals and electrostatic parameters, and special hydrogen bond functions do not improve the fit of CVFF to experimental data.

6.6.2 Potentials

The reliability of atomistic MC and MD simulation techniques that have impacted on areas ranging from drug design to crystal growth depends on the use of appropriate interatomic energies and forces. These interactions are generally described using either analytic potential energy expressions or semiempirical electronic structure methods, or obtained from a

Table 6.1 Original parameters for Tersoff EIP [5] for carbon-based systems.

$A = 1393.6\,\text{eV}$	$B = 346.74\,\text{eV}$
$\lambda_1 = 3.4879\,\text{Å}^{-1}$	$\lambda_2 = 2.2119\,\text{Å}^{-1}$
$\lambda_3 = 0.0000\,\text{Å}^{-1}$	$n = 0.727\,51$
$c = 38\,049.0$	$\beta = 1.5724 \times 10^{-7}$
$d = 4.3484$	$h = -0.570\,58$
$R = 1.95\,\text{Å}$	$D = 0.15\,\text{Å}$

Source: Modified from Tersoff [5].

Table 6.2 Original parameters for Brenner EIP [6] for solid-state carbon structures.

$A = 10\,953.544\,162\,170\,\text{eV}$	$B_1 = 12\,388.791\,977\,98\,\text{eV}$
$B_2 = 17.567\,406\,450\,9\,\text{eV}$	$B_3 = 30.714\,932\,080\,65\,\text{eV}$
$\alpha = 4.746\,539\,060\,\text{Å}^{-1}$	$\lambda_1 = 4.720\,452\,312\,7\,\text{Å}^{-1}$
$\lambda_2 = 1.433\,213\,249\,9\,\text{Å}^{-1}$	$\lambda_3 = 1.382\,691\,250\,6\,\text{Å}^{-1}$
$Q = 0.313\,460\,296\,083\,3\,\text{Å}$	$R = 2\,\text{Å}$
$D = 1.7\,\text{Å}$	$T_0 = -0.008\,096\,75$
$\beta_0 = 0.7073$	$\beta_1 = 5.6774$
$\beta_2 = 24.0970$	$\beta_3 = 57.5918$
$\beta_4 = 71.8829$	$\beta_5 = 36.2789$

Source: Modified from Brenner [6].

total-energy first-principles calculation. While the latter approach is not subject to errors that can arise from the assumed functional forms and parameter fitting usually required in the first two methods, there remain clear advantages to classical potentials for large systems and long simulation times. In the following sections, we will describe the bond order potentials, namely Tersoff model [5] and Brenner model [6] which is also known as Reactive Empirical Bond Order Potential (REBO potential). These are not based on a traditional many-body expansion of potential energy in bond lengths and angles instead, they use a parameterized bond order function to introduce many-body effects and chemical bonding into a pair potential. The Tersoff and Brenner Empirical Inter-atomic Potentials (EIPs) are convenient, short-range, bond-order, empirical potentials which are often used in MD simulations and other calculations to model different properties of carbon-based materials. The convenience of these potentials comes from their rather simple, analytical forms and the short-range of atomic interactions. For carbon-based systems, the Tersoff model has nine adjustable parameters which are listed in Table 6.1, which were originally fit to cohesive energies of various carbon systems, the lattice constant of diamond, and the bulk modulus of diamond. The Brenner EIP is based directly on the Tersoff EIP but has additional terms and parameters which allow it to better describe various chemical reactions in hydrocarbons and include nonlocal effects. These parameters are listed in Table 6.2.

6.6.2.1 Tersoff Model

The analytical form for the pair-potential, V_{ij}, of the Tersoff model [5] is given by the following functions with corresponding parameters listed in Table 6.1:

$$V_{ij} = f_{ij}^C (a_{ij} f_{ij}^R - b_{ij} f_{ij}^A) \tag{6.20}$$

$$f_{ij}^R = A e^{-\lambda_1 r_{ij}} \tag{6.21}$$

$$f_{ij}^A = B e^{-\lambda_2 r_{ij}} \tag{6.22}$$

where

r_{ij} = distance between atoms i and j;
f_{ij}^A and f_{ij}^R = competing attractive and repulsive pair-wise terms;
f_{ij}^C = cut-off term which ensures only nearest neighbor interactions;
a_{ij} = is a range-limiting term on the repulsive potential that is typically set equal to one;

The bond angle term, b_{ij}, depends on the local coordination of atoms around atom i and the angle between atoms i, j, and k:

$$b_{ij} = (1 + \beta^n \xi_{ij}^n)^{\frac{-1}{2n}} \tag{6.23}$$

$$\xi_{ij} = \sum_{k \# i,j} f_{ik}^C g_{ijk} e^{\lambda_3^3 (r_{ij} - r_{ik})^3} \tag{6.24}$$

$$g_{ijk} = 1 + \frac{c^2}{d^2} - \frac{c^2}{d^2 + (h - \cos[\theta_{ijk}])^2} \tag{6.25}$$

where θ_{ijk} is the angle between atoms i, j, k. This bond angle term allows the Tersoff model to describe the strong covalent bonding that occurs in carbon systems, which cannot be represented by purely central potentials. This angle-dependent term also allows for description of carbon systems that bond in different geometries, such as tetrahedrally bonded diamond and 120° tri-bonded graphene.

6.6.2.2 Brenner Model

The Brenner potential [6] for solid-state carbon structures is given by the following functions with corresponding parameters listed in Table 6.2:

$$V_{ij} = f_{ij}^C (f_{ij}^R - \overline{b_{ij}} f_{ij}^A) \tag{6.26}$$

$$f_{ij}^R = \left(1 + \frac{Q}{r_{ij}}\right) A e^{-\alpha r_{ij}} \tag{6.27}$$

$$f_{ij}^A = \sum_{n=1}^{3} B_n e^{-\lambda_n r_{ij}} \tag{6.28}$$

where many of the terms are similar to the Tersoff model described above and the bond angle term, $\overline{b_{ij}}$, is given by

$$\overline{b_{ij}} = \frac{1}{2}(b_{ij}^{\sigma-\pi} + b_{ji}^{\sigma-\pi}) + \Pi_{ij}^{\mathrm{RC}} + b_{ij}^{\mathrm{DH}} \tag{6.29}$$

$$b_{ij}^{\sigma-\pi} = \left(1 + \sum_{k \# i,j} f_{ik}^C g_{ijk}\right)^{-1/2} \tag{6.30}$$

$$g_{ijk} = \sum_{i=0}^{5} \beta_i \cos^i[\theta_{ijk}] \tag{6.31}$$

Here, $b_{ij}^{\sigma-\pi}$ depends on the local coordination of atoms around atom i and the angle between atoms i, j, and k, θ_{ijk}. The coefficients, β_i, in the bond-bending spline function, g_{ijk}, fit to experimental data for graphite and diamond and are also listed in Table 6.2. The term, Π_{ij}^{RC}, accounts for various radical energetics, such as vacancies, which are not considered here; thus, this term is taken to be zero. The term, b_{ij}^{DH}, is a dihedral bending function that depends on the local conjugation and is zero for diamond but important for describing graphene and SWCNTs. This dihedral function involves third nearest-neighbor atoms and is given by

$$b_{ij}^{\mathrm{DH}} = \frac{T_0}{2} \sum_{k,l\#i,j} f_{ik}^C f_{jl}^C (1 - \cos^2[\Theta_{\mathrm{ijkl}}]) \tag{6.32}$$

where T_0 is a parameter, f_{ij}^C is the cut-off function, and Θ_{ijkl} is the dihedral angle of four atoms identified by the indices, i, j, k, and l, and is given by

$$\cos[\Theta_{ijkl}] = \vec{\eta}_{jik} \cdot \vec{\eta}_{ijl} \tag{6.33}$$

$$\vec{\eta}_{jik} = \frac{\vec{r}_{ji} \times \vec{r}_{ik}}{|\vec{r}_{ji}||\vec{r}_{ik}| \sin[\theta_{ijk}]} \tag{6.34}$$

where $\vec{\eta}_{jik}$ and $\vec{\eta}_{ijl}$ are unit vectors normal to the triangles formed by the atoms given by the subscripts, $\vec{r}_{ij}$ is the vector from atom i to atom j, and θ_{ijk} is the angle between atoms i, j, and k. In flat graphene, the dihedral angle, Θ_{ijkl}, is either 0 or π, and the dihedral term is subsequently zero. Bending of the graphene layer leads to a contribution from this term.

Some of the main differences when compared to the Tersoff EIP are the Brenner EIP and includes two additional exponential terms with corresponding adjustable parameters in the attractive pair-wise term, it includes a screened Coulomb term in the repulsive pair-wise term, it uses a fifth-order polynomial spline between bond orders for diamond and graphite, and it includes a dihedral bending term for bond energies which plays a role in SWCNTs and graphene.

6.6.2.3 Morse Potential

The Morse potential [7], named after physicist Philip M. Morse, is a convenient model for the potential energy of a diatomic molecule. It is better approximation for the vibrational structure of the molecule than the quantum harmonic oscillator because it explicitly includes the effects of bond breaking, such as the existence of unbound states. It also accounts for the an-harmonicity of real bonds. The Morse potential can also be used to model other interactions such as the interaction between an atom and a surface. Figure 6.1 shows a comparison between a simple harmonic potential and more advanced Morse potential. Unlike the energy levels of the harmonic oscillator potential, which are evenly spaced by $\hbar \times \omega$ ($\hbar = h/2\pi$), where "h" is Planck's constant, the Morse potential level spacing decreases as the energy approaches the dissociation energy. The dissociation energy D_e is larger than the true energy required for dissociation D_0 due to the zero point energy of the lowest ($v = 0$) vibrational level. The Morse potential energy function is of the form:

$$V(r) = D_e(1 - e^{-a(r-r_e)})^2 \tag{6.35}$$

Here, r is the distance between the atoms, r_e is the equilibrium bond distance, D_e is the well depth (defined relative to the dissociated atoms), and a controls the "width" of the potential

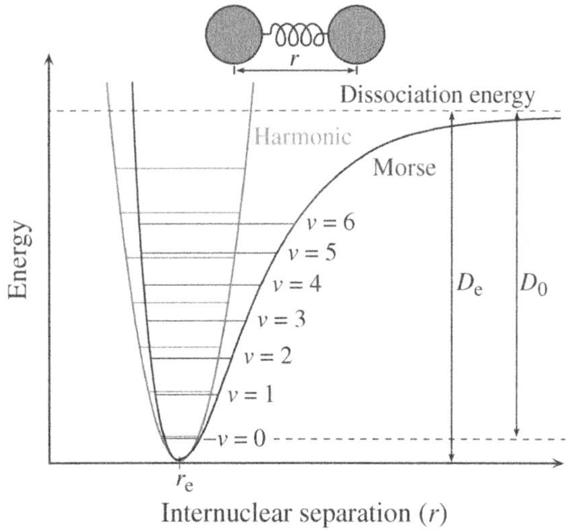

Figure 6.1 Comparison of harmonic potential and Morse potential.

(the smaller a, the larger the well). The dissociation energy of the bond can be calculated by subtracting the zero point energy $E(0)$ from the depth of the well. The force constant of the bond can be found by Taylor expansion of $V(r)$ around $r = r_e$ to the second derivative of the potential energy function, from which it can be shown that the parameter, a, is given by Eq. (6.36):

$$a = (k_e/2D_e)^{1/2} \tag{6.36}$$

where

k_e = force constant at the minimum of the well.

The zero of potential energy is arbitrary and the equation for the Morse potential can be rewritten in any number of ways by adding or subtracting a constant value.

6.6.2.4 Lennard–Jones Potential

An important concept in simulation methodology is that a significant limitation on the computation of interaction energies or forces between particles leads to an extraordinary reduction of the simulation time. To understand this concept, we consider the interaction energies between particles or potential curves. The Lennard–Jones potential [8] U_{LJ} is expressed as follows:

$$U_{LJ} = 4\varepsilon\,[(\sigma/r)^{12} - (\sigma/r)^{6}] \tag{6.37}$$

This potential is usually used as a model potential for rare gases such as Ar molecule. Here, ε is the well depth, σ is the quantity corresponding to the particle diameter and r is the separation between particles/molecules. Figure 6.2 shows the curve of the Lennard–Jones potential, in which U_{LJ} and r, are nondimensionalized by ε and σ. It shows a steep potential barrier in the range of $r \leq \sigma$, which induces such a significant repulsive interaction that particles are prevented from significantly overlapping, and an attractive interaction in the range of $r \geq \sigma$, which rapidly decreases to zero. These characteristics of the potential curve indicate that the interaction energy after a distance of approximately $r = 3\sigma$ can be assumed to be negligible. Hence, particle interaction energies or forces do not need to be calculated in the range of

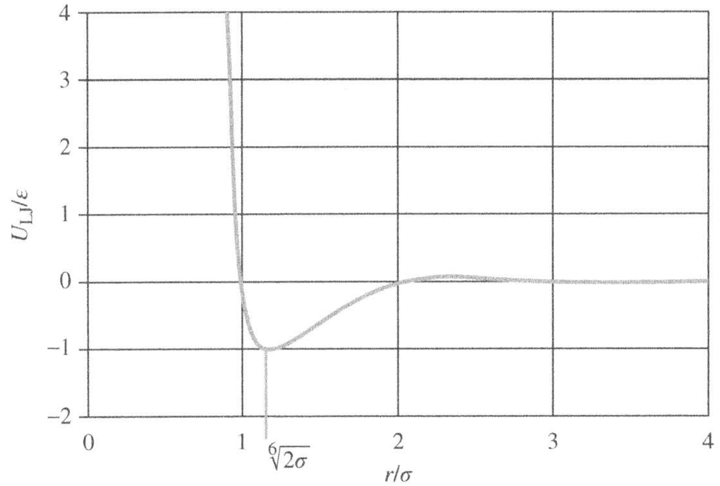

Figure 6.2 Lennard-Jones potential.

$r > 3\sigma$ in actual simulations. The distance for cutting off the calculation of energies or forces is known as the cutoff distance or cutoff radius, denoted by r_{coff}.

6.6.3 Ensemble

Suppose we have a macroscopic pure liquid sample, which might consist of 10^{23} particles, and we want to try and model some simple thermodynamic properties like the pressure p, the internal energy U, or the Gibbs energy G. At room temperature, the individual particles making up the sample will be in motion so at first sight, we ought to try and solve the equations of motion for these particles. In view of the large number of particles present, such an approach would be folly. Just to try and specify the initial positions and momentum of so many particles is not possible, and in any case, such a calculation will give us too much information. For the sake of argument, suppose that the container is a cube. Figure 6.3 shows a two-dimensional slice through the cube where the size of the particles has been exaggerated by a factor of approximately 10^{10}. The pressure exerted by a gas on a container wall depends on the rate at which the particles collide with the wall. It is not necessary to know which particle underwent a particular collision. What we need to know about is the root mean square speed of the particles, their standard deviation about the mean, the temperature, and so on. In statistical thermodynamics, we do not enquire about the behavior of the individual particles that make up a macroscopic sample; we just enquire about their average properties. Classical thermodynamics is essentially particle-free. All that really matters to such a thermodynamicist is bulk properties such as the number of particles N, the temperature T, and the volume of the container V. This information is shown in the right-hand box in Figure 6.3. Rather than worry about the time development of the particles in the left-hand box in Figure 6.3, we can make a very large number of copies of the system on the right-hand side. We then calculate average values over this large number of replications and, according to the ergodic theorem, the average value we calculate is exactly the same as the time average we would calculate by studying the time evolution of the original system. The two are the same. All the cells in the ensemble are not exact replicas at the molecular level. We just ensure that

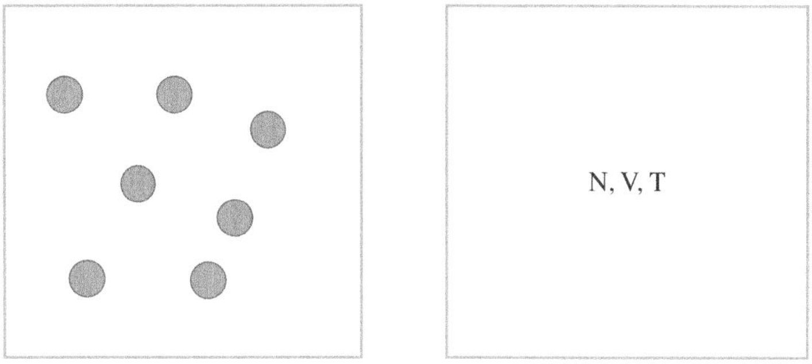

(a) (b)

Figure 6.3 A box of particles.

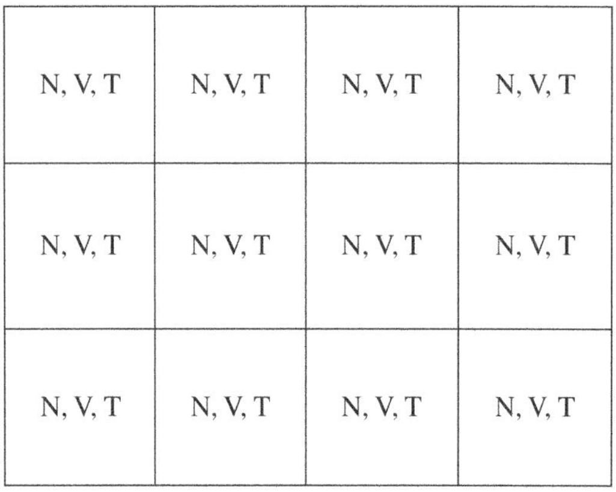

Figure 6.4 A canonical ensemble.

each cell has a certain number of thermodynamic properties that are the same. There is no mention of molecular properties at this stage.

Figure 6.4 shows an ensemble of cells all with the same values of N, V, and T. This array of cells is said to form a canonical ensemble. There are three other important ensembles in the theory of statistical thermodynamics, and they are classified according to what is kept constant in each cell. Apart from the canonical ensemble, where N, V, and T are kept constant, we have the following ensembles:

(i) The microcanonical ensemble where N and the total energy E and V are kept constant in each cell. In fact, this is a very simple ensemble because energy cannot flow from one cell to another.
(ii) In an isothermal-isobaric ensemble, N, T, and the pressure p are kept constant.
(iii) Finally, we have the grand canonical ensemble, where V, T, and the chemical potential are kept constant. The grand canonical ensemble is a fascinating one because the number of particles is allowed to fluctuate.

6.6.4 Thermostat

An equilibration procedure may be necessary in order to obtain the desired system temperature T. In the example of a liquid, the temperature T_{cal}, which is calculated from averaging the assigned velocities of particles, may differ significantly from the desired system temperature T. This may be due to the energy exchange between the kinetic and the potential energies. Hence, an equilibration procedure is frequently necessary before starting the main loop in a simulation program.

6.6.4.1 Andersen's Method

According to Andersen [9], the temperatures calculated from the translational and angular velocities of particles are denoted by $T_{cal}^{(t)}$ and $T_{cal}^{(r)}$, respectively, and written as follows:

$$T_{\mathrm{cal}}^{(t)} = \frac{1}{3N}\sum_{i=1}^{N}\frac{mv_i^2}{k},\ T_{\mathrm{cal}}^{(r)} = \frac{1}{3N}\sum_{i=1}^{N}\frac{I\omega_i^2}{k} \tag{6.38}$$

where N is the total number of particles, assumed to be $N \gg 1$. $T_{\mathrm{cal}}^{(t)}$ and $T_{\mathrm{cal}}^{(r)}$ calculated from v_i and ω_i ($i = 1, 2,\ldots, N$) are generally not equal to the desired temperature T. This equilibration procedure adjusts temperatures calculated from the translational and angular velocities of particles to T during the simulation by using the method of scaling the translational and angular velocities of each particle. If $T_{\mathrm{cal}}^{(t)\mathrm{ave}}$ and $T_{\mathrm{cal}}^{(r)\mathrm{ave}}$ denote the averaged values of $T_{\mathrm{cal}}^{(t)}$ and $T_{\mathrm{cal}}^{(r)}$ taken, for example, over 50 time steps, then the scaling factors $c_0^{(t)}$ and $c_0^{(r)}$ are determined as follows:

$$c_0^{(t)} = \sqrt{\frac{T}{T_{\mathrm{cal}}^{(t)\mathrm{ave}}}},\ c_0^{(r)} = \sqrt{\frac{T}{T_{\mathrm{cal}}^{(r)\mathrm{ave}}}} \tag{6.39}$$

With the scaling factors determined, the translational and angular velocities of all particles in a system are scaled as follows:

$$\boldsymbol{v}_i' = c_0^{(t)}\boldsymbol{v}_i\ ,\ \ \omega_i' = c_0^{(r)}\omega_i \qquad (i = 1, 2, \ldots, N) \tag{6.40}$$

This treatment yields the desired system temperature T. In this example, the scaling procedure would be conducted at every 50 time steps, but in practice an appropriate time interval must be adopted for each simulation case. The abovementioned equilibration procedure is repeated to give rise to the desired system temperature with sufficient accuracy.

6.6.4.2 Berendsen Thermostat

The Berendsen thermostat [10] is an algorithm to rescale the velocities of particles in MD simulations to control the simulation temperature. In this scheme, the system is weakly coupled to a heat bath with some temperature. The thermostat suppresses fluctuations of the kinetic energy of the system and therefore cannot produce trajectories consistent with the canonical ensemble. The temperature of the system is corrected such that the deviation exponentially decays with some time constant τ.

To maintain the temperature, the system is coupled to an external heat bath with fixed temperature T_0. The velocities are scaled at each step, such that the rate of change of temperature is proportional to the difference in temperature:

$$\frac{\mathrm{d}T(t)}{\mathrm{d}t} = \frac{1}{\tau}(T_0 - T(t)) \tag{6.41}$$

where τ is the coupling parameter which determines how tightly the bath and the system are coupled together. This method gives an exponential decay of the system toward the desired temperature. The change in temperature between successive time steps is

$$\Delta T = \frac{\delta t}{\tau}(T_0 - T(t)) \tag{6.42}$$

Thus, the scaling factor for the velocities is

$$\lambda^2 = 1 + \frac{\delta t}{\tau}\left(\frac{T_0}{T\left(t - \frac{\delta t}{2}\right)} - 1\right) \tag{6.43}$$

The term $T\,(t - \frac{\delta t}{2})$ is due to the fact that the leap-frog algorithm is used for the time integration. In practice, τ is used as an empirical parameter to adjust the strength of the coupling. Its value has to be chosen with care. In the limit $\tau \to \infty$, the Berendsen thermostat is inactive, and the run is sampling a microcanonical ensemble. The temperature fluctuations will grow until they reach the appropriate value of a microcanonical ensemble. However, they will never reach the appropriate value for a canonical ensemble. On the other hand, too small values of τ will cause unrealistically low-temperature fluctuations. If τ is chosen the same as the time step δt, the Berendsen thermostat is nothing else than the simple velocity scaling. Values of $\tau \approx 0.1$ ps are typically used in MD simulations of condensed-phase systems. The ensemble generated when using the Berendsen thermostat is not a canonical ensemble.

Though the thermostat does not generate a correct canonical ensemble (especially for small systems), for large systems of the order of hundreds or thousands of atoms/molecules, the approximation yields roughly correct results for most calculated properties. The scheme is widely used due to the efficiency with which it relaxes a system to some target (bath) temperature. In many instances, systems are initially equilibrated using the Berendsen scheme, while properties are calculated using the widely known Nosé–Hoover thermostat, which correctly generates trajectories consistent with a canonical ensemble.

6.6.4.3 Nosé–Hoover Thermostat

The Berendsen thermostat is extremely efficient for relaxing a system to the target temperature, but once our system has reached equilibrium, it might be more important to probe a correct canonical ensemble. The extended system method was originally introduced by Nose [11] and subsequently developed by Hoover [12]. The idea of the method proposed by Nose was to reduce the effect of an external system, acting as heat reservoir, to an additional degree of freedom. This heat reservoir controls the temperature of the given system, i.e. the temperature fluctuates around target value. The idea is to consider the heat bath as an integral part of the system by addition of an artificial variable, $\tilde{s}$, associated with a "mass" $Q > 0$ as well as a velocity $\dot{\tilde{s}}$.

The magnitude of Q determines the coupling between the reservoir and the real system and so influences the temperature fluctuations. The artificial variable $\tilde{s}$ plays the role of a time-scaling parameter. More precisely the timescale in the extended system is stretched by the factor, $\tilde{s}$:

$$\mathrm{d}\tilde{t} = \tilde{s}\mathrm{d}t \tag{6.44}$$

The atomic coordinates are identical in both systems. This leads to

$$\tilde{r} = r, \dot{\tilde{r}} = \tilde{s}^{-1}\dot{r}\ , \tilde{s} = s \qquad \text{and} \qquad \dot{\tilde{s}} = \tilde{s}^{-1}\dot{s} \tag{6.45}$$

The Lagrangian for the extended system is chosen to be

$$\mathcal{L} = \sum_i \frac{m_i}{2}\tilde{s}^2\dot{\tilde{r}}_i^2 - U(\tilde{r}) + \frac{1}{2}Q\dot{\tilde{s}}^2 - gk_bT_0 \ln \tilde{s} \tag{6.46}$$

The first two terms of the Lagrangian represent the kinetic energy minus the potential energy of the real system. The additional terms are the kinetic energy of $\tilde{s}$ and the potential, which is chosen to ensure that the algorithm produces a canonical ensemble where $g = N_{df}$ in real time sampling (Nose–Hoover formalism) and $g = N_{df} + 1$ for virtual-time sampling (Nose-formalism). This leads to the Nose equations of motion:

$$\ddot{\tilde{r}}_i = \frac{\tilde{F}_i}{m_i\tilde{s}^2} - \frac{2\dot{\tilde{s}}\dot{\tilde{r}}_i}{\tilde{s}} \tag{6.47}$$

$$\ddot{\tilde{s}} = \frac{1}{Q\tilde{s}}\left(\sum_i m_i\tilde{s}^2\dot{\tilde{r}}_i^{\,2} - gk_bT_0\right) \tag{6.48}$$

These equations sample a microcanonical ensemble in the extended system $(\tilde{r},\tilde{p},\tilde{t})$. However, the energy of the real system is not constant. Accompanying the fluctuations of $\tilde{s}$, heat transfers occur between the system and a heat bath, which regulate the system temperature. It can be shown, that the equations of motion sample a canonical ensemble in the real system. The Nose equations of motion are smooth, deterministic, and time-reversible. However, because the time-evolution of the variable $\tilde{s}$ is described by a second-order equation, heat may flow in and out of the system in an oscillatory fashion, leading to nearly periodic temperature fluctuations. The stretched timescale of the Nose equations is not very intuitive, and the sampling of a trajectory at uneven time intervals is rather impractical for the investigation of dynamical properties of a system. However, as shown by Nose and Hoover, the Nose equations of motion can be reformulated in terms of real system variables. The transformation is achieved through

$$s = \tilde{s},\quad \dot{s} = \tilde{s}\dot{\tilde{s}},\quad \ddot{s} = \tilde{s}^2\ddot{\tilde{s}} + \tilde{s}\dot{\tilde{s}}^2, r = \tilde{r},\quad \dot{r} = \tilde{s}\dot{\tilde{r}}\ ,\quad \ddot{r} = \tilde{s}^2\ddot{\tilde{r}} + \tilde{s}\dot{\tilde{r}}^2 \tag{6.49}$$

and with substituting

$$\gamma = \frac{\dot{s}}{s} \tag{6.50}$$

the Lagrangian equations of motion can be written as

$$\ddot{r}_i = \frac{\boldsymbol{F}_i}{m_i} - \gamma\boldsymbol{r}_i\ \ddot{\boldsymbol{r}}_i = \frac{\boldsymbol{F}_i}{m_i} - \gamma\boldsymbol{r}_i \tag{6.51}$$

$$\dot{\gamma} = \frac{-k_BN_{df}}{Q}T(t)\left(\frac{g}{N_{df}}\frac{T_0}{T(t)} - 1\right) \tag{6.52}$$

In both algorithms, some care must be taken in the choice of the fictious mass Q and extended system energy E_e. On the one hand, too large values of Q (loose coupling) may cause a poor temperature control (Nose–Hoover thermostat with $Q \to \infty$ is MD which generates a microcanonical ensemble). Although any finite (positive) mass is sufficient to guarantee in principle the generation of a canonical ensemble, if Q is too large, the canonical distribution will only be obtained after very long simulation times. On the other hand, too small values (tight coupling) may cause high-frequency temperature oscillations. The variable $\tilde{s}$ may oscillate at a very high frequency. It will tend to be off-resonance with the characteristic frequencies of the real system, and effectively decouple from the physical degrees of freedom (slow exchange of kinetic energy). As a more intuitive choice for the coupling strength, the Nose equations of motion can be expressed as

$$\dot{\gamma} = \frac{-1}{\tau_{\mathrm{NH}}}\left(\frac{g}{N_{df}}\frac{T_0}{T(t)} - 1\right) \tag{6.53}$$

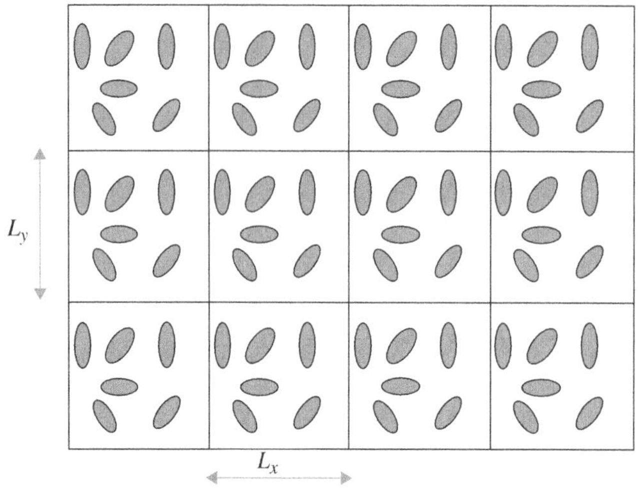

Figure 6.5 Periodic boundary conditions.

with the effective relaxation time

$$\tau_{NH}^2 = \frac{Q}{N_{df} k_B T_0} \tag{6.54}$$

The relaxation time can be estimated when calculating the frequency of the oscillations for small deviations $\delta\tilde{s}$ from the average $\langle\tilde{s}\rangle$.

6.6.5 Boundary Conditions

A system of one-mol-order size, being composed of about 6×10^{23} particles, never needs to be directly treated in molecular simulations for thermodynamic equilibrium. The use of the periodic boundary condition, explained below, enables us to treat only a relatively small system of about 100–10 000 particles in order to obtain such reasonable results as to explain the corresponding experimental data accurately.

6.6.5.1 Periodic Boundary Condition

Figure 6.5 schematically illustrates the concept of the periodic boundary condition for a two-dimensional system composed of spherocylinder particles. The central square is a simulation region and the surrounding squares are virtual simulation boxes, which are made by replicating the main simulation box. As Figure 6.5 shows, the origin of the *xy*-coordinate system is taken at the center of the simulation region, and the dimensions of the simulation region in the *x*-and *y*-directions are denoted by L_x and L_y. The two specific procedures are necessary in treating the periodic boundary condition.

First is the treatment of outgoing particles crossing the boundary surfaces of the simulation region and second is the calculation of interaction energies or forces with virtual particles being in the replicated simulation boxes. As shown in Figure 6.5, a particle crossing and exiting the left boundary surface has to enter from the right virtual box. When the interaction energy or force of particle *i* with other particles, for example, particle *j* has to be calculated, an appropriate particle *j* has to be chosen as an object from real and virtual particles *j*. This may be done in such a way that the distance between particle *i* and particle *j* is minimal.

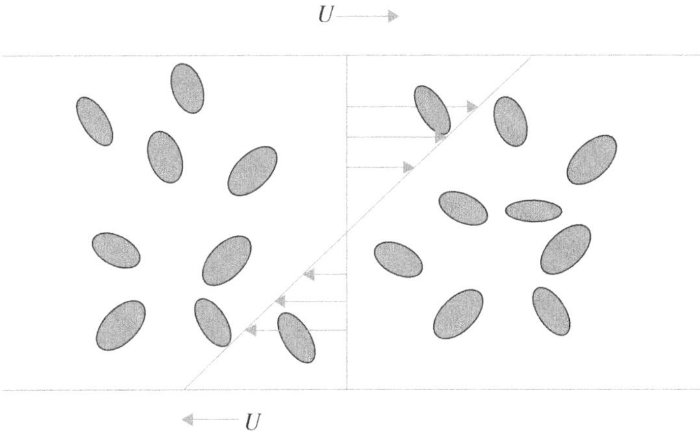

Figure 6.6 Simple shear flow.

6.6.5.2 Lees–Edwards Boundary Condition

The periodic boundary condition is quite useful for molecular simulations of a system in thermodynamic equilibrium, but this boundary condition is unsuitable for nonequilibrium situations. In treating the dynamic properties of a system in nonequilibrium, the most basic and important flow is a simple shear flow, as shown in Figure 6.6. The velocity profile, linearly varying from $-U$ at the lower surface to U at the upper one, can be generated by sliding the lower and upper walls in the left and right directions with the velocity U, respectively. This flow field is called the "simple shear flow." In generating such a simple shear flow in actual molecular simulations, the upper and lower replicated simulation boxes, shown in Figure 6.5, are made to slide in different directions with a certain constant speed. This sliding periodic boundary condition is called the "Lees–Edwards boundary condition" [13]. Figure 6.7 schematically depicts the concept of this boundary condition. The replicated boxes in the upper and lower layers slide in each direction by the distance ΔX. If particles move out of the simulation box by crossing the boundary surface normal to the x-axis, as shown in Figure 6.7, they are made to come into the simulation box through the opposite boundary surface, which is exactly the same procedure as the periodic boundary condition.

The important treatment in the Lees–Edwards boundary condition concerns the particles crossing the boundary surfaces normal to the y-axis. The same treatment of the periodic boundary condition is applied to the y-coordinate of such particles, but the x-coordinate should be shifted from x to $(x - \Delta X)$ in the case of Figure 6.7. In addition, the x-component of velocity, v_x of these particles needs to be modified to $(v_x - U)$, but the y-component v_y can be used without modification. For the case of evaluating interaction energies or forces, similar procedures have to be conducted for the particles interacting with virtual particles which are in the replicated simulation boxes in the upper or lower layers.

6.7 Molecular Dynamics Methodology

If the mass of molecule i is denoted by m_i, and the force acting on molecule i by the ambient molecules and an external field denoted by f_i, then the motion of a particle is described by

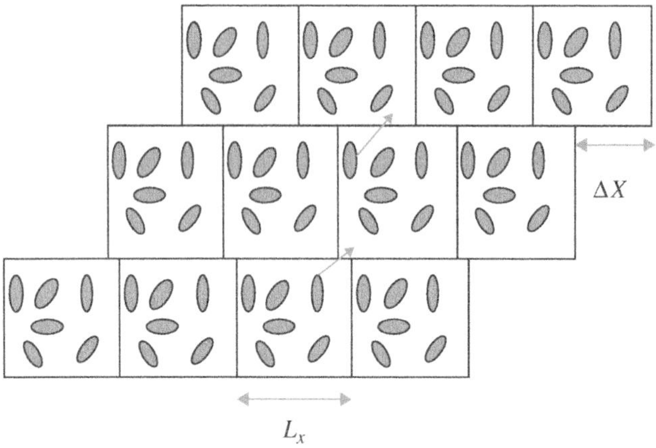

Figure 6.7 Lees–Edwards boundary condition.

Newton's equation of motion:

$$m_i \frac{d^2 r_i}{dt^2} = f_i \tag{6.55}$$

If a system is composed of N molecules, there are N sets of similar equations, and the motion of N molecules interacts through forces acting among the molecules. Differential equations such as Eq. (6.55) are unsuitable for solving the set of N equations of motion on a computer. Computers readily solve simple equations, such as algebraic ones, but are quite poor at intuitive solving procedures such as a trial and error approach to find solutions. Hence, Eq. (6.55) will be transformed into an algebraic equation. To do so, the second-order differential term in Eq. (6.55) must be expressed as an algebraic expression, using the following Taylor series expansion:

$$x(t+h) = x(t) + h\frac{dx(t)}{dt} + \frac{1}{2!}h^2\frac{d^2x(t)}{dt^2} + \frac{1}{3!}h^3\frac{d^3x(t)}{dt^3} + \ldots \tag{6.56}$$

Eq. (6.56) implies that x at time $(t+h)$ can be expressed as the sum of x itself, the first-order differential, the second-order differential, and so on, multiplied by a constant for each term. If x does not significantly change with time, the higher-order differential terms can be neglected for a sufficiently small value of the time interval h. In order to approximate the second-order differential term in Eq. (6.55) as an algebraic expression, another form of the Taylor series expansion is necessary:

$$x(t-h) = x(t) - h\frac{dx(t)}{dt} + \frac{1}{2!}h^2\frac{d^2x(t)}{dt^2} - \frac{1}{3!}h^3\frac{d^3x(t)}{dt^3} + \ldots \tag{6.57}$$

If the first-order differential term is eliminated from Eqs. (6.56) and (6.57), the second-order differential term can be solved as

$$\frac{d^2x(t)}{dt^2} = \frac{x(t+h) - 2x(t) + x(t-h)}{h^2} + O(h^2) \tag{6.58}$$

The last term on the right-hand side of this equation implies the accuracy of the approximation, and, in this case, terms higher than h^2 are neglected. If the second-order differential is approximated as

$$\frac{d^2x(t)}{dt^2} = \frac{x(t+h) - 2x(t) + x(t-h)}{h^2} \tag{6.59}$$

This expression is called the "central difference approximation." With this approximation and the notation $r_i = (x_i, y_i, z_i)$ for the molecular position and $f_i = (f_{xi}, f_{yi}, f_{zi})$ for the force acting on particle i, the equation of the x-component of Newton's equation of motion can be written as

$$x_i(t+h) = 2x_i(t) - x_i(t-h) + \frac{h^2}{m_i} f_{xi}(t) \tag{6.60}$$

Similar equations are satisfied for the other components. Since Eq. (6.60) is a simple algebraic equation, the molecular position at the next time step can be evaluated using the present and previous positions and the present force [14]. If a system is composed of N molecules, there are $3N$ algebraic equations for specifying the motion of molecules; these numerous equations are solved on a computer, where the motion of the molecules in a system can be pursued with the time variable. Eq. (6.60) does not require the velocity terms for determining the molecular position at the next time step. This scheme is called the "Verlet method." The velocity, if required, can be evaluated from the central difference approximation as

$$\boldsymbol{v}_i(t) = \frac{\boldsymbol{r}_i(t+h) - \boldsymbol{r}_i(t-h)}{2h} \tag{6.61}$$

This approximation can be derived by eliminating the second-order differential terms in Equations (6.56) and (6.57). It has already been noted that the velocities are unnecessary for evaluating the position at the next time step; however, a scheme using the positions and velocities simultaneously may be more desirable in order to keep the system temperature constant. If we take into account that the first and second-order differentials of the position are equal to the velocity and acceleration, respectively, the neglect of differential terms equal to or higher than third-order in Eq. (6.56) leads to the following equation:

$$\boldsymbol{r}_i(t+h) = \boldsymbol{r}_i(t) + h\boldsymbol{v}_i(t) + \frac{h^2}{2m_i} \boldsymbol{f}_i(t) \tag{6.62}$$

This equation determines the position of the molecules, but the velocity term arises on the right-hand side, so that another equation is necessary for specifying the velocity. The first-order differential of the velocity is equal to the acceleration:

$$\boldsymbol{v}_i(t+h) = \boldsymbol{v}_i(t) + \frac{h}{m_i} \boldsymbol{f}_i(t) \tag{6.63}$$

In order to improve accuracy, the force term in Eq. (6.63) is slightly modified and the following equation is obtained:

$$\boldsymbol{v}_i(t+h) = \boldsymbol{v}_i(t) + \frac{h}{2m_i} (\boldsymbol{f}_i(t) + \boldsymbol{f}_i(t+h)) \tag{6.64}$$

The scheme of using Eqs. (6.62) and (6.64) for determining the motion of molecules is called the "velocity Verlet method." It is well known that the velocity Verlet method is significantly superior in regard to the stability and accuracy of a simulation. Consider another representative scheme. Noting that the first-order differential of the position is the velocity and that of the velocity is the acceleration, the application of the central difference approximation to these first-order differentials leads to the following equations:

$$\boldsymbol{r}_i(t+h) = \boldsymbol{r}_i(t) + h\boldsymbol{v}_i(t+h/2) \tag{6.65}$$

$$\boldsymbol{v}_i(t+h/2) = \boldsymbol{v}_i(t-h/2) + \frac{h}{m_i} \boldsymbol{f}_i(t) \tag{6.66}$$

The scheme of pursuing the positions and velocities of the molecules with Eqs. (6.65) and (6.66) is called the "leapfrog method." This name arises from the evaluation of the positions and forces, and then the velocities, by using time steps in a leapfrog manner. This method is also a significantly superior scheme in regard to stability and accuracy, comparable to the velocity Verlet method. The MD method is applicable to both equilibrium and nonequilibrium physical phenomena, which makes it a powerful computational tool that can be used to simulate many physical phenomena if computing power is sufficient. The main procedure for conducting the MD simulation using the velocity Verlet method in the following steps:

(i) Specify the initial position and velocity of all molecules.
(ii) Calculate the forces acting on molecules.
(iii) Evaluate the positions of all molecules at the next time step from Eq. (6.62).
(iv) Evaluate the velocities of all molecules at the next time step from Eq. (6.64).
(v) Repeat the procedures from step 2.

In the above procedure, positions and velocities will be evaluated at every time interval h in the MD simulation. The method of specifying the initial positions and velocities is as follows.

6.7.1 Initial Positions

In order to develop a simulation program, it is necessary to have an overview of the general methodology, which should include the assignment of the initial configuration and velocities, the treatment of boundary conditions, and techniques for reducing computation time. An appropriate initial configuration has to be set with careful consideration given to the physical property of interest so that the essential phenomena can be grasped. For example, if nonspherical molecules or particles are known to incline in a preferred direction, there may be some advantages to using a parallelepiped rectangular simulation region rather than a cubic one. The periodic boundary condition is a representative model to manage the boundary of a simulation region. It is almost always used for systems in thermodynamic equilibrium. On the other hand, for investigating the dynamic properties of a system, the simple shear flow is frequently treated and in this case the Lees–Edwards boundary condition is available. Techniques for reducing computation time become very important in large-scale three-dimensional simulations, and methods of tracking particle neighbors, such as the cell index method are indispensable.

6.7.1.1 Spherical Systems

Setting an initial configuration of particles is an indispensable procedure for both MD and MC methods. Although it is possible to assign randomly the initial position of particles in a simulation region, a regular configuration, such as a simple cubic lattice or a face-centered cubic lattice, is handled in a more straightforward manner. The random allocation suffers from the problem of the undesirable overlap of particles and from possible difficulties in achieving high-packing fractions. Lattice assignments are almost free from the overlap problem and can achieve high packing fractions. However, the lattice packing may be too perfect for some simulations, requiring the adjustment of a small random perturbation. In the following paragraphs, the method of setting the initial configuration in a regular lattice formation for a two-dimensional configuration has been discussed, which will be followed by a three-dimensional configuration.

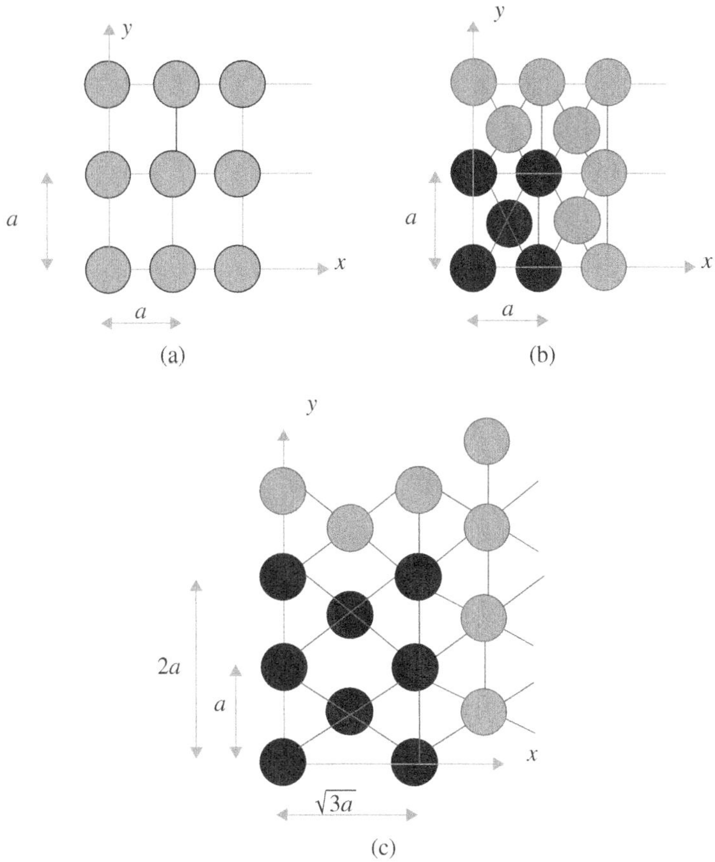

Figure 6.8 Initial conditions for a two-dimensional system.

Figure 6.8 shows several lattice systems that may be used to assign an initial configuration for a two-dimensional system. A basic lattice form is expanded to fill the whole simulation region and particles are then located at each lattice point. Figure 6.8a, the simplest lattice model, may be suitable for a gaseous system. However, even if the particle-particle distance "a" is equal to the particle diameter, a high-packing fraction cannot be obtained by using this simple lattice model. Hence, it is inappropriate for the simulations of a liquid or solid system. Since there is only one particle in the unit cell shown in Figure 6.8a, a system with total particle number N $(=Q^2)$ can be generated by replicating the unit cell $(Q-1)$ times in each direction to make a square simulation region of side length $L = Qa$. So for the use of this lattice system as the initial configuration, the particle number N has to be taken from $N = 1, 4, 9, 16, 25, 36, 49, 64, 81, 100$, and so on. The number density of particles n is given by $n = N/L^2$, and the area fraction is given by $\Phi_S = \pi\,(d/2)^2\,N/L^2$, where d is the particle diameter. In practice, the number of particles N and the area fraction Φ_S are first chosen; then the values of Q and L are evaluated, from which the value of "a" can be determined. With these values, the initial configuration of particles can be assigned according to the simple lattice system shown in Figure 6.8a.

The lattice system shown in Figure 6.8b can yield a higher-packing fraction and therefore may be applicable for an initial configuration of a gaseous or liquid state, but it has limited application to a solid state. Since there are two particles in the unit cell of this lattice, a system with total particle number $N = 2Q^2$ of particles can be generated by replicating the unit cell $(Q-1)$ times in each direction. In this case, the simulation region is also a square of side length $L = Qa$, and the possible value of N is taken from 2, 8, 18, 32, 50, 72, 98, 128, 162, 200, and so on. The number density of particles n is given by $n = N/L^2$, and the area fraction Φ_s is given by $\Phi_s = \pi\,(d/2)^2\,N/L^2$. Figure 6.8c shows the most compact lattice for a two-dimensional system. This lattice model may also be applicable to a solid system. If the dark particles are assumed to constitute the unit lattice, it follows that there are four particles in this unit lattice. Hence, by replicating the unit lattice $(Q-1)$ times in each direction, the simulation region becomes a rectangle of side lengths $L_x = 3^{1/2}aQ$ and $L_y = 2aQ$, with a total number of particles $N = 4Q^2$, where the possible value of N is taken from 4, 16, 36, 64, 100, 144, 196, 256, 324, 400, and so on. The particle number density n is given by $n = N/L_xL_y$, and the area fraction Φ_s is given by $\Phi_s = \pi\,(d/2)^2\,N/L_xL_y$. The actual assignment of the abovementioned quantities for simulations is similar to that for Figure 6.8a.

Figure 6.9 shows several lattice models for a three-dimensional system. Figure 6.9a is the simple cubic lattice model, which is suitable as an initial configuration mainly for a gaseous or liquid system. Since there is only one particle in the unit cell, the number of particles in a system is given by $N = Q^3$ by replicating the unit cell $(Q-1)$ times in each direction. In this case, the possible value of N is taken from $N = 1, 8, 27, 64, 125, 216, 343, 512, 729, 1000$, and so on. The simulation region is a cube of side length $L = Qa$. The number density n and the volumetric fraction Φ_V are given by $n = N/L^3$ and $\Phi_V = 4\pi\,(d/2)^3\,N/3L^3$, respectively. The face-centered cubic lattice model shown in Figure 6.9b is one of the close-packed lattices, and therefore may be applicable as an initial configuration of a solid state. Since there are four particles in the unit cell, the total number of particles in the simulation region is given by $N = 4Q^3$ by replicating the unit cell $(Q-1)$ times in each direction. In this case, the total number of particles is taken from $N = 4, 32, 108, 256, 500, 864, 1372$, and so on. The number density and the volumetric fraction are given by $n = N/L^3$ and $\Phi_V = 4\pi\,(d/2)^3\,N/3L^3$, respectively. As in a two-dimensional system, for the actual assignment of the abovementioned quantities, the particle number N and the volumetric fraction Φ_V are first chosen, then Q and L are evaluated, and finally the lattice distance "a" is determined.

For a gaseous or liquid system, the simple lattice models shown in Figures 6.8a and 6.9a are applicable in a straightforward manner for developing a simulation program. In contrast, for the case of a solid system, the choice of an appropriate lattice used for the initial configuration of particles is usually determined by the known physical properties of the solid.

6.7.1.2 Nonspherical Systems

For a nonspherical particle system, the orientation of the particles must be assigned in addition to their position so that the technique for setting the initial configuration is a little more difficult than that for a spherical particle system. For this purpose, a versatile technique whereby a wide range of initial configurations can be assigned is desirable. If particle-particle interactions are large enough to induce the cluster formation of particles in a preferred direction, then an appropriately large initial configuration has to be adopted in order for the simulation to capture such characteristic aggregate structures. We consider the example of a system composed of spherocylinder particles, as shown in Figure 6.10, with a magnetic moment at the particle center normal to the long particle axis. The spherocylinder is a cylinder with hemisphere caps at both the ends. An ensemble of these particles can be expected to aggregate to form raft-like clusters with the magnetic moments inclining in the applied

(a)

(b)

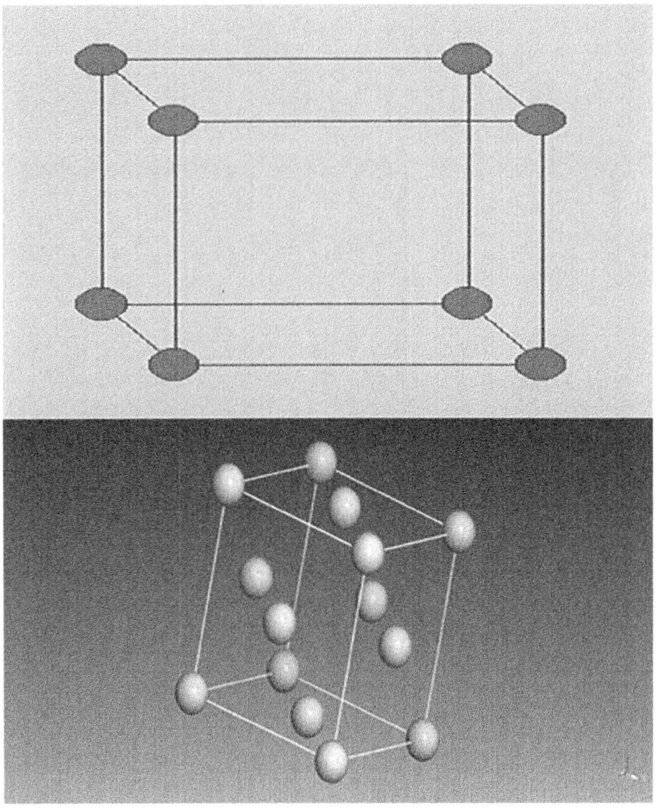

Figure 6.9 Initial positions for a three-dimensional system.

magnetic field direction. Hence, a simulation region with sufficient length in the direction of the cluster formation has to be taken in order for the simulation particles to aggregate in a reasonable manner.

The spherocylinder can be characterized by the ratio of the particle length l to the diameter d of the cylindrical part, known as the aspect ratio $r_p = l/d$. For Figure 6.10 where $r_p = 3$, the particles are placed in contact with three and nine rows in the x- and y-directions, respectively, leading to a configuration of 27 particles in a square region in the xy-plane. Extending this configuration to 18 layers in the z-direction yields an initial configuration of spherocylinder particles with a simulation region $(L_x, L_y, L_z) = (3r_pd, 3r_pd, 6r_pd)$ with total number of particles $N = 486$; if four rows are arranged in the x-direction, then a simulation region larger than the present case can be adopted with a simulation region $(L_x, L_y, L_z) = (4r_pd, 4r_pd, 8r_pd)$.

If the particle–particle distances are expanded equally in each direction to yield a desired volumetric fraction of particles Φ_V, then this expanded system may be used as an initial configuration for simulations. Such an expansion with a factor α of particle-particle distances gives rise to the system volume $V = 54r_p^3d^3\alpha^3$. The volumetric fraction Φ_V is related to the system volume as $\Phi_V = NV_p/V$, in which V_p is the volume of a spherocylinder particle, expressed as $V_p = \pi d^3(3r_p-1)/12$. From these expressions, the expansion ratio α can be obtained as

$$\alpha = \frac{1}{r_p}\left(\frac{3\pi(3r_p - 1)}{4\Phi_V}\right)^{\frac{1}{3}} \tag{6.67}$$

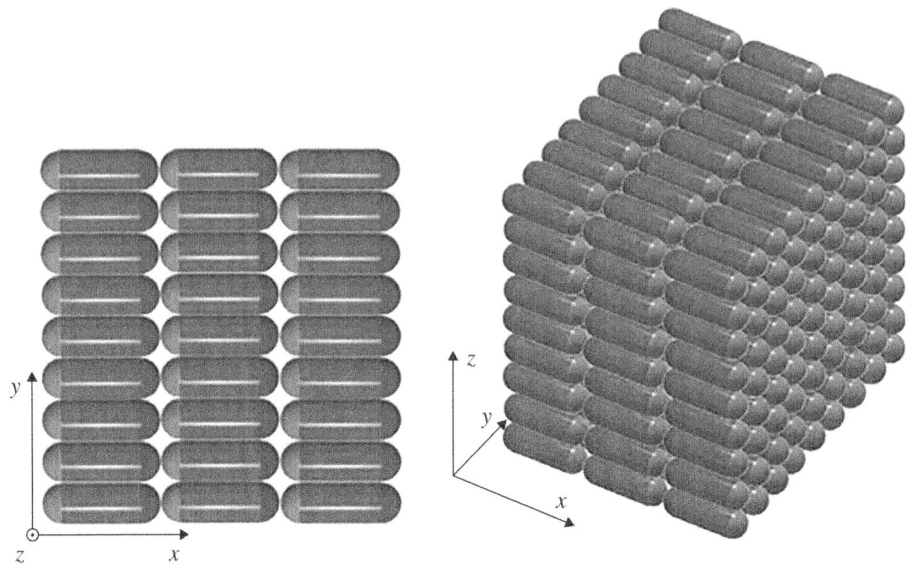

Figure 6.10 Initial conditions for spherocylinder particles.

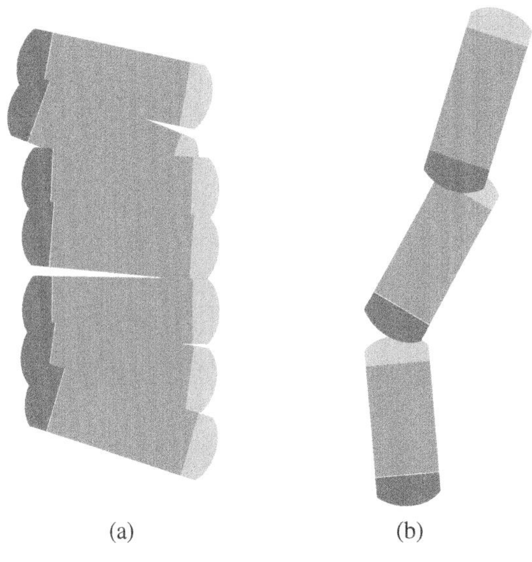

Figure 6.11 Aggregation for spherocylinder particles in (a) short axis direction and (b) long axis direction.

This initial configuration is applicable for a system in which particles are expected to aggregate in the direction of the particle short axis, as shown in Figure 6.11a. If particles are expected to aggregate in the direction of the particle long axis, as shown in Figure 6.11b, it is straightforward to follow a similar procedure with the spherocylinder particles aligned in the z-direction in Figure 6.10.

The main procedure for setting the initial configuration is summarized below:

(i) Consider an appropriate initial configuration, with sufficient consideration given to the physical phenomenon of interest.
(ii) Set a nearly close-packed situation as an initial configuration.
(iii) Calculate the total number of particles N.

(iv) Evaluate the expansion ratio α from Eq. (6.67) to give rise to the desired volumetric fraction Φ_V.
(v) Expand particle–particle distances equally by the factor α.
(vi) Perturb the particle positions by small distances using random numbers in order to destroy the regularity of the initial configuration; otherwise, all particle-particle interactions may be zero and therefore the particles may not move with time.

6.7.2 Initial Velocities

There is a difference in the method of assigning the initial velocities for spherical and nonspherical systems. For nonspherical systems, we have to assign angular velocities in addition to translational velocities. Methods for assigning the initial velocities for these systems have been discussed below.

6.7.2.1 Spherical Systems

In the MD method, the motion of particles is described by pursuing their position and velocity over time, so these factors have to be specified as an initial condition. If the system of interest is in thermodynamic equilibrium with temperature T, the particle velocities are described by the following Maxwellian distribution [15]:

$$f(\boldsymbol{v}_i) = \left(\frac{m}{2\pi kT}\right)^{\frac{3}{2}} \exp\left\{-\frac{m}{2kT}(\boldsymbol{v}_{ix}^2 + \boldsymbol{v}_{iy}^2 + \boldsymbol{v}_{iz}^2\right\} \tag{6.68}$$

in which k is Boltzmann's constant, m, is the mass of particles, and $v_i = (v_{ix}, v_{iy}, v_{iz})$ is the velocity vector of particle i. Since the Maxwellian distribution f is the probability density distribution function, the probability of particle i being found in the infinitesimal velocity range between v_i and $(v_i + \mathrm{d}v_i)$ becomes $f(v_i)\mathrm{d}v_i$. Characteristics of this function can be understood more straightforwardly by treating the distribution function f_x as the x-velocity component. The probability density distribution function $\chi(\mathrm{v}_i)$ for the speed $\mathrm{v}_i = (\mathrm{v}_{ix}^2 + \mathrm{v}_{iy}^2 + \mathrm{v}_{iz}^2)$ of particle i can be derived from Eq. (6.68) as

$$\chi(\boldsymbol{v}_i) = 4\pi\left(\frac{m}{2\pi kT}\right)^{\frac{3}{2}} \boldsymbol{v}_i^2 \exp\left\{-\frac{m}{2kT}\boldsymbol{v}_i^2\right\} \tag{6.69}$$

For a given system temperature T, the initial velocities of particles for simulations can be assigned according to the probability density function in Eq. (6.68) or Eq. (6.69). In order to set the initial velocities of particles in MD simulations, or to generate random displacements in BD and DPD simulations, it is necessary to generate random numbers according to a particular probability distribution. The probability distributions of interest here are the Gaussian distribution (also known as the normal distribution), and the Maxwell–Boltzmann distribution (or Maxwellian distribution). Since the velocity of particles theoretically has the Maxwellian velocity distribution for thermodynamic equilibrium, the initial velocity of particles in simulations must have such a velocity distribution. The method of setting the initial velocity of particles according to the Maxwellian distribution has been discussed in the following paragraph. It has been assumed that the stochastic variable x, such as the particle velocity or a random displacement, obeys the following normal distribution $\rho(x)$:

$$\rho(x) = \frac{1}{(2\pi)^{1/2}\sigma} \exp\left\{-\frac{(x-\overline{x})^2}{2\sigma^2}\right\} \tag{6.70}$$

in which σ^2 is the variance and $\overline{x}$ is the average of the stochastic variable x. In order to generate the stochastic variable x according to this normal distribution, the following equations are used together with a uniform random number sequence ranging from zero to unity:

$$x = \overline{x} + (-2\sigma^2 \ln R_1)^{1/2} \cos(2\pi R_2) \quad \text{or} \quad x = \overline{x} + (-2\sigma^2 \ln R_1)^{1/2} \sin(2\pi R_2) \tag{6.71}$$

According to either equation of Eq. (6.71), the required number of values of the stochastic variable are generated using a series of random numbers, such as R_1 and R_2, taken from a uniform random number sequence. In this way, the initial velocities of particles and random displacements can be assigned. The technique in Eq. (6.71) is called the Box–Muller method [16].

For generating a uniform random number sequence, there is an arithmetic method and a machine-generated method. The arithmetic method is reproducible, and the same random number sequence can be obtained at any time in the simulations. In contrast, the machine-generated method is generally not a reproducible sequence, and a different sequence of random numbers is generated each time a simulation is run. For the case of the Maxwellian velocity distribution, the velocity components of particle i can be assigned using the random numbers $R_1, R_2, \ldots, R_6$ taken from a uniform random number sequence as

$$\begin{aligned} v_{ix} &= \left\{-2\left(\frac{kT}{m}\right)\ln R_1\right\}^{1/2} \cos(2\pi R_2) \\ v_{iy} &= \left\{-2\left(\frac{kT}{m}\right)\ln R_3\right\}^{1/2} \cos(2\pi R_4) \\ v_{iz} &= \left\{-2\left(\frac{kT}{m}\right)\ln R_5\right\}^{1/2} \cos(2\pi R_6) \end{aligned} \tag{6.72}$$

In this way, all the initial velocity components can be assigned using random numbers. Each particle requires a new, that is, a different, set of random numbers.

The temperature which is evaluated from the initial particle velocities assigned by the abovementioned method is approximately equal to the desired system temperature, but may not necessarily be satisfactory. Hence, an equilibration procedure is usually necessary before starting the main loop in an actual simulation program. This will be explained in the section that follows.

6.7.2.2 Nonspherical Systems

For a nonspherical particle system, the initial angular velocities need to be assigned in addition to the translational velocities. Similar to the translational velocity $v = (v_x, v_y, v_z)$ discussed above, the angular velocity $\omega = (\omega_x, \omega_y, \omega_z)$ is also governed by the Maxwellian distribution $f_\omega(\omega)$. The expression for $f_\omega(\omega)$ is

$$f_\omega(\omega) = \left(\frac{I}{2\pi kT}\right)^{\frac{3}{2}} \exp\left\{-\frac{I}{2kT}(\omega_x^2 + \omega_y^2 + \omega_z^2)\right\} \tag{6.73}$$

where I is the inertia moment of a particle. The characteristics of the exponential function in Eq. (6.68) or Eq. (6.73) demonstrate that the probability of particles appearing with larger translational and angular velocities increases with the system temperature. The method for setting the initial translational velocities using uniform random numbers, explained in the previous subsection, is applicable to the present angular velocity case. Here, m and (v_{ix}, v_{iy}, v_{iz}) in Eq. (6.72) are replaced by I and $(\omega_{ix}, \omega_{iy}, \omega_{iz})$. New uniform random numbers need to be used for each particle.

6.8 Molecular Potential Energy Surface

The complete mathematical description of a molecule, including both quantum mechanical and relativistic effects, is a formidable problem, due to the small scales and large velocities. However, for this discussion, these intricacies are ignored and the focus is on general concepts because molecular mechanics and dynamics are based on empirical data that implicitly incorporate all the relativistic and quantum effects. Since no complete relativistic quantum mechanical theory is suitable for the description of molecules, we start with the nonrelativistic, time-independent form of the Schrödinger [17] description:

$$H\,\psi\,(R,r) = E\,\psi\,(R,r) \tag{6.74}$$

where H is the Hamiltonian for the system, ψ is the wave function, and E is the energy. In general, ψ is a function of the coordinates of the nuclei (R) and of the electrons (r).

Although the Eq. (6.74) is quite general, it is too complex for any practical use, so approximations are made. Noting that the electrons are several thousands of times lighter than the nuclei and therefore move much faster, Born and Oppenheimer [18] proposed what is known as the Born–Oppenheimer approximation: the motion of the electrons can be decoupled from that of the nuclei, giving two separate equations. The first equation describes the electronic motion or the potential energy surface:

$$H\,\psi\,(r;R) = E\,\psi\,(r;R) \tag{6.75}$$

This depends only parametrically on the positions of the nuclei. Note that this equation defines energy $E(R)$, which is a function of only the coordinates of the nuclei. This energy is usually called the potential energy surface. The second equation then describes the motion of the nuclei on this potential energy surface $E(R)$:

$$H\,\phi\,(R) = E\,\phi\,(R) \tag{6.76}$$

Solving Eq. (6.76) is important if we are interested in the structure or time evolution of a model. Equation (6.76) is the Schrödinger equation for the motion of the nuclei on the potential energy surface. In principle, Eq. (6.75) can be solved for the potential energy E, and then Eq. (6.76) could be solved. However, the effort required to solve Eq. (6.75) is extremely large, so usually an empirical fit to the potential energy surface, commonly called a force field is used. Since the nuclei are relatively heavy objects, quantum mechanical effects are often insignificant, in which case (6.76) can be replaced by Newton's equation of motion:

$$-\frac{\mathrm{d}V}{\mathrm{d}R} = m\,\frac{d^2R}{\mathrm{d}t^2} \tag{6.77}$$

The solution of Eq. (6.77) using an empirical fit to the potential energy surface $E(R)$ is called MD. Molecular mechanics ignores the time evolution of the system and instead focuses on finding particular geometries and their associated energies or other static properties. This includes finding equilibrium structures, transition states, relative energies, and harmonic vibrational frequencies.

Materials' properties result from quantum mechanical interactions described by solutions of a Schrodinger equation. Solving the full many-body problem was first addressed by Hohenberg, Kohn, and Sham (HKS) [19, 20] who formulated a theory known as "Density functional theory" (DFT), where electrons are replaced by effective electrons with the same total density moving in the potential generated by the other electrons and ion cores. Density functional theory (DFT) is a quantum mechanical modeling method used in physics and chemistry to investigate the electronic structure (principally the ground state) of many-body systems, in particular atoms, molecules, and the condensed phases. With this theory, the properties of a many-electron system can be determined by using functionals, i.e. functions of another function, which in this case is the spatially dependent electron density. Hence, the name DFT comes from the use of functionals of the electron density. DFT is among the most popular and versatile methods available in condensed-matter physics, computational physics, and computational chemistry. DFT was put on a firm theoretical footing by the two Hohenberg–Kohn theorems (H–K). The original H–K theorems held only for nondegenerate ground states in the absence of a magnetic field, although they have since been generalized to encompass these.

The first H–K theorem demonstrates that the ground state properties of a many-electron system are uniquely determined by an electron density that depends on only three spatial coordinates. It lays the groundwork for reducing the many-body problem of N electrons with $3N$ spatial coordinates to three spatial coordinates, through the use of functionals of the electron density. This theorem can be extended to the time-dependent domain to develop time-dependent density functional theory (TDDFT), which can be used to describe excited states. The second H–K theorem defines energy functional for the system and proves that the correct ground state electron density minimizes this energy functional.

Within the framework of Kohn–Sham DFT (KS DFT), the intractable many-body problem of interacting electrons in a static external potential is reduced to a tractable problem of noninteracting electrons moving in an effective potential. The effective potential includes the external potential and the effects of the Coulomb interactions between the electrons. Modeling the interactions becomes the difficulty within KS DFT. The simplest approximation is the local-density approximation (LDA), which is based upon exact exchange energy for a uniform electron gas, which can be obtained from the Thomas–Fermi model, and from fits to the correlation energy for a uniform electron gas. Noninteracting systems are relatively easy to solve as the wave function can be represented as a Slater determinant of orbitals.

In a system of N electrons, the total density is expressed as a sum of the orbitals ψ_i as

$$\rho(r) = \sum_{i=1}^{N} |\psi_i(r)|^2 \tag{6.78}$$

where ψ_i are solutions of the Kohn–Sham equations:

$$\left(-\frac{\hbar^2}{2m}\nabla^2 + V_{\text{eff}}\right)\psi_i(r) = \varepsilon_i \psi_i(r) \tag{6.79}$$

where

$\hbar$ = reduced Planck's constant;

V_{eff} = effective potential due to Coulomb and exchange-correlation contributions;

ε_i = orbital energy of the corresponding Kohn–Sham orbital, ψ_i.

However, the HKS theory, usually called DFT does not provide an algorithm of determining this functional. The solution came from the LDA which assumes that the functional depends only on the local electron density. LDA is an accurate approximation for systems with slowly varying charge densities, like many metals; however, it has some limitations in systems far from equilibrium and for nonuniform charge densities. Development of efficient algorithms has made DFT the primary method for calculating properties such as band structures, cohesive energies, and activation barriers. However, DFT is inefficient if the atoms are allowed to move as, for example, during the geometrical optimization of a molecule.

References

1 Cornell, W.D., Cieplak, P., Bayly, C.I. et al. (1995). A second generation force field for the simulation of proteins, nucleic acids, and organic molecules. *Journal of American Chemical Society* 117: 5179–5197.

2 Sun, H. (1998). COMPASS: an ab initio force-field optimized for condensed-phase applications overview with details on alkane and benzene compounds. *Journal of Physical Chemistry* 5647 (98): 7338–7364.

3 Brooks, B.R., Bruccoleri, R.E., Olafson, B.D. et al. (1983). CHARMM: A program for macromolecular energy, minimization, and dynamics calculations. *Journal of Computational Chemistry* 4 (2): 187–217.

4 Dauber-Osguthorpe, P., Roberts, V.A., Osguthorpe, D.J. et al. (1988). Structure and energetics of ligand binding to proteins: Escherichia coli dihydrofolate reductase-trimethoprim a drug-receptor system. *Proteins: Structure, Function, and Genetics* 4: 31–47.

5 Tersoff, J. (1988). New empirical approach for the structure and energy of covalent systems. *Physical Review B* 37: 6991.

6 Brenner, D.W. (1990). Empirical potential for hydrocarbons for use in simulating the chemical vapor deposition of diamond films. *Physical Review B* 42 (15): 9458.

7 Morse, P.M. (1929). Diatomic molecules according to the wave mechanics II. Vibrational levels. *Physical Review* 34: 57–64.

8 Lennard-Jones, J.E. (1924). On the determination of molecular fields. *Proceedings of Royal Society of London A* 106 (738): 463–477.

9 Andersen H.C. (1980), "Molecular dynamics simulations at constant pressure and/or temperature", The *Journal of Chemical Physics*, Vol. 72 (4), pp. 2384.

10 Berendsen, H.J.C., Postma, J.P.M., van Gunsteren, W.F. et al. (1984). Molecular-dynamics with coupling to an external bath. *Journal of Chemical Physics* 81 (8): 3684–3690.

11 Nosé, S. (1984). A unified formulation of the constant temperature molecular-dynamics methods. *Journal of Chemical Physics* 81 (1): 511–519.

12 Hoover, W.G. (1985). Canonical dynamics: Equilibrium phase-space distributions. *Physical Review A.* 31 (3): 1695–1697.

13 Lees, A.W. and Edwards, S.F. (1972). The computer study of transport processes under extreme conditions. *Journal of Physics C: Solid State Physics* 5: 1921.

14 Satoh, A. (2011). *Introduction to Practice of Molecular Simulation*, 1e. London: Elsevier (Chapter 2).

15 Maxwell, J.C. (1860). Illustrations of the dynamical theory of gases. Part I. On the motions and collisions of perfectly elastic spheres. *The London, Edinburgh, and Dublin Philosophical Magazine and Journal of Science* 19 (4): 19–32.

16 Box, G.E.P. and Muller, M.E. (1958). A note on the generation of random normal deviates. *The Annals of Mathematical Statistics* 29 (2): 610–611.

17 Schrödinger, E. (1926). An undulatory theory of the mechanics of atoms and molecules. *Physical Review* 28 (6): 1049–1070.

18 Born, M. and Oppenheimer, J.R. (1927). Zur Quantentheorie der Molekeln. *Annalen der Physik* 389 (20): 457–484.

19 Hohenberg, P. and Kohn, W. (1964). Inhomogeneous electron gas. *Physical Review* 136 (3B): B864–B871.

20 Kohn, W. and Sham, L.J. (1965). Self-consistent equations including exchange and correlation effects. *Physical Review* 140 (4A): A1133–A1138.

7

Molecular Dynamics Simulation-Case Studies

7.1 Carbon Nanofiber–Reinforced Polymer Composites

In the last two decades [1], significant research has been devoted to vapor grown carbon nanofibers (VGCNFs) which are produced by catalytic exposure of gaseous hydrocarbons to high temperatures often resulting in highly ordered oblique graphene layers arranged as stacked truncated cones surrounded by turbostratic carbon. This arrangement is less ordered and, therefore, more defect prone compared to the graphitic inner phase. Although the fabrication and structural aspects of this class of nanofibers have received sufficient attention, very few studies have directly and indirectly addressed their mechanical behavior as unequivocal mechanical property measurement methods are still lacking at this scale. VGCNFs have been used to reinforce a variety of polymers, including polypropylene (PP), polycarbonate, nylon, poly(ether sulfone), poly(ethylene terephthalate), poly(phenylene sulfide), acrylonitrile-butadiene-styrene (ABS), and epoxy. CNFs are known to have wide ranging morphologies from structures with a disordered bamboo-like structure to highly graphitized "cup stacked" structures as given by Endo et al. [1] where the conical shells of nanofiber are nested within each other.

CNFs have transport and mechanical properties that approach the theoretical values of single-crystal graphite, similar to the fullerenes, but they can be made in high volumes at low cost, ultimately lower than that of conventional carbon fibers. In equivalent production volumes, CNFs are projected to have a cost comparable to E-glass on a per-pound basis, yet they possess properties that far exceed those of glass and are equal to, or exceed, those of much more costly commercial carbon fiber. Graphite fiber was firstly reported by Hughes and Chambers [2]. Initially, CNF was regarded as an impediment because it blocked the reactors, reduced the heat transfer coefficients, and even deactivated the catalysts employed in these cases including hydrogenation reactions, steam reforming processes, and Fisher–Tropsch synthesis. CNFs did not actually attract much attention of the academic community until the 1980s, when their extraordinary physical and chemical properties were found. Their potential applications as catalysts, catalyst supports, gas storage materials, electrodes for fuel cell, and polymer additives have been searched since then. CNFs have been produced mainly by chemical vapor deposition (CVD), from which the CNF diameters vary from 2 to 100 nm, with their lengths from 5 to 100 μm. In spite of the enormous progress in the synthesis and application, fundamental studies on CNFs obviously lag behind. There are mainly three types of CNFs: the Fishbone or herringbone carbon nanofibers (f-CNFs), their graphene layers, stacking obliquely with respect to the fiber axis; the platelet CNFs, their graphene layers being perpendicular to the axis; and the ribbon CNFs, their graphene layers paralleling to

Mechanics of Particle- and Fiber-Reinforced Polymer Nanocomposites: From Nanoscale to Continuum Simulations, First Edition. Sumit Sharma.

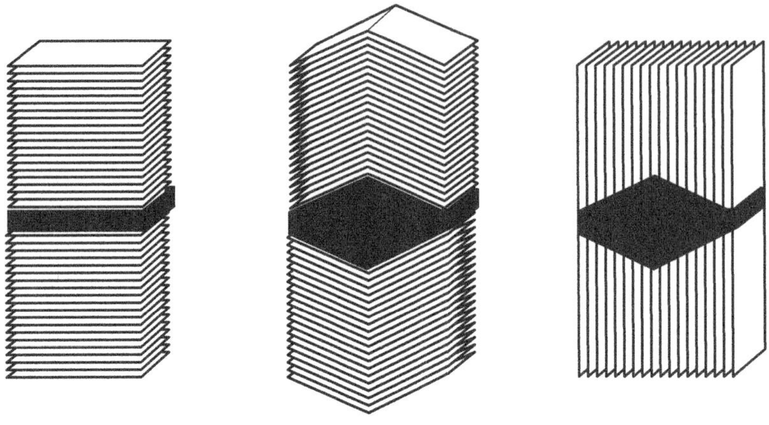

Figure 7.1 Schematic representation for CNFs: (a) platelet-type CNF; (b) fishbone-type CNF; (c) ribbon-type CNF.

the growth axis [3]. Figure 7.1 schematically presents these three types of CNF structures based on their HRTEM photos. These morphological structures only provide stacking modes of the graphene sheets, which help to account for the growth mechanism of CNFs. However, these structures are still insufficient as a deeper insight is necessary. It is of great importance to establish a model depicting the structures of CNFs at the atomic scale. The f-CNF is supposedly the most complicated one among the three CNF types, and its model has not been discussed systematically.

Electron energy loss spectra (EELS) on f-CNFs have revealed typical features of sp^2 carbon confirming their "graphitic" nature. Theoretical studies indicated some relationships between the electronic structure and the topological nature of graphitic cones of which the f-CNFs were made up. Modeling for boron nitride cones was also performed, and an ordered stacking mode of cones was claimed. It was reported by Rodriguez [4] that CNF consisted of a quasi-cylindrical arrangement which displayed a hollow or amorphous core surrounding the axis, while the outer portion presented an extremely well-defined arrangement of paralleling graphite layers. People noticed, especially for f-CNFs that they were stacked by lampshade-type or closed graphitic cones oriented in the growth direction, and the formation of curved carbon surface was determined by the occurrence of four square or pentagonal carbon rings (ring defects) in the hexagonal network.

The present study focuses on modeling of CNFs and evaluating the effect of their reinforcement in PP matrix. Molecular dynamics (MD) simulation has been used to study the effect of CNF volume fraction (V_f) and aspect ratio (l/d) on the mechanical properties of CNF/PP composites. Results have also been obtained for damping.

Mall et al. [5] tested an instrumented indentation technique on three types of carbon nanotube/nanofiber reinforced composites to investigate its applicability for measuring mechanical properties (tensile modulus and hardness). There was good agreement in the measured elastic modulus between the instrumented indentation and uniaxial tension tests for the case of a composite with a harder epoxy matrix material. In contrast, there was considerable difference in elastic modulus between the two tests in the case of a composite with a softer polystyrene matrix material. The instrumented indentation test was also utilized in the case of a patterned nanotube array-reinforced epoxy matrix composite. This clearly showed the modulus of the array composite improved considerably compared to that of the neat epoxy resin.

Minaie et al. [6] manufactured hierarchical polymer composites defined as carbon nanofiber/fiber-reinforced polymer composites using multiscale-reinforcement fabrics (MRFs), and they were characterized for their mechanical properties. Compared to the base composite (not containing CNF), the hierarchical composites containing the functionalized CNF showed an increase in interlaminar shear strength (ILSS) and compressive strength. Panels containing amine-functionalized CNF had the highest increase in properties: 12% in ILSS and 13% in compressive strength. The study establishes that it is possible to obtain a significant increase in ILSS and compressive strength by adding CNF only to the carbon fiber layers. Tibbetts and Beetz [7] grew large diameter vapor grown carbon fibers (VGCFs) and directly measured their tensile properties. The effect of fiber diameter on tensile properties was studied for the diameter range of 6.5–31.5 μm. The study indicated that the tensile properties decrease with increasing VGCF diameter. For a 7.5 μm diameter VGCF, the tensile strength and tensile modulus were 2.92 and 237 GPa, respectively. However, for a 31.5 μm in diameter VGCF, the tensile strength and tensile modulus decreased to 0.73 and 120 GPa, respectively.

Patton et al. [8] estimated the lower limit of the tensile modulus and tensile strength of Pyrograf III VGCNF based on the rule of mixtures and experimentally measured mechanical properties of 15.5 vol% VGCNF reinforced epoxy. Their calculations revealed that the lower limit of tensile modulus and tensile strength are in the range of 88–166 and 1.7–3.38 GPa, respectively. Direct measurement of mechanical properties of nanofibers and nanotubes is technically difficult because of their small diameter. Therefore, there are limited experimental data on the mechanical properties of nanofillers. Endo et al. [9] reported the mechanical tensile properties of sub-micron vapor grown carbon fiber (s-VGCF). The results showed that the tensile strength of graphitized s-VGCF increases with decreasing the fiber diameter. The tensile strength of s-VGCF having diameters of 100 and 300 nm were 2.2 and 1.77 GPa, respectively. The tensile strength and tensile modulus of s-VGCF were reported to be higher than that of general grade CF. On average, the tensile strength and tensile modulus of the s-VGCF are 2 and 200 GPa, respectively. Zhu et al. [10] evaluated carbon nanofillers with different surface functional groups and aspect ratios, including carboxyl carbon nanotubes, un-functionalized CNFs, glycidyloxypropyl-trimethoxysilane carbon nanotubes (GPS-CNTs), and nanofibers for their potential for increasing the interlaminar fracture toughness of an S2-glass fiber/epoxy composite. The fillers were added in the matrix of the fiber reinforced plies, in the resin interlayer between plies, or in both regions. Comparisons were made based on mode I and mode II interlaminar fracture toughness. For composites made with CNTs dispersed in the matrix, fracture toughness was largely unaffected except for a slight increase seen with long GPS-CNTs. However, adding a CNF or CNT modified resin interlayer significantly increased the fracture toughness, with the highest improvement over the baseline material achieved by adding long GPS-CNTs in the interlayer (79% and 91% for mode I and mode II onset toughness, respectively). Important material parameters identified for improving interlaminar fracture toughness are the nanofiller aspect ratio and concentration at the fracture plane.

Sansoz and Gu [11] used classical MD simulations to study the effects of cone angle on mechanical properties and failure mechanisms in thermally treated cup-stacked CNF. The authors reported a 22-fold reduction in elastic modulus and 4-fold decrease in tensile strength of the cup-stacked CNF with a wide range of cone angles between 19.2° and 180°. Their results show significant elastic stiffening for intermediate angles between 38.9° and 112.9°, as well as a minimum in tensile strength at a critical cone angle, due to the competition between weak van der Waals forces between layers and strong strengthening mechanisms from surface bonds introduced during thermal treatment. Different failure modes in CNF subjected to tensile deformation were also predicted as a function of cone angle. Their study constitutes an important step toward understanding the origin of strength dispersions observed experimentally in CNF and suggests that the design of high-strength CNF can be optimized structurally by appropriately tuning the cone angle.

Past experimental studies have shown significant statistical variations in the elastic modulus of CNF-reinforced PP composites. Limited information is available at the molecular scale about the interaction between CNF and PP matrix. Computational studies complement experiment by providing easy manipulation, analysis, and insights on the molecular level. The extent to which mechanical reinforcement can be achieved depends on several factors, including uniformity of dispersion, degree of alignment of CNFs, and the strength of polymer–CNF interfacial bonding. Since it is difficult to control and measure many of these properties experimentally, computational modeling can provide some crucial insights. For this reason, theoretical and computational methods have been widely applied to study polymer/CNF composites.

Due to the difficulties in the experimental characterization of nanotubes, computer simulation has been regarded as a powerful tool for modeling the properties of nanofibers. Among the available modeling techniques, MD simulation has been used most extensively. First principles methods are able to generate reasonably accurate data of structures and energies relevant to polymer nanocomposite systems, but they can only be used to study small systems over short times due to their computational expense. In contrast, molecular simulation methods such as molecular mechanics (MM) and MD that are based on analytic force fields are computationally cheaper compared to first principles methods. They can therefore be used to study larger molecular systems for longer times. As described below, MD can also be used to obtain macroscopic properties such as the tensile modulus.

The present study is undertaken to investigate changes in the static mechanical properties under the influence of varying compositions (by volume) of pristine CNFs in PP by employing MD technique without any additional component or any surface modification in the composite. Elastic modulus has been evaluated using Accelrys Materials Studio 5.5. CNF composition in PP was varied by volume from 0% to 16%. A number of simulations were also performed for a fixed CNF volume fraction and varying aspect ratio. Aspect ratio of CNFs was varied from $l/d = 5$ to $l/d = 100$. Variation of damping with V_f and l/d has been studied. These properties would help us in mechanical characterization of PP-based composites and the usefulness of CNFs as reinforcement from a comprehensive applications point of view.

7.1.1 Molecular Modeling of CNF and CNF/PP Composites

In all molecular modeling calculations, including the energy minimization process using the steepest descent method, a commercially available molecular simulation engine Materials Studio version 5.5 (MS 5.5) was used. Constant temperature (NVT ensemble) and constant number of particles, constant pressure, and constant temperature i.e. (NPT ensemble) both were used to apply the dynamics on to CNTs. Intermolecular and intramolecular atomic interactions were described by using COMPASS (condensed phase optimized molecular potential for atomistic simulation studies), the first and only ab initio force field that

Table 7.1 Physical and mechanical properties of CNF.

Property	Pyrograph III® PR-24-PS
Ultimate strength, GPa	7.0
Tensile modulus, GPa	600
Diameter, μm	0.060–0.2
Length, μm	30–100
Density, g/cm^3	1.95
Electrical resistivity, μΩ-cm	55
Loss factor (η)	0.0015 (Finegan et al. [12])

Source: Based on Zhou et al. [13].

Table 7.2 Properties of polypropylene (PP).

Property	Polypropylene
Storage modulus, GPa	1.23 (Finegan et al. [12])
Loss factor, η	0.06 (Finegan et al. [12])
Density, g/cm^3	0.900 (Alam et al. [14])
Poisson's ratio, ν^m	0.30

enables an accurate and simultaneous prediction of various gas-phase and condensed phase properties of organic and inorganic materials. The simulation in each case was performed with an interval of 1 fs (femto second) with total time of 50 ps (pico seconds). Table 7.1 lists the properties of CNF and Table 7.2 shows the properties of PP.

7.1.2 Modeling of CNFs

Long sheets of plane carbon atoms are drawn in Materials Studio 5.5, using partial double bonds to join each other forming hexagonal rings to the adjacent atoms. Figure 7.2 shows one of the plane sheets. An imaginary wedge of 60° angle is drawn on the modeled plane grapheme sheet, replacing all the carbon atoms falling on it with a single partial double bond. The stable configuration of this molecular model is achieved by performing energy minimization using the steepest descent algorithm available in "Geometry Optimization" menu in "Forcite" module of Materials Studio. After approximately 5000 iterations of geometry optimization, the plane sheet is transformed into a cone as shown in the Figure 7.3. The cone, thus formed is then taken as the base for modeling of CNFs, creating the second, third, fourth, and other cones by following the same procedure.

7.1.3 Modeling of CNF–PP Composites

The cones modeled in Section 7.1.2 are then cut and stacked one over the other having an offset distance between them of 3.5 Å, which is close to the value of 3.4 Å, reported by Zhu et al. [10]. The schematic in Figure 7.4 displays six independent cones stacked over each other

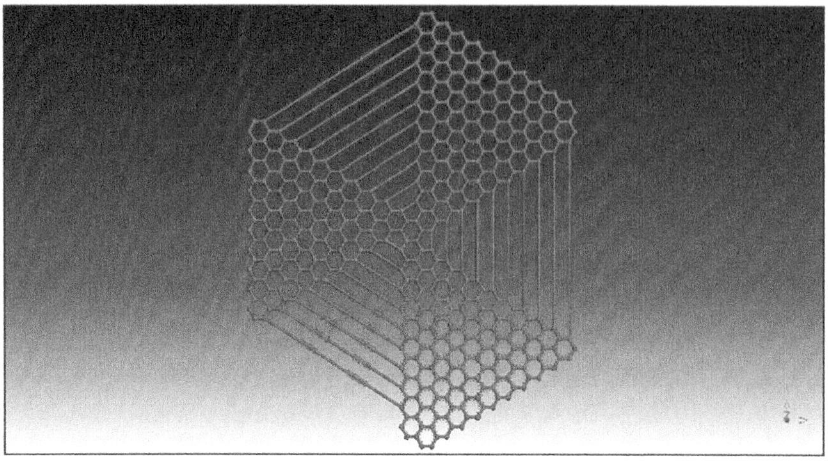

Figure 7.2 Plane sheet of carbon atoms.

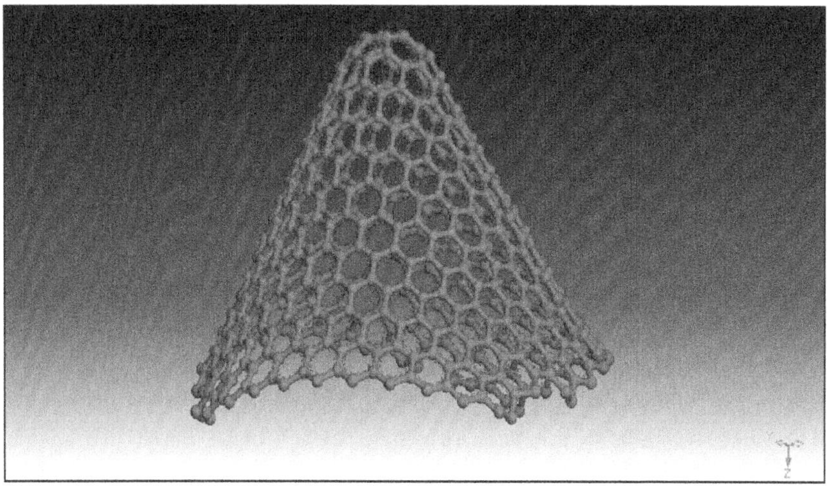

Figure 7.3 Carbon nanocone after energy minimization.

having length $L = 27$ Å, outer and inner diameter of 15.324 and 6 Å, respectively. These cones are then inserted into simulation cells of different size. PP is then packed outside the modeled CNFs of varying lengths.

PP, also known as polypropylene, is a thermoplastic polymer, made by the chemical industry and used in a wide variety of applications, including packaging, textiles (e.g. ropes, thermal underwear, and carpets), stationery, plastic parts and reusable containers of various types, laboratory equipment, loudspeakers, automotive components, and polymer banknotes. In this study, we have constructed several chains of PP, each containing seven units of propylene as shown in Figure 7.5. The simulation procedure consists of geometry optimization, dynamics, and forcite mechanical properties calculation. For studying the effect of varying volume fraction (V_f), the aspect ratio of CNF ($l/d = 10$) was kept constant. The size of the simulation cell varies in transverse directions and was kept constant in the CNF direction. Cells of different size were simulated using Materials Studio 5.5. Volume

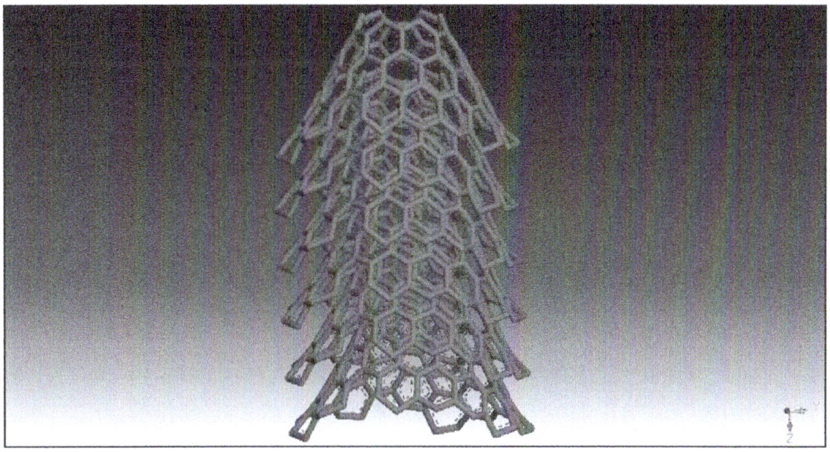

Figure 7.4 Molecular model of stacked cones.

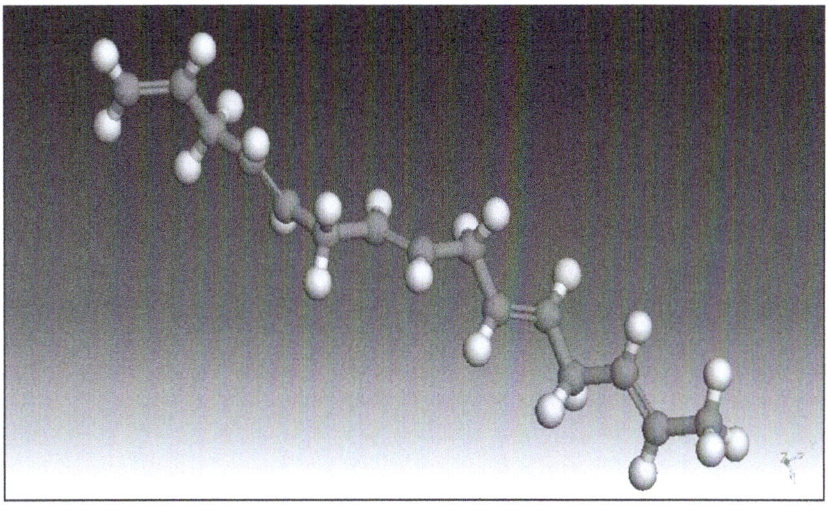

Figure 7.5 Chain of polypropylene (PP) containing seven monomer units of propylene.

fraction was varied from $V_f = 0\%$ to $V_f = 16\%$. Number of atoms in the cells varied from 6000 atoms for $V_f = 0$–17 500 atoms for $V_f = 16\%$. PP was packed outside the CNF using "Amorphous" module in Materials Studio 5.5. One of the simulation cells in which CNF is surrounded by PP is shown in Figure 7.6. For studying the effect of variation of the CNF aspect ratio (l/d), the volume fraction (V_f) was kept constant at 4%. Aspect ratio varies from $l/d = 5$ to $l/d = 60$. Aspect ratio (l/d) was varied by changing the length of CNF and keeping the diameter constant. By keeping the diameter of CNF same, we have not changed the basic properties of CNF.

A similar procedure was adopted for studying the variation of thermal conductivity and damping of CNF-reinforced PP composites, with V_f and l/d. After completing the construction of models, the simulation procedure consists of geometry optimization, dynamics, and forcite mechanical properties calculation. The geometry optimization procedure is necessary because the initial configuration is generally in a higher state of energy. To bring it into a state

Figure 7.6 CNF packed inside PP matrix.

of equilibrium, we have conducted the geometry optimization with the steepest descent algorithm. The number of iterations was taken as 5000.

A frequent activity in MD simulation is the optimization or minimization (with respect to potential energy) of the system being examined. For instance, it is often desirable to optimize a structure after it has been sketched, since sketching often creates the molecule in a high energy configuration and starting a simulation from such an un-optimized structure can lead to erroneous results. There are a number of optimization techniques available in Materials Studio viz., steepest descent method, conjugate gradient, and Newton–Raphson method.

In the steepest descent method, the line search direction is defined along the direction of the local downhill gradient. Each line search produces a new direction that is perpendicular to the previous gradient; however, the directions oscillate along the way to the minimum. This inefficient behavior is characteristic of steepest descents, especially on energy surfaces having narrow valleys. Convergence is slow near the minimum because the gradient approaches zero, but the method is extremely robust, even for systems that are far from being harmonic. It is the method most likely to generate the true lowest energy structure, regardless of what the function is or where the process begins. Therefore, the steepest descent method is often used when the gradients are large and the configurations are far from the minimum. This is commonly the case for initial relaxation of poorly refined crystallographic data or for graphically built models.

Once an energy expression and, if necessary, an optimized structure have been defined for the system of interest, a dynamic simulation can be run. The basis of this simulation is the classical equations of motion which are modified, where appropriate, to deal with the effects of temperature and pressure on the system. The main product of a dynamics run is a trajectory file that records the atomic configuration, atomic velocities, and other information in a sequence of time steps which can be analyzed subsequently. In this study, NVT ensemble was used i.e. constant number of atoms, volume, and temperature. The time step has been taken as 1 fs and the total simulation time as 100 ps. Andersen thermostat has been used with a repulsive cutoff of 6 Å.

Through this dynamic run, the system is allowed to evolve with time in order to study the interactions between the atoms. In all the simulations, periodic boundary conditions were applied to the simulation box so as to make it as a representative volume element imitating the bulk phase of the nanocomposite. The equations of motion of the polymer and nanofiber atoms were integrated by the velocity-Verlet integrator. The dynamic properties of pure PP have been computed using "Forcite" module.

A constant strain of 0.005 has been applied to the simulation cell. Before calculating the mechanical properties, geometry optimization has been carried out to achieve the minimum energy configuration after the dynamics run. The Forcite mechanical properties calculations use the "Constant strain" approach. The process starts by removing symmetry from the system, followed by an optional reoptimization of the structure, where the cell parameters can be varied. Optimization at this stage is always advised as incorrect results can be obtained if the structure is far from its lowest energy configuration. For each configuration, a number of strains are applied, resulting in a strained structure. The resulting structure is then optimized, keeping the cell parameters fixed. Elastic moduli have been predicted using COMPASS forcefield in Forcite module of Materials Studio 5.5. The number of strains has been taken as six with maximum strain of 0.005. The number of iterations has been taken as 10.

7.1.4 Damping in CNF–PP Composites

With increasing interest in alternative and efficient energy, there is rapidly growing demand in the use of fiber composite materials due to their high strength/stiffness-to-weight ratios in a wide variety of structural applications such as aerospace structures or wind turbine blades. Those composite structures are often subjected to dynamic loadings and exposed to undesirable mechanical vibrations from operating environments. The consequential trade-off for such high specific strength/stiffness ratios includes poor acoustic and vibrational damping capacity, which leads to several issues with structural vibrations, noise attenuation, and fatigue for composite structures placed in service with dynamic loadings.

Damping, which is one of the manifestations of viscoelastic characteristics, is an obviously important material property for noise and vibration mitigation in structures. Many solutions have been proposed such as embedded damping layers or active polymers with external control systems to improve energy dissipation capability in the composites and therefore mitigate the unwanted vibrations encountered in service. Yet all these solutions lead to a reduced structural performance and often incur significant penalties in the form of additional weight or external power requirements. Numerous studies have been performed to improve the energy dissipation capability in composites with the addition of nanofillers including CNFs. It was successfully demonstrated that the energy dissipation capability of composites and multi-scale composites can be improved due to the combination of the CNFs' high surface area and stick–slip friction.

In this study, damping properties of CNF/PP composites have been expressed analytically from the theory of short fiber composites. The correspondence principle – "A linear elastostatic analysis can be converted to a vibratory linear visco-elastic analysis by replacing the elastic moduli with the corresponding complex moduli," has been used to define the dynamic properties. The complex modulus has a real and an imaginary part. The damping loss factor can be expressed as the ratio of imaginary to the real part of the complex modulus. Tandon and Weng [15] derived explicit expressions for the elastic constants of a short fiber composite using the Mori–Tanaka approach. These equations have been solved using a mathematical tool such as Matlab. By writing a program in Matlab, we can solve the equations and can study the effect of various parameters such as aspect ratio and fiber volume fraction on the modulus and damping loss factor. In Materials Studio 5.5, the energy dissipated can be obtained after

Table 7.3 Comparison for variation of E_{11} with CNF aspect ratio (l/d), between this study and other researchers.

CNF aspect ratio (l/d)	Young's modulus (E_{11}), GPa (this study)	Young's modulus (E_{11}), GPa
5	185	180–245 (Chasiotis et al. [17])
10	200	
15	215	
20	228	
25	238	
30	246	
35	256	245 (Tibbetts [18])
40	260	
45	264	
50	268	
55	271	
60	273	

completion of mechanical properties calculation task. Dividing this dissipated energy by the energy obtained before the unloading cycle begins, gives the measure of loss factor (η).

Results of MD have been compared with the results obtained from Mori–Tanaka model. In the approach suggested by Mori and Tanaka [16], each inclusion behaves as an isolated one, subject to the averaged stress fields acting on it from all the other inclusions.

7.1.5 Results and Discussion

This section has been divided in two parts. In Section 7.1.5.1, the results obtained for elastic moduli have been discussed, and in Section 7.1.5.2, the damping results have been explained.

7.1.5.1 Elastic Moduli

Table 7.3 shows the comparison for variation of longitudinal elastic modulus (E_{11}) with CNF aspect ratio (l/d), between this study and other researchers. From the Table 7.3, it can be inferred that the results of this study are in agreement with Chasiotis et al. [17] and Tibbetts and McHugh [18]. Figure 7.7 shows the variation of Young's modulus (E_{11}) of CNF with aspect ratio (l/d). As length of the CNF increases, the efficiency of load transfer is increased dramatically. But for larger lengths, the increase in van der Waals forces causes the modulus to increase at a slower rate.

Figure 7.8 shows the variation of Young's modulus, E_{11} and E_{22}, with percentage of the CNF in PP. It shows that E_{11} increases with increase in CNF volume fraction. The percentage increase in E_{11} with an increase in CNF volume fraction is shown in Table 7.4. From Figure 7.8 and Table 7.4, it can be observed that E_{11} increases sharply until CNF volume fraction is 2%. Thereafter, the rise is minimal due to the slippery nature of CNFs which does not assist in further improvement of mechanical properties. Figure 7.8 also shows that the increase in transverse modulus (E_{22}) is negligible in comparison to the rise in E_{11}. This is due

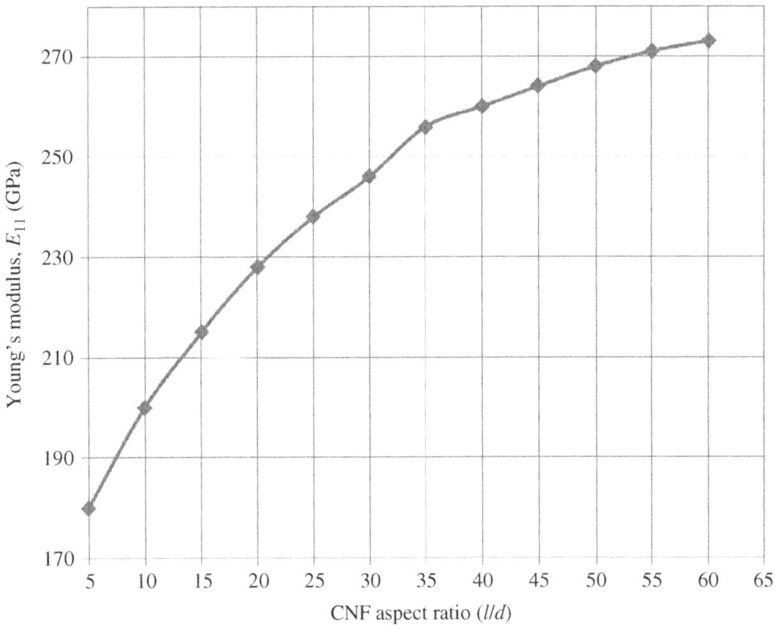

Figure 7.7 Variation of Young's modulus (E_{11}) of CNF with aspect ratio (l/d).

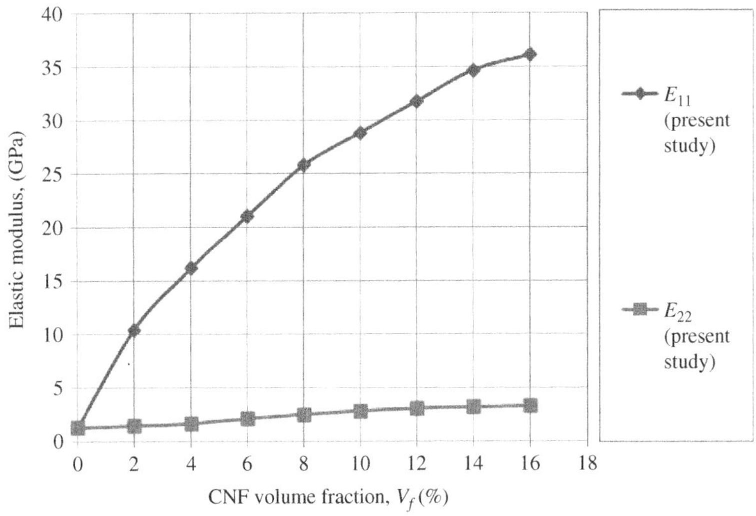

Figure 7.8 Variation of Young's modulus, E_{11} and E_{22}, with percentage of CNF in PP for fixed aspect ratio ($l/d = 10$).

to the fact that the fibers are aligned in the longitudinal direction. As percentage of CNF in PP increases, volume of the fibers is increasing. Since CNFs have high modulus in the longitudinal direction, the composite has improved properties in the said direction. In transverse direction, the properties of the composite are matrix dominated and hence, lower values of E_{22}.

Table 7.4 Percentage variation in longitudinal modulus (E_{11}) with increasing CNF volume fraction.

CNF (vol%)	Percentage difference in longitudinal modulus (E_{11}) w.r.t previous value	Percentage increase in E_{11}	
0	—	From $V_f = 0\%$ to $V_f = 6\%$	From $V_f = 6\%$ to $V_f = 16\%$
2	748.78		
4	54.69		
6	30.15		
8	22.64	1608.94	71.74
10	11.75		
12	10.23		
14	9.19		
16	4.09		

Table 7.5 Percentage increase in shear moduli with increasing CNF volume fraction.

	Increase in shear moduli (%)		
CNF range (vol%)	G_{23}	G_{31}	G_{12}
0–6	234.78	204.35	191.30
6–16	31.34	30.00	27.27

Figure 7.9 shows the comparison of relative Young's modulus of CNF–reinforced PP composites obtained from present study with other researchers, for fixed aspect ratio ($l/d = 10$). It is apparent that for all of the VGCNF/PP composites studies, there is a linear increase in relative Young's modulus with increasing VGCNF content. Figure 7.10 shows the variation of shear moduli (G_{23}, G_{31}, and G_{12}) with an increase in CNF volume fraction. With an increase in the CNF volume fraction from 0 to 0.06, the percent increase in transverse shear modulus G_{23} is approximately 183%, whereas the percent increase reduces to merely 10% when CNF volume fraction increases from 0.06 to 0.16. Also, rise in all the shear moduli is negligible in comparison to the rise in E_{11}. This is again due to the fact that in every shear mode, the properties of the composite are matrix dominated and hence lower values of all shear moduli. Table 7.5 shows the percentage increase in shear moduli with increasing CNF volume fraction.

Figure 7.11 shows the variation of Young's modulus E_{11} and E_{22}, with aspect ratio (l/d) of CNF–reinforced PP composite with volume fraction ($V_f = 0.04$). With increase in CNF aspect ratio (l/d) until $l/d = 50$, both the longitudinal and transverse elastic modulus increase rapidly. Thereafter, the increase is smaller. When the length of the CNFs increases, the efficiency of load transfer is increased dramatically. But for larger lengths, the increase in van der Waals forces causes the modulus to increase at a slower rate, as can be observed from Figure 7.11. Figure 7.12 shows the variation of shear modulus with aspect ratio (l/d) of CNF–reinforced PP composite with fixed volume fraction ($V_f = 0.04$). It can be inferred

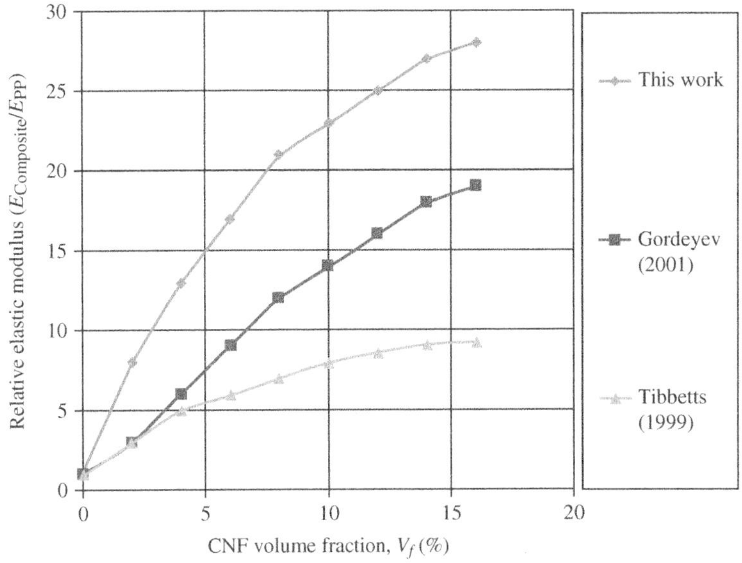

Figure 7.9 Comparison of relative Young's modulus of CNF/PP composites obtained from present study with other researchers, for fixed aspect ratio ($l/d = 10$).

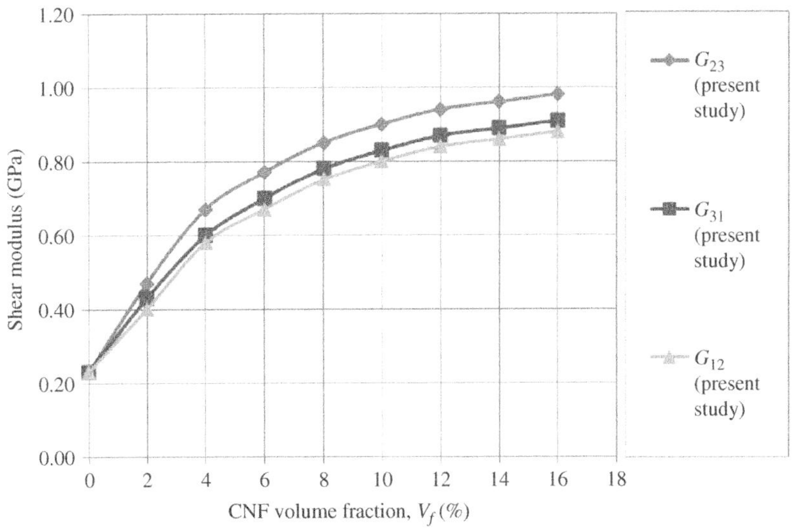

Figure 7.10 Variation of shear modulus, G_{23}, G_{31}, and G_{12}, with percentage of CNF in PP for fixed aspect ratio ($l/d = 10$).

that the increase in shear moduli is significantly smaller in comparison to the increase in longitudinal and transverse elastic moduli. All the shear moduli increase until $l/d = 50$. After that, the rise is minimal, which may be due to the increase in van der Waals force for higher aspect ratios. Figure 7.12 also shows that the increase in G_{31} and G_{12} is smaller than the increase in G_{23}.

Table 7.6 shows the percentage increase in elastic moduli with an increase in the CNF aspect ratio (l/d) for fixed volume fraction ($V_f = 0.04$). From Figure 7.11 and Table 7.6, it

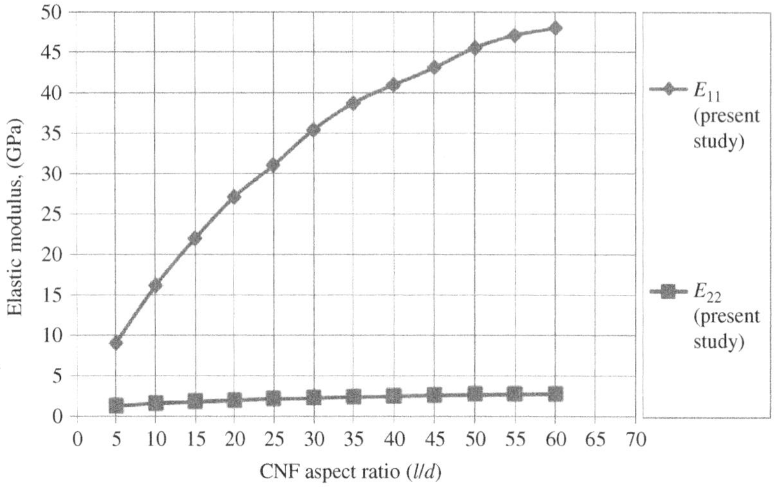

Figure 7.11 Variation of Young's modulus, E_{11} and E_{22}, with CNF aspect ratio (l/d) for fixed volume fraction ($V_f = 0.04$).

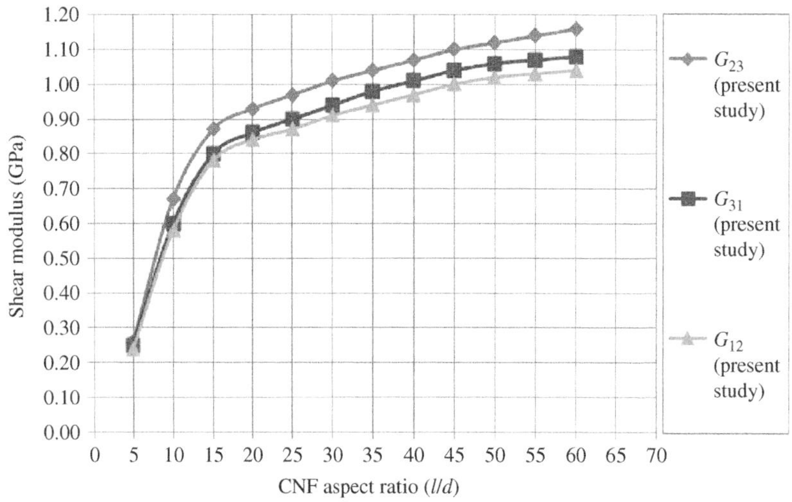

Figure 7.12 Variation of shear modulus with CNF aspect ratio (l/d) with fixed volume fraction ($V_f = 0.04$).

can be concluded that E_{11} and E_{22} increases sharply from $l/d = 5$ to $l/d = 50$. A possible reason for this is the presence of shear stress concentration at numerous fiber ends in a short fiber composite. However, for $l/d > 50$, both the moduli increase slowly. This is because, after $l/d = 50$, higher fiber axial stress occurs which causes the moduli to increase at a slower rate. Table 7.7 shows the percentage increase in shear moduli with an increase in the CNF aspect ratio (l/d). It shows that the percentage increase in all the shear moduli is greater when the aspect ratio varies from $l/d = 5$ to $l/d = 50$. This is probably, due to the fact that at low fiber aspect ratios, number of fibers in the composite will be more for the same volume fraction. More the number of discontinuous fibers more are the number of fiber-ends available for stress concentration. Thus, at low values of l/d, the stress concentration at numerous fiber

Table 7.6 Percentage increase in elastic modulus with increase in CNF aspect ratio (*l*/*d*).

CNF aspect ratio (*l*/*d*)	Percentage increase in E_{11}	Percentage increase in E_{22}
5–50	403.42	106.92
50–60	5.34	4.08

Table 7.7 Percentage increase in shear moduli with increase in CNF aspect ratio (*l*/*d*).

CNF aspect ratio (*l*/*d*)	Percentage increase in G_{23}	Percentage increase in G_{31}	Percentage increase in G_{12}
5–50	330.77	325.00	324.00
50–60	3.57	1.96	1.89

ends leads to rise in the values of all the shear moduli. As the fiber aspect ratio increases, shear moduli become constant as shown in Figure 7.12.

7.1.5.2 Damping

In Materials Studio 5.5, the energy dissipated can be obtained after completion of mechanical properties calculation task. From "Forcite" mechanical properties task, we get two values of energy. Energy in frame 1 gives the energy during loading, whereas the energy in frame 2 gives the energy during unloading cycle. Area between loading and unloading curves gives us the energy dissipated. Dividing this dissipated energy by the energy obtained before the unloading cycle begins, gives the measure of loss factor (η). Results of MD have been compared with the results obtained from Mori–Tanaka model.

Figure 7.13 shows the variation of longitudinal loss factor (η_{11}) with percentage of CNF (by volume) in PP. It shows that η_{11} decreases with an increase in the CNF volume fraction. The percent decrease in η_{11} with an increase in the CNF volume fraction is shown in Table 7.8. From Figure 7.13 and Table 7.8, it can be observed that η_{11} decreases sharply until CNF volume fraction is 2%. Thereafter, the fall steadies due to the curvy and the slippery nature of CNFs which does not assist in further decline of damping properties. It can also be observed that the difference between Mori–Tanaka model and MD model goes on increasing as the volume fraction increases.

Figure 7.14 shows the variation of η_{11} with aspect ratio (l/d) of armchair CNF-reinforced PP composite with fixed volume fraction ($V_f = 0.04$). With an increase in the CNF aspect ratio (l/d) until $l/d = 50$, the longitudinal loss factor decreases rapidly. Thereafter, the decrease is smaller. When the length of the CNFs increases, the efficiency of load transfer is increased dramatically. But for larger lengths, the increase in van der Waals force causes the loss factor to decrease at a slower rate. Mori–Tanaka model gives higher values of loss factor (η_{11}) than MD. Finegan et al. [12] predicted damping values of CNF/PP composites using dynamic mechanical analyzer (DMA). Their predicted values do not show the measured decrease in damping with increasing fiber volume fraction. It was also observed that composites having

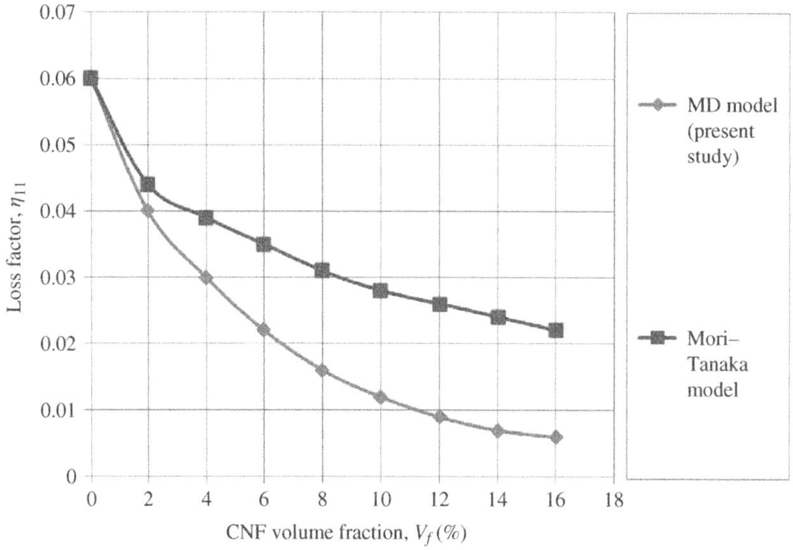

Figure 7.13 Variation of loss factor (η_{11}) with percentage of CNF in PP for fixed aspect ratio ($l/d = 10$).

Table 7.8 Percentage variation in loss factor (η_{11}) with increasing CNF volume fraction.

CNF (vol%)	Percentage difference in loss factor (η_{11}) w.r.t previous value using MD	Percentage difference between Mori–Tanaka model and MD
0	—	—
2	33.33	9.09
4	27.27	23.08
6	26.67	37.14
8	25.00	48.39
10	25.00	57.14
12	25.00	65.38
14	22.22	70.83
16	14.28	72.73

very low fiber aspect ratios should have higher damping than those having high fiber aspect ratios.

Table 7.9 shows the percentage decrease in η_{11} with an increase in the CNF aspect ratio (l/d) for fixed volume fraction ($V_f = 0.04$). From Figure 7.14 and Table 7.9, it can be concluded that η_{11} decreases sharply from $l/d = 5$ to $l/d = 50$. A possible reason for this is the presence of shear stress concentration at numerous fiber ends in a short fiber composite. However, for $l/d > 50$, η_{11} decreases slowly. This is because, after $l/d = 50$, higher fiber axial stress occurs which causes the loss factor to decrease at a slower rate. It can also be observed that at low aspect

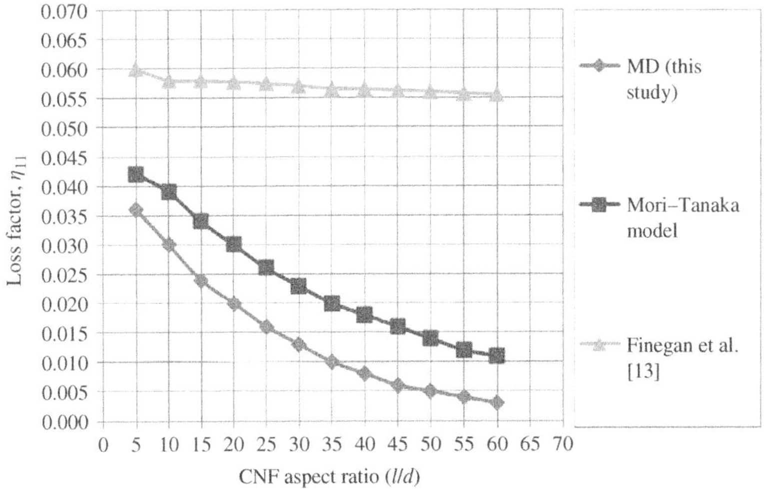

Figure 7.14 Variation of loss factor (η_{11}) with CNF aspect ratio (l/d) for fixed volume fraction ($V_f = 0.04$).

Table 7.9 Percentage decrease in loss factor (η_{11}) with increase in CNF aspect ratio (l/d).

CNF aspect ratio (l/d)	Percentage decrease in η_{11}	Percentage difference between Mori–Tanaka model and MD model at		
		$l/d = 5$	$l/d = 50$	$l/d = 60$
5–50	86.11	14.28	64.28	72.72

ratios, Mori–Tanaka model and MD give almost similar results. At $l/d = 5$, the percentage difference between them is only 14.28%.

There are two possible mechanisms that could be responsible for damping in CNF-reinforced polymer composites at molecular level: (i) energy dissipation caused by interfacial sliding at the nanofiber–polymer interface and (ii) energy dissipation caused by interfacial stick–slip sliding at the nanofiber–nanofiber interface. When a normal tensile stress is applied to a composite, it starts elongating. As a result of the applied stress, the resin starts applying a shear stress on the nanofiber, thus causing the load to be transferred to nanofibers. Consequently, normal strain starts appearing in nanofibers and they start elongating accordingly. When the applied stress is small, the nanofiber remains bonded to the matrix (sticking phase). Both the resin and the nanofiber move together during this phase, and the strains are equal in both epoxy resin and nanofiber. As the applied stress is increased, the shear stress on CNF increases. At a certain value of shear stress, called the critical shear stress, the nanofiber debonds from the resin. When the shear stress on the nanofiber increases beyond this value (as a result of increased applied stress), the epoxy starts flowing over the surface of the nanofiber. The strain in the nanofiber remains constant at its maximum level, while the strain in the epoxy increases (slipping phase). In this phase, there is no transfer of load between CNF and matrix, and because of this energy dissipation due to slippage occurs which results in structural damping.

With greater stiffness, CNFs have an increased capability to resist applied loading, which could lead to more slippage and energy dissipation through friction. In other words, the increase in contact area dramatically increases the potential for energy dissipation due to interfacial friction. Nevertheless, it can be concluded that, by taking advantage of the large interfacial contact area between CNFs and resins, as well as the high stiffness and low density properties of CNFs, high performance in energy dissipation and structural damping can be achieved. It can be concluded that the elastic modulus increases with the addition of CNFs, and there is good interfacial bonding between CNF and PP, which can be observed from appreciable increment of Young's modulus and decreasing tendency of loss factor.

From the results of this study, elastic modulus which is a measure of the stiffness of an elastic material and is a quantity used to characterize materials, increases with increase in volume of CNFs. Considering the effect of volume studies, it is safe to state that 5% CNF composition in pure PP is the most suitable composition to enhance the mechanical properties to a sufficient extent for any mechanical load and pressure related application. Plastic deformation in amorphous polymers occurs due to nucleation and propagation of shear bands. In the unreinforced polymer matrix, shear bands propagate unchecked as there are no barriers for their movement. On the other hand, the presence of CNFs in the composites could offer resistance for the propagation of shear bands. The reasons for the enhancement of elastic modulus in polymer CNF nanocomposites are good mechanical interlocking and the presence of obstacles to the motion of shear bands.

7.1.6 Conclusions

The study presents results of MD simulation for elastic modulus obtained by using Materials Studio 5.5. The results indicate unambiguously that static mechanical properties of pure PP are greatly enhanced by composing these with less than 10% of CNF. On the basis of the findings of this study of elastic properties of CNF–PP composites, it is safe to conclude that a concentration around 2% of CNF is a good compromise to form composites suitable for enhancing both static as well as dynamic properties. The main findings of the study are summarized as below:

(1) With only 2% addition by volume of CNF in PP, E_{11} increases 748%. Thereafter, the increase is at a lower rate. Increase in E_{22} is much less in comparison to the increase in E_{11}.
(2) Until $l/d = 50$, both longitudinal and transverse elastic moduli increase rapidly. Thereafter, the increase is smaller.
(3) Loss factor (η_{11}) decreases with increase in CNF volume fraction. η_{11} decreases sharply until CNF volume fraction is 2%. Thereafter, the fall steadies due to the curvy and slippery nature of CNFs which does not assist in further decline of damping.
(4) With an increase in the CNF aspect ratio (l/d) until $l/d = 60$, the longitudinal loss factor (η_{11}) decreases rapidly. Thereafter, the decrease is smaller. For larger lengths, the increase in van der Waals force causes the loss factor to decrease at a slower rate.

7.2 Silica Nanoparticle/Hydroxyapatite Fiber Reinforced bis-GMA/TEGDMA Composites

Dental composites, generally referred to as resin-based composites have gained considerable attention around the world because of their application in dentistry as a restorative material. These are composed of epoxy resins such as bis-phenol A-glycidyl methacrylate (bis-GMA) and other di-methacrylate monomers viz. tri-ethylene glycol di-methacrylate (TEGDMA) reinforced with inorganic nanofillers of diameter 5–100 nm [19] vis-à-vis silica nanoparticles and a photo-initiator. A photo-initiator system is composed of camphorquinone (CQ), ethyl 4-dimethylaminobenzoate (EDMAB), and 4-(2-hydroxytetradecyl) oxy phenyl-phenyliodonium hexafluoroantimonate (SarcatTM CD1012) [20]. The mechanical properties of dental composites can be tailored by making a change in the concentration of different constituents. Dental composite has an expected life span of five years [21]. It has established itself as a versatile material with its astonishing mechanical properties as the gamut of its diametric tensile strengths is 45–70 MPa [19] and flexural strength ranging from 100 to 140 MPa [22].

Satapathy et al. [19] revealed that the compressive strength of nanocomposites incorporated with nano silica reflected a decreasing trend with decrease in the silica particle size with a gamut of 187–273 MPa. This phenomenon was attributed to the size scale combination of nanofillers in-turn effecting their distribution and dispersion in the matrix. The possible application of nanotechnology can help in revealing novel materials possessing strong and defensible filler-resin network for bearing effective stress leading to ameliorated thermo-mechanical, fatigue, fracture, and wear resistance. Certain issues like micro-leakage, decrease in strength over a period of time after being in contact with water, and poor shrinkage and many more aspects being reported by several researchers can be avoided using nanotechnology. It has been observed that poor polymerization shrinkage led to micro-crack initiation [23].To avoid this, epoxy–polyol matrices were included in dental composites. The ring opening reaction happening during the curing stage led to a reduction in the shrinkage percentage but apart from that it also increased the water sorption capacity leading to nucleation and propagation of cracks during hydration [24].

Dental composites must have high wear resistance, enhanced fracture toughness, and high strength along with lower hardness and stiffness to avoid any failure of the antagonistic cusp. This makes the selection of the constituents of a dental composite really difficult [25]. A significant improvement in polishability and water absorption property of dental composites was reported after the introduction of nanofibers. It also led to an increase in the strength and toughness of the filler-resin network [26–30]. Once the microcrack was initiated in the matrix, the reinforced nanofillers remained intact across the crack planes thus supporting the applied load. Tian et al. [31, 32] reported that by varying the mass fraction of nano fibrillar silicate (FS) in bis-GMA, dental resin promised superior mechanical properties. FS belongs to the class of hydrated magnesium/aluminum silicate and consists of nanoscale single crystals. It is very good to explore this realm because of its abundance in nature. The FS mineral commonly found in nature is attapulgite/palygorskite. A remarkable improvement in the mechanical properties of dental composites was seen after the addition of small mass fraction of FS. Higher mass fraction (greater than 4%) showed adverse effect on the mechanical properties of dental composites. This fact was attributed to the agglomeration of FS particles at higher mass fractions [21].

Saeb et al. [33] synthesized and modified Fe_3O_4 magnetic nanoparticles (MNPs) with chitosan as a natural surface modifier to prevent their aggregation and make them potent to contribute to ring opening of epoxy. The cure behavior and kinetics of epoxy nanocomposites filled with bare and chitosan-modified Fe_3O_4 were analyzed using isothermal and nonisothermal differential scanning calorimetry (DSC). The kinetic models applied to DSC experimental data suggested positive effect of Chitosan@ Fe_3O_4 MNPs on cure reaction between epoxy and polyamidoamine hardener. Computational studies demonstrated that curing of epoxy resin with aminoamide hardener has been governed by the interaction between epoxy and Chitosan@ Fe_3O_4.

In another work, Saeb et al. [34] synthesized magnetic nanoparticles of Fe_3O_4 via precipitation method, functionalized with β-cyclodextrin hydroxyl-rich precursor, and characterized by FTIR, TEM, and magnetic hysteresis loop techniques. Epoxy nanocomposites containing bare and β-cyclodextrin-modified Fe_3O_4 nanoparticles were prepared, and their cure behavior/kinetics was studied by isoconversional models applied on dynamic DSC data. Adhesion energy between cure moieties and surface of β-cyclodextrin was quantified using MD simulation. Activation energy of epoxy resin increased from 56 to 62 kJ/mol upon introduction of β-cyclodextrin-modified Fe_3O_4, which suggests retarded crosslinking compared to the blank system while a highly cured network was identified for the resin containing β-cyclodextrin-modified Fe_3O_4 particles due to additional chemical reaction of hydroxyl functional groups attached to Fe_3O_4 surface with cure moieties. MD simulation confirmed either quantitatively or qualitatively the strong adhesion between cure moieties and β-cyclodextrin surface.

Bahlakeh et al. [35] studied the effect of different types of iron oxides, which naturally exist on steel substrate, on the interfacial interaction between an epoxy coating and a carbon steel substrate at the molecular/atomic level by employing MD simulations and quantum mechanics (QM) calculations. Three types of iron oxide, that is, ferrous oxide (FeO), ferric oxide (Fe_2O_3, hematite), and ferrous ferric oxide (Fe_3O_4, magnetite), were considered for modeling, and their binding energies were calculated and compared by altering the concentration of hydroxide groups on the surface. It was found that aminoamide-cured epoxy was adsorbed on all oxide substrates through a mechanism of electrostatic and donor–acceptor interactions, with binding energies of −113.6, −1035.9, and −304.4 kcal/mol, respectively, assigned to FeO, Fe_2O_3, and Fe_3O_4 iron oxides. In the case of a hydroxylated surface, aminoamide-conjugated epoxy adhesive was found in the hydrogen bond interface as well, which was evidence of strengthened binding at a surface populated by hydroxide groups. Moreover, theoretical explorations showed that the type of covalent linkage between the curing molecule and epoxy resin governs the extent of cured epoxy adhesion to the surface of iron oxide.

Most of the above studies dealt with studying the effects of large size, composition, and content of reinforcement on the properties of composites. But in practice, the dispersion of nanofillers in the resin is very difficult and leads to agglomeration, coalescence, and clustering of nanofillers. This results in weakening of the real nanofiller effect on the mechanical properties of dental composites. MD simulation provides an insight into the behavior of the material at atomic level. There are very limited studies of bis-GMA and TEGDMA dental composites, using MD because of their complexity. The authors have used MD simulations for exploring the properties of various nanomaterials [36–46]. Thus, in this work MD simulations have been used to predict the mechanical and dynamic properties of blends of dental resin and composites. The aim of this work is to explore the influence of inorganic nanoparticles such as silicon-dioxide (SiO_2) on the properties of bis-GMA and TEGDMA dental composites at the nanoscale using MD. The mechanical properties of SiO_2 nanoparticle-reinforced bis-GMA/TEGDMA composites have been predicted using MD simulation.

7.2.1 Molecular Dynamics Methodology

All the MD simulations in this study have been performed using Biovia Materials Studio [47]. Molecular models of both unfilled polymer blends and filled resin composites were prepared. These were then used for the prediction of nanomaterial properties.

7.2.1.1 Molecular Models of Unfilled Polymers

The amorphous module of Biovia Materials Studio [47] was used for constructing the monomers of bis-GMA and TEGDMA as shown in Figure 7.15a,b. The number of monomers of bis-GMA and/or TEGDMA in each chain was taken as 15. The number of monomers in each chain was selected based on the essential condition of maintaining the polymer characteristics and also to imitate the photo-polymerization effect [48]. The number of polymer chains, for both resin blends as well as composite materials, was selected in such a way that the conversion of double bonds was around 60–70% [49]. A range of blends of resin with different mass fractions of bis-GMA and TEGDMA, such as, 80/20, 70/30, 60/40, 50/50, 40/60, 30/70, and 20/80 were constructed to study the structure–property effects of organic components. One such bis-GMA/TEGDMA blend with mass fraction of 20/80 has been shown in Figure 7.16.

7.2.1.2 Molecular Models of Filled Polymer Composites

Silica nanoparticles of 20 Å diameter were added to the 20/80 bis-GMA/TEGDMA polymer blend in various wt% of 3, 5, 7, 9, and 11 for studying the effect of loading fraction on the mechanical properties of nanocomposites. For eliminating the effect of free radicals, a modified silica nanoparticle was created with hydrogen atoms and OH groups added to it. Figure 7.17 shows the equilibrium structure of one of the resin blends with weight percentage of silica nanoparticles as 5%. In this study, a single silica nanoparticle per unit cell has been used. It has been shown that the results of eight nanofillers were similar to the results obtained using a single nanofiller [50]. Though the computational resources have limited the number of nanofillers and polymer chains in a typical system but the mechanical properties which agree with the experimental results can still be obtained as has been shown by [48]. Periodic boundary conditions have been used throughout this study. Thus, the silica nanoparticles were uniformly dispersed in the resin blend to represent a bulk composite. The silica nanoparticle reinforced resin blends were also reinforced with hydroxyapatite fibers (HAFs), which have anti-bacterial properties. One such hybrid composite containing silica nanoparticle, hydroxyapatite (HA) and resin blends has been shown in Figure 7.18. A number of similar unit cells containing varying weight percentage of HA (0, 2, 4, 6, 8, 10, and 12%) were modeled for studying the effect of HAFs on the mechanical properties of silica nanoparticle-reinforced resin blends.

7.2.1.3 MD Methodology

COMPASS II forcefield has been used in this study for describing intermolecular and intramolecular interactions for both filled and un-filled dental materials. This forcefield has been widely used for simultaneous prediction of gas and condensed phase properties of a variety of polymers and organic and inorganic molecules [51]. The monomers of bis-GMA and TEGDMA were optimized through the steepest descent algorithm. This resulted in

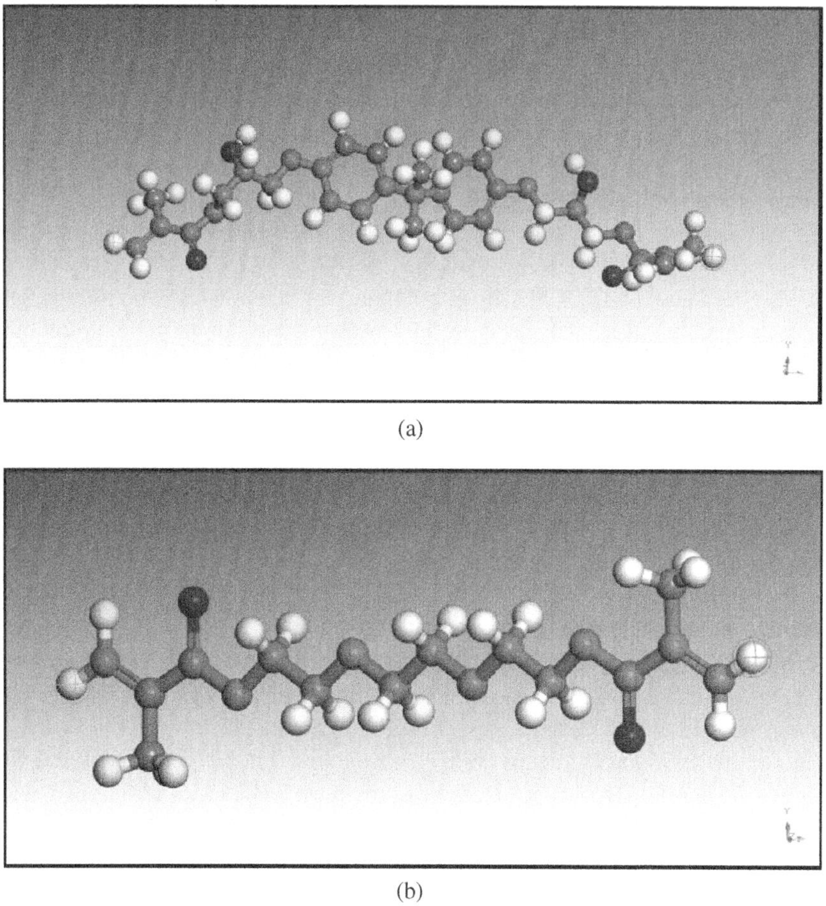

Figure 7.15 Molecular models of (a) bis-GMA and (b) TEGDMA. Source: Based on Dassault Syste'mes BIOVIA [47].

the minimization of energy. The criterion of convergence was kept as 0.001 kcal/(mol Å). The polymers were then created using the Amorphous Cell module. The resin blends were created by random distribution of bis-GMA and TEGDMA monomers and the polymers in a cubic representative volume element (RVE). The length of the RVE's was varied from 40 to 60 Å. The number of atoms in the silica nanoparticle was 504. It also contained 106 hydroxyl groups that were used for eliminating the effect of the free radicals on the surface of the silica nanoparticle.

Long simulation runs performed using numerical integration resulted in the problem of energy drifting [52]. In computer simulations of mechanical systems, energy drift is the gradual change in the total energy of a closed system over time. According to the laws of mechanics, the energy should be a constant of motion and should not change. However, in simulations the energy might fluctuate on a short time scale and increase or decrease on a very long time scale due to numerical integration artifacts that arise with the use of a finite time step (Δt). This is somewhat similar to the flying ice cube problem, whereby numerical errors in handling equi-partition of energy can change vibrational energy into translational energy. More specifically, the energy tends to increase exponentially; its increase

Figure 7.16 Equilibrium molecular structure of bis-GMA/TEGDMA 20/80 resin blend.

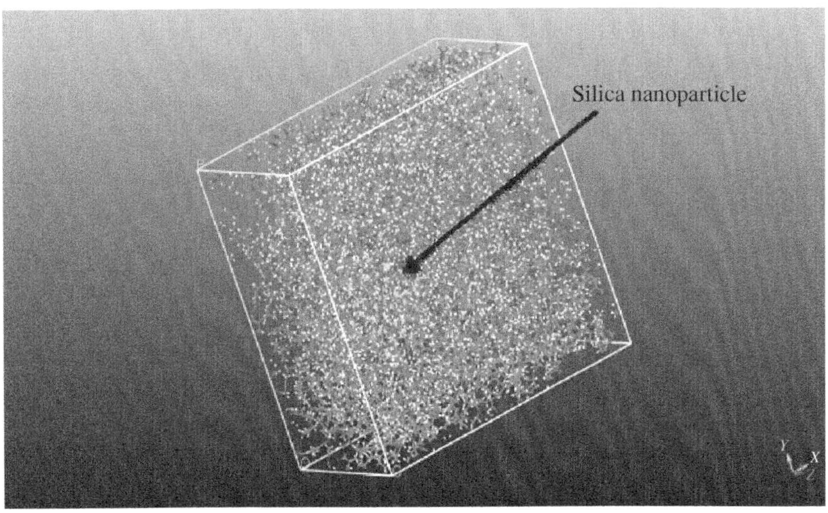

Figure 7.17 Bis-GMA/TEGDMA resin blend containing 5 wt% of silica nanoparticle with total number of atoms as 17 504.

can be understood intuitively because each step introduces a small perturbation to the true velocity which results in a second-order increase in the energy. Energy drift – usually damping – is substantial for numerical integration schemes that are not symplectic, such as the Runge–Kutta family. Symplectic integrators usually used in MD, such as the Verlet integrator family, exhibit increases in energy over very long time scales, though the error remains

Figure 7.18 Bis-GMA/TEGDMA resin blend containing 8 wt% of silica nanoparticles along with 8 wt% of hydroxyapatite fiber with total number of atoms as 16 894.

roughly constant. These integrators do not in fact reproduce the actual Hamiltonian mechanics of the system; instead, they reproduce a closely related "shadow" Hamiltonian whose value they conserve many orders of magnitude more closely. The accuracy of the energy conservation for the true Hamiltonian is dependent on the time step. Hence for equilibration process, several short simulation runs were conducted for the resin blends as well as the composites. The sequence of the MD simulation has been given below.

(a) The resin blends as well as the composite materials were first attached to a barostat having high pressures. This resulted in the acceleration of the equilibration process [53].
(b) The systems were then equilibrated at a temperature of 298 K and at a pressure of 950 bar for 180 ps.
(c) Further equilibration was performed at a temperature of 298 K and at a pressure of 475 bar for 180 ps.
(d) Finally, the systems were equilibrated at the same temperature, but the pressure was reduced to 1 bar for 900 ps.
(e) The ensemble was then changed to NVT with a temperature of 298 K and the simulation time was 900 ps.

The step (e) was repeated several times till the conversion of total energy and temperature. Forcite module was used for conducting the MD simulations. Nose–Hoover thermostat [54] with a relaxation time of 1 ps was used for maintaining the temperature constant. The Berendsen barostat [55] with a time-constant of 0.1 ps was used for coupling the pressure. The quality of calculations was set to ultra-fine for achieving higher accuracy in the simulations. The van der Waals interaction cutoff distance was taken as 18 Å using the atom-based method. Ewald summation [56] was used for the long-range electrostatic interactions. The mechanical properties were then calculated using the NVT ensemble at 298 K for another 900 ps.

7.2.2 Results and Discussion

7.2.2.1 Chain Configuration

It has been reported that the chain configuration significantly effects the mechanical reinforcement of composites [57]. However, the effect of addition of nanoparticles on the polymer matrix has not been explored in detail. Radius of gyration (R_g) provides an insight of the effect of adding nanoparticles in the polymer matrix. It is defined as an average squared distance between monomers in given conformation and the molecule's center of mass and is given by Eq. (7.1).

$$R_g = \sqrt{\frac{\left\langle \sum_{i=1}^{N} (r_i - r_{cm})^2 m_i \right\rangle}{\sum_{i=1}^{N} m_i}} \tag{7.1}$$

where

N=number of atoms in a chain

m_i=mass of ith atom

r_i=position of ith atom

$r_{cm}=\dfrac{\left\langle \sum_{i=1}^{N} m_i r_i \right\rangle}{\sum_{i=1}^{N} m_i}$ = centre of mass of the chain

The predicted R_g of the polymer chains as a function of TEGDMA weight percentage has been shown in Figure 7.19. It could be observed from Figure 7.19 that R_g in various blends of resin lie in the range of 16–17 Å and thus was approximately independent of bis-GMA/TEGDMA weight percentage. Thus, the interaction between the organic components had no influence on the chain configuration. Also, Figure 7.20 depicts the R_g of the polymer chains and silica nanoparticles as a function of nanoparticle weight percentage. The radii of gyration of the polymer chains around the silica nanoparticle were found to be much smaller in comparison to that of the polymer chains in resin blends. This indicated that the silica nanoparticles could easily disturb the behavior of chains of polymer. The average value of R_g for the chains of polymers was found to decrease from 16 to 10 Å as the weight percentage of silica nanoparticle was increased from 0% to 11%. Whereas, average value of R_g of silica nanoparticles remained almost constant at 8 Å, independent of the nanoparticle weight percentage. Similarly, the average value of R_g for the chains of polymers was found to decrease from 16 to 12 Å as the weight percentage of HAFs was increased from 0% to 12%. Whereas, average value of R_g of HAFs remained almost constant at 10 Å, independent of the HAFs weight percentage. This has been shown in Figure 7.21. These results were found to be in agreement with previous studies [58, 59].

7.2.2.2 Effect of Hydrogen Bonding

The mechanical performance of blends of polymers and composites has been found to be strongly influenced by the hydrogen bond interaction [60]. Intermolecular hydrogen bonds

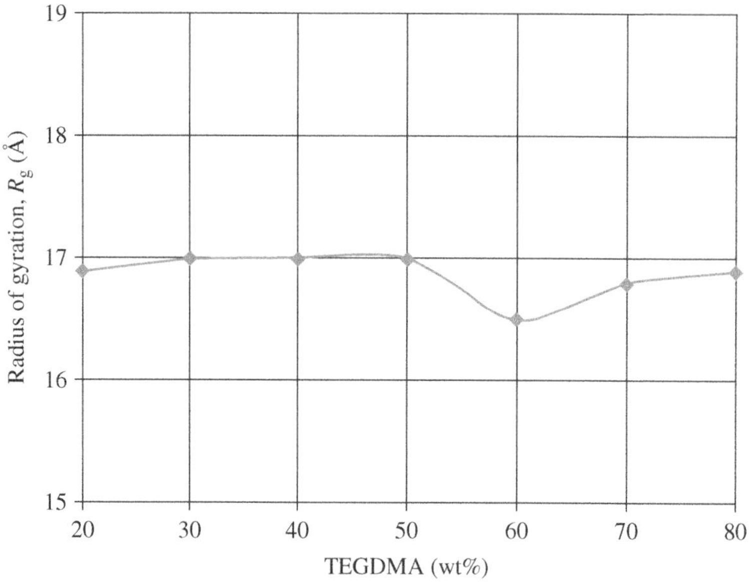

Figure 7.19 Variation of radius of gyration (R_g) of the polymer chains with different weight percentages of TEGDMA.

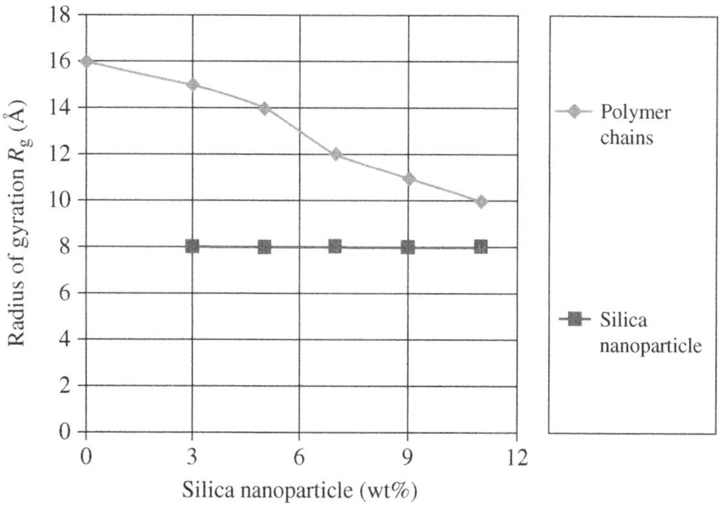

Figure 7.20 Variation of radius of gyration (R_g) of the polymer chains and silica nanoparticle with different weight percentages of nanoparticles.

could be formed by the hydroxyl groups present in bis-GMA and silica nanoparticles in the dental composites. Thus, the change in hydrogen bonds in the dental composites has been critically examined in this study. For studying the effect of hydrogen bonding on dental composites, a geometric criterion has been used where the hydrogen bond formation was described by two variables, the bond orientation and the atomic distance. A combination of O and H (donor) and acceptor O in O—H···O configuration has been defined as a hydrogen bond only if the angle of bond in O—H···O was greater than 120° and the spacing between

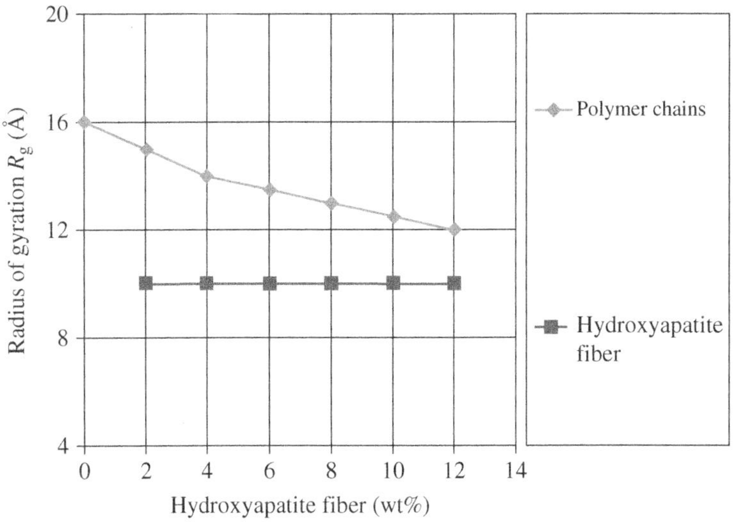

Figure 7.21 Variation of radius of gyration (R_g) of the polymer chains and hydroxyapatite fiber with different weight percentages of hydroxyapatite.

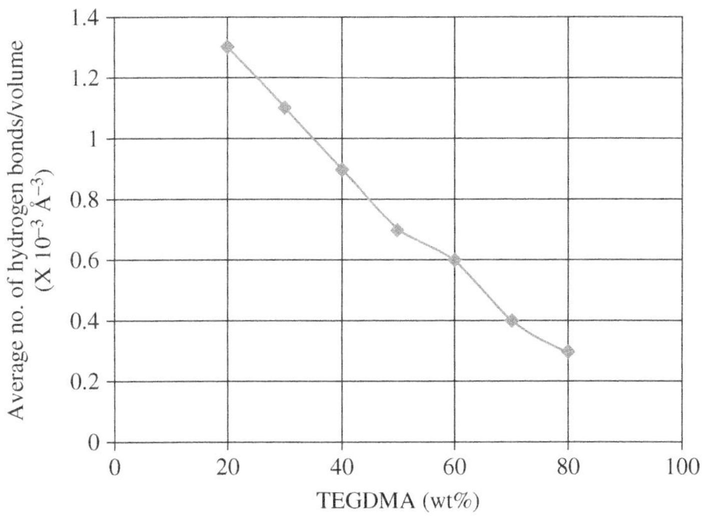

Figure 7.22 Variation of average volumetric hydrogen bonds in resin blends with TEGDMA weight percentage.

H and acceptor O was less than 2.5 Å. Using the above criterion, the average number of hydrogen bonds per unit volume in the blends and composite materials have been predicted and have been shown in Figures 7.22 and 7.23.

Figure 7.22 shows the variation of average volumetric hydrogen bonds in resin blends with TEGDMA weight percentage. As the weight percentage of TEGDMA in dental resin blends was increased from 20% to 80%, the hydrogen bonds per unit volume were found to decrease from 1.3×10^{-3} to 0.3×10^{-3} Å^{-3}, which could be ascribed to the fact that there was reduction in hydroxyl groups in bis-GMA with an increase in weight percentage of TEGDMA.

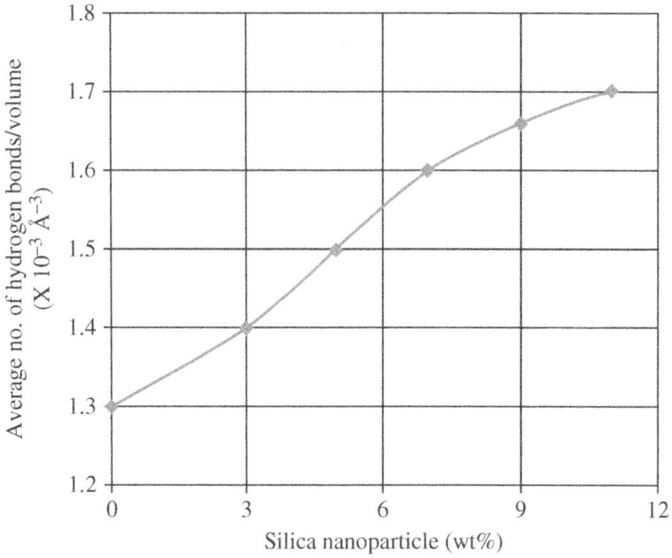

Figure 7.23 Variation of average volumetric hydrogen bonds in composites with silica nanoparticle weight percentage.

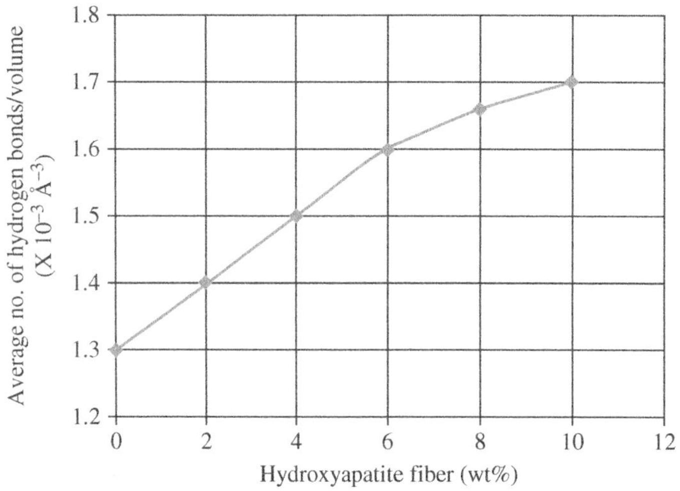

Figure 7.24 Variation of average volumetric hydrogen bonds in composites with hydroxyapatite weight percentage.

Figure 7.23 shows the variation of average volumetric hydrogen bonds in composites with silica nanoparticle weight percentage. It was observed that there was an increase in the hydrogen bonds from 1.3×10^{-3} to 1.7×10^{-3} Å^{-3}, when the silica nanoparticle weight percentage was increased from 0 to 11 wt%. Similarly, Figure 7.24 shows that there was an increase in the hydrogen bonds from 1.3×10^{-3} to 1.5×10^{-3} Å^{-3}, when the hydroxyapatite weight percentage was increased from 0–12 wt%.

7.2.2.3 Prediction of Mechanical Properties

In an atomistic calculation, the internal stress tensor could be obtained using the so-called virial expression given by Swenson [61], which has been shown in Eq. (7.2).

$$\sigma = -\frac{1}{V_0}\left[\left(\sum_{i=1}^{n} m_i(v_i v_i^T)\right) + \left(\sum_{i<j} r_{ij} f_{ij}^T\right)\right] \tag{7.2}$$

where index i runs over all particles 1 through N; m_i, v_i, and f_i denote the mass, velocity, and force acting on particle i; and V_0 denotes the (un-deformed) system volume. The application of stress to a body results in a change in the relative positions of particles within the body, expressed quantitatively via the strain tensor:

$$\varepsilon_{ij} = \begin{bmatrix} \varepsilon_{11} & \varepsilon_{12} & \varepsilon_{13} \\ \varepsilon_{21} & \varepsilon_{22} & \varepsilon_{23} \\ \varepsilon_{31} & \varepsilon_{32} & \varepsilon_{33} \end{bmatrix} \tag{7.3}$$

The elastic stiffness coefficients, relating the various components of stress and strain have been defined by Eq. (7.4).

$$C_{lmnk} = \left.\frac{\partial \sigma_{lm}}{\partial \varepsilon_{nk}}\right|_{T,\ z_{nk}} = \left.\frac{1}{V_0}\frac{\partial^2 A}{\partial \varepsilon_{lm}\,\partial \varepsilon_{nk}}\right|_{T,\ z_{lm},\ z_{nk}} \tag{7.4}$$

where, A denotes the Helmholtz free energy. For small deformations, the relationship between the stresses and strains could be expressed in terms of a generalized Hooke's law:

$$\sigma_{lm} = C_{lmnk}\varepsilon_{nk} \tag{7.5}$$

In this study, a constant-strain approach has been used for prediction of mechanical properties of resin blends as well as dental composites. A small strain of 0.002 was applied to all the periodic structures. This was accomplished by uniformly stretching the RVE in the deformation direction followed by a rescaling process of fitting the coordinates of the molecules within the new dimensions. Then keeping the lattice parameters fixed, minimization of potential energy of each RVE was performed. The stiffness matrix was then obtained from the second derivative of potential energy (A) with respect to the applied strain, as given by Eq. (7.4). The mechanical properties could then be calculated using the following equations.

$$E = \mu\frac{3\lambda + 2\mu}{\lambda + \mu} \tag{7.6}$$

$$G = \mu \tag{7.7}$$

$$v = \frac{\lambda}{2(\lambda + \mu)} \tag{7.8}$$

where, the Lame's constants are:

$$\lambda = \frac{1}{3}(C_{11} + C_{22} + C_{33}) - \frac{2}{3}(C_{44} + C_{55} + C_{66}) \tag{7.9}$$

$$\mu = \frac{1}{3}(C_{44} + C_{55} + C_{66}) \tag{7.10}$$

Here,

E = Elastic modulus
G = Shear modulus
ν = Poisson's ratio

Values of the stiffness matrix, for a particular resin blend containing bis-GMA/TEGDMA in the ratio 80/20 (by weight), haves been calculated as shown in Eq. (7.11).

$$C_{ij} = \begin{bmatrix} 5.1784 & 1.9217 & 1.9277 & -0.0091 & 0.0435 & -0.0210 \\ 1.9217 & 5.6218 & 2.9607 & -0.2191 & 0.0451 & -0.0820 \\ 1.9277 & 2.9607 & 5.5721 & -0.2187 & 0.0371 & -0.0541 \\ -0.0091 & -0.2191 & -0.2187 & 1.3919 & -0.0186 & 0.0479 \\ 0.0435 & 0.0451 & 0.0371 & -0.0186 & 1.4041 & -0.0518 \\ -0.0210 & -0.0820 & -0.0541 & 0.0479 & -0.0518 & 1.4089 \end{bmatrix} \text{GPa} \tag{7.11}$$

Similarly, for a particular dental composite containing silica nanoparticles (7% by weight) and for a particular dental composite containing HAFs (8% by weight), the stiffness matrices have been predicted as shown in Eqs. (7.12) and (7.13), respectively.

$$C_{ij} = \begin{bmatrix} 5.9725 & 2.4711 & 2.1793 & 0.0391 & 0.0653 & 0.1683 \\ 2.4711 & 6.4500 & 3.1617 & 0.8145 & -0.0781 & 0.0998 \\ 2.1793 & 3.1617 & 6.2175 & 0.1871 & 0.1897 & -0.0801 \\ 0.0391 & 0.8145 & 0.1871 & 1.4623 & -0.0146 & 0.0471 \\ 0.0653 & -0.0781 & 0.1897 & -0.0146 & 1.5528 & -0.0218 \\ 0.1683 & 0.0998 & -0.0801 & 0.0471 & -0.0218 & 1.4242 \end{bmatrix} \text{GPa} \tag{7.12}$$

$$C_{ij} = \begin{bmatrix} 5.4225 & 2.1162 & 2.0187 & 0.0193 & 0.0555 & 0.1120 \\ 2.1162 & 6.0500 & 3.0471 & 0.4451 & -0.0787 & 0.0884 \\ 2.0187 & 3.0471 & 6.0172 & 0.1472 & 0.1179 & -0.0710 \\ 0.0193 & 0.4451 & 0.1472 & 1.4215 & -0.0150 & 0.0472 \\ 0.0555 & -0.0787 & 0.1179 & -0.0150 & 1.4991 & -0.0322 \\ 0.1120 & 0.0884 & -0.0710 & 0.0472 & -0.0322 & 1.4111 \end{bmatrix} \text{GPa} \tag{7.13}$$

Eqs. (7.11)–(7.13) show that the polymer blends as well as composites are not completely isotropic. However, the non-diagonal terms are much smaller in comparison to the diagonal terms. Thus, this static method was found to be feasible for predicting the properties of dental composites. Finally, the stiffness matrix for a dental composite containing 9 wt% of silica nanoparticles along with 8 wt% of HAF in a bis-GMA/TEGDMA (60/23 by weight) resin blend has also been shown in Eq. (7.14).

$$C_{ij} = \begin{bmatrix} 6.0871 & 2.7821 & 2.1793 & 0.0391 & 0.0653 & 0.1683 \\ 2.4711 & 6.9500 & 3.7745 & 0.9952 & -0.0817 & 0.1052 \\ 2.1793 & 3.7745 & 6.5714 & 0.2561 & 0.2079 & -0.0712 \\ 0.0391 & 0.9952 & 0.2561 & 1.6432 & -0.0061 & 0.0812 \\ 0.0653 & -0.0817 & 0.2079 & -0.0061 & 1.7081 & -0.0023 \\ 0.1683 & 0.1052 & -0.0712 & 0.0812 & -0.0023 & 1.5252 \end{bmatrix} \text{GPa} \tag{7.14}$$

It could be seen from Eq. (7.14) that the reinforcement effect of hydroxyapatite was significant and the silica nanoparticles also helped in improving the stiffness of the dental composites. The silica nanoparticles are known to be efficient anti-bacterial agents. Thus, not only

Table 7.10 Properties of resin blends obtained from MD simulations.

TEGDMA weight percentage	Density, ρ (g/cm^3)	Poisson's ratio, ν	Modulus of elasticity, E (GPa)	Shear modulus, G (GPa)
20	1.159	0.29	3.32	1.30
30	1.149 (1.228[a], 1.212[b])	0.28	3.30	1.29
40	1.142	0.28	3.28	1.27
50	1.139 (1.232[a])	0.31	3.25	1.25
60	1.137 (1.229[a])	0.33	3.23	1.24
70	1.133 (1.227[a], 1.204[b])	0.32	3.18	1.20
80	1.130 (1.237[a])	0.32	3.09	1.17

a) Experimental result from [63]
b) Experimental result from [64]

the stiffness was found to improve with the addition of silica nanoparticles and HAFs but also the anti-bacterial properties are also likely to be improved, as has been shown by [62]. The mechanical properties of resin blends along with a comparison of their densities obtained from this simulation study and previous experiments have been tabulated in Table 7.10. It could be inferred from Table 7.10 that the elastic and shear moduli of dental resin blends decrease slowly with an increase in the TEGDMA weight percentage. Whereas, an increase in the bis-GMA weight percentage from 20% to 80% caused the elastic moduli to increase by up to 8.2%.

Similarly, for dental composites with varying weight percentage of silica nanoparticles (0–11%), the results obtained from MD have been given in Table 7.11. It was observed that the elastic and shear moduli increased by 12.95% and 11.5%, respectively. This could be attributed to the fact that silica nanoparticles have much higher moduli in comparison to that of the polymer blend. Also, because of the enhanced resin-particle interaction due to the hydrogen bonds, the physical and mechanical properties were found to improve. The value of Poisson's ratio obtained for dental composites was found to be in the range of 0.29–31. This agreed well with the value of 0.3 found in literature [65, 66]. The results obtained for dental composites containing varying percentage of HAFs (0–12%) have been shown in Table 7.12. It could be seen from Table 7.12 that the elastic and shear moduli increased by 8.13% and 10%, respectively. The value of Poisson's ratio obtained for dental composites containing HAFs was again found to be in the range of 0.29–31. Finally, Table 7.13 shows the properties of a typical bis-GMA/TEGDMA (60/23 by weight) dental resin composite containing 9 wt% of silica nanoparticles along with 8 wt% of hydroxyapatite fiber. Comparing Tables 7.11–7.13, it could be inferred that the elastic and shear moduli of the dental resin containing silica nanoparticles along with hydroxyapatite fibers increased considerably.

7.2.2.4 Coefficient of Diffusion

One of the major problems often encountered by dentists is the water adsorption of water molecules on the dental resins which can erode the unreacted monomers. This could result in a rapid growth of bacteria and finally result in an allergic activity [67]. Thus, the evaluation of dynamics of diffusion of the dental resins is very crucial. The diffusion coefficients, extracted from the Einstein relation, as given by Eqs. (7.15) and (7.16) have been used as a measure of

Table 7.11 Variation in properties of dental resin composites with varying weight percentage of silica nanoparticles obtained from MD simulations.

Silica nanoparticle weight percentage	Density, ρ (g/cm^3)	Poisson's ratio, ν	Modulus of elasticity, E (GPa)	Shear modulus, G (GPa)
0	1.149	0.29	3.32	1.30
3	1.160	0.29	3.47	1.35
5	1.173	0.30	3.61	1.40
7	1.186	0.30	3.68	1.42
9	1.196	0.30	3.72	1.44
11	1.205	0.31	3.75	1.45

Table 7.12 Variation in properties of dental resin composites with varying weight percentage of hydroxyapatite fibers obtained from MD simulations.

Silica nanoparticle weight percentage	Density, ρ (g/cm^3)	Poisson's ratio, ν	Modulus of elasticity, E (GPa)	Shear modulus, G (GPa)
0	1.149	0.29	3.32	1.30
2	1.154	0.29	3.42	1.34
4	1.160	0.31	3.50	1.39
6	1.168	0.31	3.54	1.40
8	1.178	0.30	3.56	1.41
10	1.186	0.30	3.58	1.42
12	1.191	0.30	3.59	1.43

Table 7.13 Properties of a typical bis-GMA/TEGDMA (60/23 by weight) dental resin composite containing 9 wt% of silica nanoparticles along with 8 wt% of hydroxyapatite fiber.

Density, ρ (g/cm^3)	Poisson's ratio, ν	Modulus of elasticity, E (GPa)	Shear modulus, G (GPa)
1.200	0.30	3.84	1.46

the dynamics of diffusion.

$$\langle \Delta r\ (t)^2 \rangle = \frac{1}{M}\sum_{i=1}^{M}\left\{r_i(t) - r_i(0)\right\}^2 \tag{7.15}$$

$$D = \lim_{t\to\infty}\frac{1}{6t}\langle \Delta r\ (t)^2 \rangle \tag{7.16}$$

Here,

$\langle \Delta r(t)^2 \rangle$ = Ensemble average mean square displacement (MSD)

M = Number of molecules

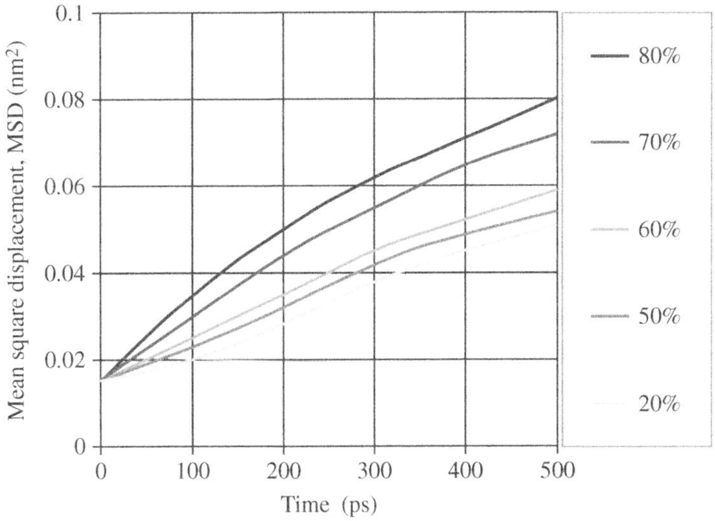

Figure 7.25 Mean square displacements (MSD) of bis-GMA/TEGDMA polymer blends from the reference positions at time *t*, at different weight percentage of TEGDMA.

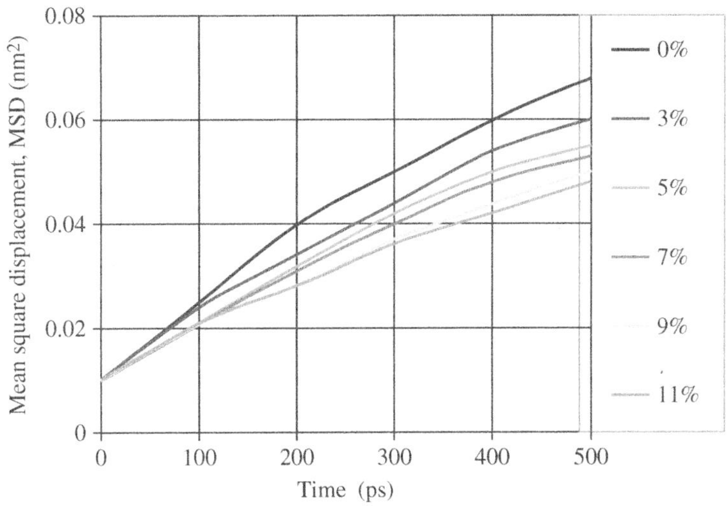

Figure 7.26 Mean square displacements (MSD) of dental resin composites from the reference positions at time *t*, with varying weight percentage of silica nanoparticles.

The MSD for dental resin and composites have been shown in Figures 7.25–7.27. Using the least squares method, the diffusion coefficients have then been predicted by fitting a straight line through MSD curves. The reference time was taken as $t = 200$–400 ps. In polymer resins, the intermolecular hydrogen bond formation was suppressed by the addition of TEGDMA, thus leading to faster diffusion of the molecules. This could be observed from Figure 7.28. Figure 7.29 shows the variation of diffusion coefficient of dental composite with varying weight percentage of silica nanoparticles. It could be inferred from Figure 7.29 that the diffusion coefficient falls approximately by 46%, from 1.15×10^{-11} to 0.62×10^{-11} m^2/s with an increase in the silica nanoparticle weight percentage from 0% to 11%. This large reduction

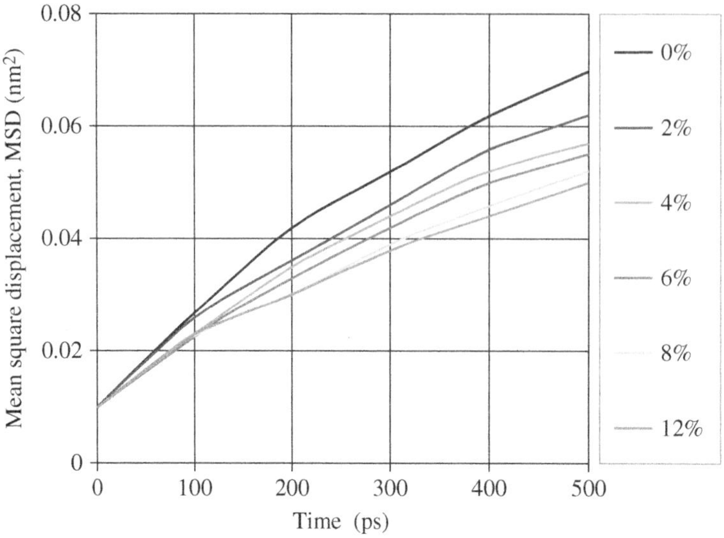

Figure 7.27 Mean square displacements (MSD) of dental resin composites from the reference positions at time *t*, with varying weight percentage of hydroxyapatite fibers.

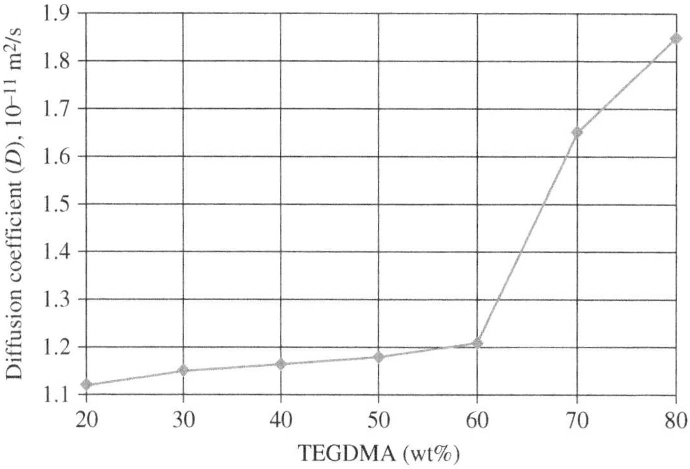

Figure 7.28 Variation of diffusion coefficient of polymer blends with varying weight percentage of TEGDMA.

in the diffusion coefficient could be attributed to the joint effect of highly compact chain configuration as well as due to an increase in hydrogen bonds, at large silica nanoparticle weight percentage. Figure 7.30 shows the variation of diffusion coefficient of dental composite with varying weight percentage of HAFs. For this case, the diffusion coefficient was found to decrease by 47.8%, from 1.15×10^{-11} to 0.6×10^{-11} m^2/s with an increase in the HAF weight percentage from 0% to 12%.

7.2.3 Conclusion

In this study, MD simulations have been performed for predicting the mechanical and dynamic properties of bis-GMA/TEGDMA polymer blends and composite materials. The

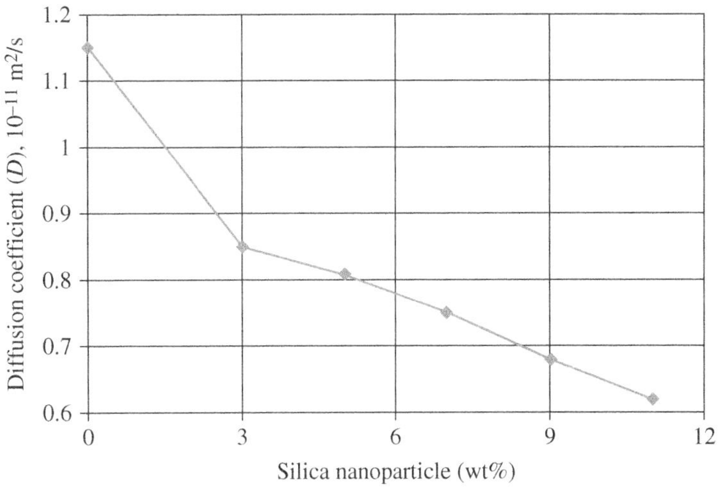

Figure 7.29 Variation of diffusion coefficient of dental composite with varying weight percentage of silica nanoparticles.

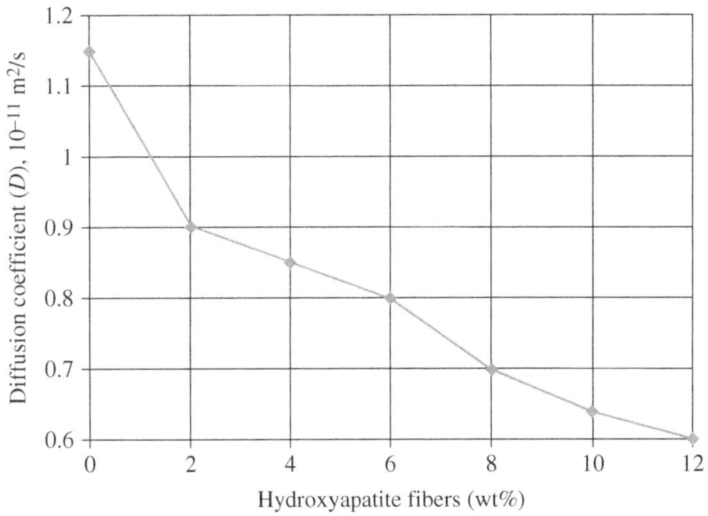

Figure 7.30 Variation of diffusion coefficient of dental composite with varying weight percentage of hydroxyapatite fibers.

mechanical reinforcement of dental polymer systems was fundamentally studied at atomic scale with the help of polymer chain configuration and hydrogen bonding. The radius of gyration of chains of polymers was found to be unaltered for different blends of polymers. But in case of composites, it was observed that there was a significant alteration in radius of gyration and with an increase in silica nanoparticle weight percentage, the polymer chains got compacted. This could be because of the perturbation of the flow behavior caused by the silica nanoparticles and HAFs. The hydrogen bonds were found to increase volumetrically with the ratio of bis-GMA or silica nanoparticles or HAFs. At high weight percentage of bis-GMA or silica nanoparticles or HAFs in the dental resin systems, the mechanical reinforcement effect was significant because of strong hydrogen bond interactions. The dynamics of the dental systems was also studied with the help of diffusion coefficients. It was found that at high weight percentage of bis-GMA or silica nanoparticles or HAFs in the dental resin systems, the diffusion dynamics behavior was restricted. This was attributed to the change in chain conformation and hydrogen bonds. These findings will be of great importance in improving the understanding of the mechanical and dynamic properties of dental materials at the atomic/nanoscale.

References

1 Endo, M., Natsuki, T., and Tantrakarn, K. (2004). Effects of carbon nanotube structures on mechanical properties. *Applied Physics A: Materials Science & Processing* 79 (1): 117–124.

2 Hughes, T.V. and Chambers C.R. (1889). Manufacture of carbon filaments. US Patent 405,480, filed 27 December 1886 and issued 30 March 1892.

3 Zhu, Y.A., Sui, Z.J., Zhao, T.J. et al. (2005). Modeling of fishbone-type carbon nanofibers: a theoretical study. *Carbon* 43 (8): 1694–1699.

4 Rodriguez, N.M. (1993). A review of catalytically grown carbon nanofibers. *Journal of Materials Research* 8 (12): 3233–3250.

5 Mall, S., Lee, H., He, P. et al. (2007). Characterization of carbon nanotube/nanofiber-reinforced polymer composites using an instrumented indentation technique. *Composites: Part B Engineering* 38 (1): 58–65.

6 Minaie, B., Rodriguez, A.J., Guzman, M.E., and Lim, C.S. (2011). Mechanical properties of carbon nanofiber/fiber-reinforced hierarchical polymer composites manufactured with multiscale-reinforcement fabrics. *Carbon* 49 (3): 937–948.

7 Tibbetts, G.G. and Beetz, C.P. (1987). Mechanical properties of vapor grown carbon fibers. *Journal of Physics D: Applied Physics* 20 (3): 292–297.

8 Patton, R.D., Pittman, C.U., Wang, L., and Hill, J.R. (1999). Vapor grown carbon fiber composites with epoxy and poly(phenylene sulfide) matrices. *Composites: Part A* 30 (9): 1081–1091.

9 Endo, M., Kim, Y.A., Hayashi, T. et al. (2001). Vapor-grown carbon fibers (VGCFs)-basic properties and their battery applications. *Carbon* 39 (9): 1287–1297.

10 Zhu, Y., Bakis, C.E., and Adair, J.H. (2012). Effects of carbon nanofiller functionalization and distribution on interlaminar fracture toughness of multi-scale reinforced polymer composites. *Carbon* 50 (3): 1316–1331.

11 Sansoz, F. and Gu, J. (2014). Role of cone angle on the mechanical behavior of cup-stacked carbon nanofibers studied by atomistic simulations. *Carbon* 66 (1): 523–529.

12 Finegan, I.C., Tibbetts, G.G., and Gibson, R.F. (2003). Modeling and characterization of damping in carbon nano-fiber/polypropylene composites. *Composites Science and Technology* 63: 1629–1635.

13 Zhou, J.H., Sui, Z.J., Li, P. et al. (2006). Structural characterization of carbon nanofibers formed from different carbon-containing gases. *Carbon* 44 (15): 3255–3262.

14 Alam, M.K., Kuriger, R.J., Anderson, D.P., and Jacobsen, R.L. (2002). Processing and characterization of aligned vapor grown carbon fiber reinforced polypropylene. *Applied Science and Manufacturing* 33 (1): 53–62.

15 Tandon, G.P. and Weng, G.J. (1984). The effect of aspect ratio of inclusions on the elastic properties of uni-directionally aligned composites. *Polymer Composites* 5: 327–333.

16 Mori, T. and Tanaka, K. (1973). Average stress in matrix and average elastic energy of materials with misfitting inclusions. *Acta Materialia* 21 (5): 571–574.

17 Chasiotis, I., Naraghi, M., and Ozkan, T. (2010). Mechanical properties of vapour grown carbon nanofibers. *Carbon* 48: 239–244.

18 Tibbetts, G.G. and McHugh, J.J. (1999). Mechanical properties of vapor-grown carbon fiber composites with thermoplastic matrices. *Journal of Materials Research* 14 (7): 2871–2880.

19 Bindu, M.G., Satapathy, B.K., Jaggi, H.S., and Ray, A.R. (2013). Size-scale effects of silica on bis-GMA/TEGDMA based nanohybrid dental restorative composites. *Composites Part B Engineering* 53 (10): 92–102.

20 Zhang, Y., Zhang, D., Qin, C., and Xu, J. (2009). Physical and mechanical properties of dental nanocomposites composed of aliphatic epoxy resin and epoxidized aromatic hyperbranched polymers. *Polymer Composites* 30 (2): 176–181.

21 Tian, M., Gao, Y., Liu, Y. et al. (2008). Fabrication and evaluation of bis-GMA/TEGDMA dental resins/composites containing nano fibrillar silicate. *Dental Materials* 24 (2): 235–243.

22 Leinfelder, K.F., Sluder, T.B., Santos, J.F.F., and Wall, J.T. (1980). Five-year clinical evaluation of anterior and posterior restorations of composite resin. *Operative Dentistry* 12 (2): 52–78.

23 Kahler, B., Kotousov, A., and Swain, M.V. (2008). On the design of dental resin – based composites: a micromechanical approach. *Acta Biomaterialia* 4 (1): 165–172.

24 Tilbrook, D.A., Clarke, R.L., Howle, N.E., and Braden, M. (2000). Photocurable epoxy-polyol matrices for use in dental composites I. *Biomaterials* 21 (17): 1743–1753.

25 Braem, M., Finger, W., Van Doren, V.E. et al. (1989). Mechanical properties and filler fraction of dental composites. *Dental Materials* 5 (5): 346–349.

26 Xu, H.H.K., Martin, T.A., Antonucci, J.M., and Eichmiller, F.C. (1999). Ceramic whisker reinforcement of dental composite resins. *Journal of Dental Research* 78 (2): 706–712.

27 Xu, H.H.K. (1999). Dental composite resins containing silica-fused ceramic single-crystalline whiskers with various filler levels. *Journal of Dental Research* 78 (7): 1304–1311.

28 Xu, H.H.K. (2000). Whisker-reinforced heat-cured dental resin composites: effects of filler level and heat-cure temperature and time. *Journal of Dental Research* 79 (6): 1392–1397.

29 Xu, H.H.K., Schumacher, G.E., Eichmiller, F.C., and Antonucci, J.M. (2000). Strengthening composite resin restorations with ceramic whisker reinforcement. *Practical Periodontics and Aesthetic Dentistry* 12 (1): 111–116.

30 Xu, H.H.K., Quinn, J.B., Smith, D.T. et al. (2003). Effects of different whiskers on the reinforcement of dental resin composites. *Dental Materials* 19 (5): 359–367.

31 Tian, M., Qu, C., Feng, Y., and Zhang, L. (2003). Structure and properties of fibrillar silicate/SBR composites by blend process. *Journal of Materials Science* 38 (24): 4917–4924.

32 Tian, M., Liang, W., Rao, G. et al. (2005). Surface modification of fibrillar silicate and its reinforcing mechanism on FS/rubber composites. *Composites Science and Technology* 65 (6–7): 1129–1138.

33 Saeb, M.R., Nonahal, M., Rastin, H. et al. (2017). Calorimetric analysis and molecular dynamics simulation of cure kinetics of epoxy/chitosan-modified Fe_3O_4 nanocomposites. *Progress in Organic Coatings* 112: 176–186.

34 Saeb, M.R., Rastin, H., Shabanian, M. et al. (2017). Cure kinetics of epoxy/β-cyclodextrin-functionalized Fe_3O_4 nanocomposites: experimental analysis mathematical modeling, and molecular dynamics simulation. *Progress in Organic Coatings* 110: 172–181.
35 Bahlakeh, G., Ghaffari, M., Saeb, M.R. et al. (2016). A close-up of the effect of iron oxide type on the interfacial interaction between epoxy and carbon steel: combined molecular dynamics simulations and quantum mechanics. *Journal of Physical Chemistry C* 120 (20): 11014–11026.
36 Sharma, S., Kumar, P., Kumar, N., and Chandra, R. (2018). Graphene/carbon nanotube reinforced nickel composites: a molecular dynamics study. *Composites: Mechanics, Computations, Applications: An International Journal* 9 (3): 247–268.
37 Sharma, S., Kumar, P., and Chandra, R. (2018). Carbon nanotube reinforced titanium composites: an experimental and molecular dynamics study. *Journal of Composite Materials* 52 (29): 4117–4123.
38 Sharma, S. and Chawla, R. (2018). A molecular dynamics study on Young's modulus and tribology of carbon nanotube reinforced styrene–butadiene rubber. *Journal of Molecular Modeling* 24 (4) https://doi.org/10.1007/s00894-018-3636-5.
39 Sharma, S. and Chawla, R. (2018). A molecular dynamics study on efficient nanocomposite formation of styrene–butadiene rubber by incorporation of graphene. *Graphene Technology* https://doi.org/10.1007/s41127-018-0018-9.
40 Sharma, S. and Chawla, R. (2017). Molecular dynamics simulation of carbon nanotube pull-out from polyethylene matrix. *Composites Science and Technology* 144 (5): 169–177.
41 Sharma, S., Chandra, R., and Kumar, P. (2016). Mechanical and thermal properties of graphene–carbon nanotube-reinforced metal matrix composites: a molecular dynamics study. *Journal of Composite Materials* 51 (23): 3299–3313.
42 Anjana, R., Sharma, S., and Bansal, A. (2016). Molecular dynamics simulation of carbon nanotube reinforced polyethylene composites. *Journal of Composite Materials* https://doi.org/10.1177/0021998316674264.
43 Sharma, S., Chandra, R., Kumar, P., and Kumar, N. (2016). Mechanical properties of carbon nanofiber reinforced polymer composites-molecular dynamics approach. *Journal of The Minerals, Metals & Materials Society* 68 (6): 1717–1727.
44 Sharma, S., Chandra, R., Kumar, P., and Kumar, N. (2016). Molecular dynamics simulation of functionalized SWCNT-polymer composites. *Journal of Composite Materials* https://doi.org/10.1177/0021998316628973.
45 Sharma, S., Chandra, R., Kumar, P., and Kumar, N. (2016). Molecular level analysis of carbon nanofiber reinforced polymer composites. *Journal of Composite Materials* 50 (13): 1787–1804.
46 Sharma, S., Chandra, R., Kumar, P., and Kumar, N. (2014). Effect of stone-Wales and vacancy defects on elastic moduli of carbon nanotubes and their composites using molecular dynamics simulation. *Computational Materials Science* 86 (4): 1–8.
47 Dassault Systèmes BIOVIA (2017). Materials Studio 7.0. San Diego: Dassault Systèmes.
48 Odegard, G.M., Clancy, T.C., and Gates, T.S. (2005). Modeling of the mechanical properties of nanoparticle/polymer composites. *Polymer* 46 (2): 553–562.
49 Turssi, C.P., Ferracane, J.L., and Vogel, K. (2005). Filler features and their effects on wear and degree of conversion of particulate dental resin composites. *Biomaterials* 26 (24): 4932–4937.

50 Dionne, P.J., Ozisik, R., and Picu, C.R. (2005). Structure and dynamics of polyethylene nanocomposites. *Macromolecules* 38 (22): 9351–9358.

51 Sun, H., Jin, Z., Yang, C. et al. (2016). COMPASS II: extended coverage for polymer and drug-like molecule databases. *Journal of Molecular Modeling* 22 (2): 1–10.

52 Guo, G.J., Zhang, Y.G., Refson, K., and Zhao, Y.J. (2002). Viscosity and stress autocorrelation function in supercooled water: a molecular dynamics study. *Molecular Physics* 100 (16): 2617–2627.

53 Li, B., Chen, Q.B., Huang, S.Y., and Liu, H.X. (2017). Developing structure and thermodynamic properties-consistent coarse-grained model for random copolymer systems. *Polymer* 123 (8): 107–120.

54 Samoletov, A.A., Dettmann, C.P., and Chaplain, M.A.J. (2007). Thermostats for slow configurational modes. *Journal of Statistical Physics* 128 (6): 1321–1336.

55 Berendsen, H.J.C., Postma, J.P.M., van Gunsteren, W.F. et al. (1984). Molecular dynamics with coupling to an external bath. *The Journal of Chemical Physics* 81 (8): 3684–3690.

56 Ewald, P.P. (1921). Die Berechnung optischer und elektrostatischer Gitterpotentiale. *Annals of Physics* 369: 253–287.

57 Allegra, G., Raos, G., and Vacatello, M. (2008). Theories and simulations of polymer-based nanocomposites: from chain statistics to reinforcement. *Progress in Polymer Science* 33 (7): 683–731.

58 Zhang, J., Lou, J., Ilias, S. et al. (2008). Thermal properties of poly(lactic acid) fumed silica nanocomposites: experiments and molecular dynamics simulations. *Polymer* 49 (9): 2381–2386.

59 Nusser, K., Neueder, S., Schneider, G.J. et al. (2010). Conformations of silica-poly(ethylene–propylene) nanocomposites. *Macromolecules* 43 (23): 9837–9847.

60 Peng, C.C., Gopfert, A., Drechsler, M., and Abetz, V. (2005). Smart silica-rubber nanocomposites in virtue of hydrogen bonding interaction. *Polymers for Advanced Technologies* 16 (11–12): 770–782.

61 Swenson, R.J. (1983). Comments on virial theorems for bounded systems. *American Journal of Physics* 51 (10): 940–942.

62 Kawashita, M., Tsuneyama, S., Miyaji, F. et al. (2000). Antibacterial silver-containing silica glass prepared by sol–gel method. *Biomaterials* 21 (4): 393–398.

63 Dewaele, M., Truffier-Boutry, D., Devaux, J., and Leloup, G. (2006). Volume contraction in photocured dental resins: the shrinkage-conversion relationship revisited. *Dental Materials* 22 (4): 359–365.

64 Ge, J., Trujillo, M., and Stansbury, J. (2005). Synthesis and photopolymerization of low shrinkage methacrylate monomers containing bulky substituent groups. *Dental Materials* 21 (12): 1163–1169.

65 El-Safty, S., Akhtar, R., Silikas, N., and Watts, D.C. (2012). Nanomechanical properties of dental resin-composites. *Dental Materials* 28 (12): 1292–1300.

66 Tang, L.C., Wan, Y.J., Peng, K. et al. (2013). Fracture toughness and electrical conductivity of epoxy composites filled with carbon nanotubes and spherical particles. *Composites Part A Applied Science and Manufacturing* 45 (2): 95–101.

67 Sideridou, I., Achilias, D.S., Spyroudi, C., and Karabela, M. (2004). Water sorption characteristics of light-cured dental resins and composites based on bis-EMA/PCDMA. *Biomaterials* 25 (2): 367–376.

8

Coupling of Scales-Continuum Mechanics and Molecular Dynamics

8.1 Introduction

Significant advancement in science and technology has resulted in the evolution of nanotechnology. At the nanoscale, the properties of nanomaterials have been found to be size-dependent. The small nano-sized materials have presented significant challenges to the researchers in predicting their mechanical properties such as Young's modulus, tensile/compressive strength, buckling resistance, etc. Accurate prediction of mechanical properties at the nanoscale is very difficult to achieve because of limitations of experimental techniques currently available. Various researchers are now using computational techniques for accurate modeling of nanomaterials. These computational techniques based on physical models not only provide the molecular details of nanomaterials for designers but also act as guidance tools for experimentalists. It is still very difficult to identify an efficient computational technique for a particular nanomaterial.

The theoretical modeling approaches can be grouped in two categories: (i) atomistic modeling and (ii) continuum mechanics modeling. In atomistic modeling, the material is treated to be composed of atoms/molecules. In this modeling approach, the techniques such as classical molecular dynamics (MD) [1, 2], tight-binding molecular dynamics (TBMD) [3], and density-functional theory [4] are included. Any problem dealing with atomic/molecular motions can be solved using these techniques. The major problem with atomistic modeling techniques is the requirement of large computational power to deal with thousands of atoms/molecules. This limits their application to systems containing only a small number of molecules/atoms. Also, the atomistic modeling techniques are applicable for studies of phenomena that are relatively short-lived, from pico- to nano-seconds.

The second approach of continuum mechanics modeling consists of neglecting the characteristics of a nanomaterial and treating it as a continuum shell. For example, Tersoff [5] predicted the energies of fullerenes based on the deformation of a graphite sheet having a planar structure, by treating the graphite sheet as an elastic continuum. Yakobson et al. [6] developed a continuum shell model of the fullerenes. But this approach fails to take into account the forces acting on each atom.

Mechanics of Particle- and Fiber-Reinforced Polymer Nanocomposites: From Nanoscale to Continuum Simulations, First Edition. Sumit Sharma.

Therefore, a new modeling technique is required that makes an analysis of the nanomaterial at an atomistic scale but is not very puzzling in time scales. Such a modeling approach is termed as "Multi-scale Modeling Approach." This approach helps us in designing novel nanodevices and multi-scale simulations of nano-systems. In the following sections, the basis of this approach has been established, which has been followed by a few computational examples. An example of carbon nanotube (CNT) will be taken up for showing the approach of multi-scale modeling. The concept stems from the fact that there are various similarities between a continuum space frame structure and a molecular model of CNT. A CNT is an elongated fullerene, named after the architect, R. Buckmister Fuller. It consists of carbon atoms bonded together through covalent bonding. These bonds have characteristic length and angles in a 3-D space. Thus, the deformation of a CNT can be simulated using the classical structural mechanics.

8.2 Structural Mechanics Review

Structural mechanics approach is useful for finding the stresses, strains, and displacements of a structure under the given loading conditions. There are various analysis techniques available out of which the stiffness matrix method is mostly used. It can be applied readily for the analysis of structures of any geometry and can be used to solve problems associated with linear elasticity, plasticity, buckling, and dynamical problems. A review of the stiffness matrix method has been performed below, since it is relevant to multi-scale modeling approach.

For an element in a space frame as shown in Figure 8.1, the equilibration equation given by Weaver and Gere [7] can be written as:

$$\mathbf{Ku} = \mathbf{f} \tag{8.1}$$

where

$$\mathbf{u} = [u_{xi}, u_{yi}, u_{zi}, \theta_{xi}, \theta_{yi}, \theta_{zi}, u_{xj}, u_{yj}, u_{zj}, \theta_{xj}, \theta_{yi}, \theta_{zj}]^{\mathrm{T}} \tag{8.2}$$

$$\mathbf{f} = [f_{xi}, f_{yi}, f_{zi}, m_{xi}, m_{yi}, m_{zi}, f_{xj}, f_{yj}, f_{zj}, m_{xj}, m_{yi}, m_{zj}]^{\mathrm{T}} \tag{8.3}$$

are the nodal displacement and force vectors, respectively. $\mathbf{K}$ is the element stiffness matrix. The matrix $\mathbf{K}$ consists of the following sub-matrices:

$$\mathbf{K} = \begin{bmatrix} K_{ii} & K_{ij} \\ K_{ij}^T & K_{jj} \end{bmatrix} \tag{8.4}$$

where

$$K_{ii} = \begin{bmatrix} \frac{EA}{L} & 0 & 0 & 0 & 0 & 0 \\ 0 & \frac{12EI_x}{L^3} & 0 & 0 & 0 & \frac{6EI_x}{L^2} \\ 0 & 0 & \frac{12EI_y}{L^3} & 0 & \frac{-6EI_y}{L^2} & 0 \\ 0 & 0 & 0 & \frac{GJ}{L} & 0 & 0 \\ 0 & 0 & \frac{-6EI_y}{L^2} & 0 & \frac{4EI_y}{L} & 0 \\ 0 & \frac{6EI_x}{L^2} & 0 & 0 & 0 & \frac{4EI_x}{L} \end{bmatrix} \tag{8.5}$$

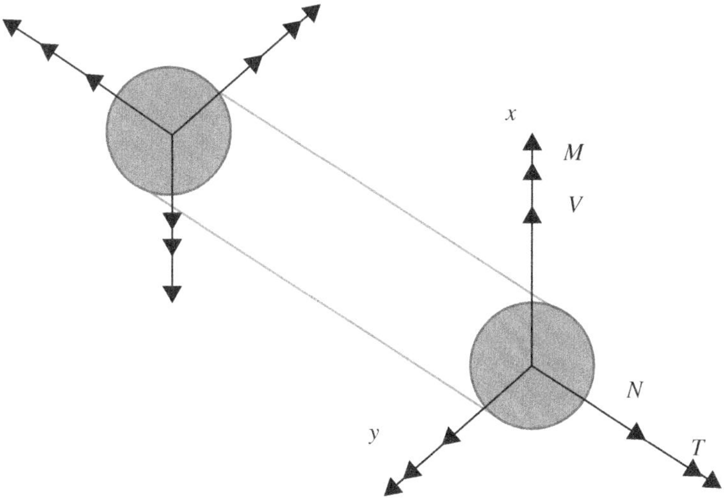

Figure 8.1 A beam element in a space frame.

$$K_{ij} = \begin{bmatrix} \frac{-EA}{L} & 0 & 0 & 0 & 0 & 0 \\ 0 & \frac{-12EI_x}{L^3} & 0 & 0 & 0 & \frac{6EI_x}{L^2} \\ 0 & 0 & \frac{-12EI_y}{L^3} & 0 & \frac{-6EI_y}{L^2} & 0 \\ 0 & 0 & 0 & \frac{-GJ}{L} & 0 & 0 \\ 0 & 0 & \frac{6EI_y}{L^2} & 0 & \frac{2EI_y}{L} & 0 \\ 0 & \frac{-6EI_x}{L^2} & 0 & 0 & 0 & \frac{2EI_x}{L} \end{bmatrix} \tag{8.6}$$

$$K_{jj} = \begin{bmatrix} \frac{EA}{L} & 0 & 0 & 0 & 0 & 0 \\ 0 & \frac{12EI_x}{L^3} & 0 & 0 & 0 & \frac{-6EI_x}{L^2} \\ 0 & 0 & \frac{12EI_y}{L^3} & 0 & \frac{6EI_y}{L^2} & 0 \\ 0 & 0 & 0 & \frac{GJ}{L} & 0 & 0 \\ 0 & 0 & \frac{6EI_y}{L^2} & 0 & \frac{4EI_y}{L} & 0 \\ 0 & \frac{-6EI_x}{L^2} & 0 & 0 & 0 & \frac{4EI_x}{L} \end{bmatrix} \tag{8.7}$$

From Eqs. (8.5)–(8.7), it can be seen that even when the length of the element is known, there are four parameters that still need to be determined viz. the tensile resistance EA, the flexural rigidity EI_x and EI_y, and the torsional stiffness GJ. The above equations must be specified for each element in the space frame structure and should be transferred from a local to a global coordinate system. A system of simultaneous linear equations can be formulated as per the conditions of equilibrium. Finally, by taking into account the boundary constraints, the system of equations can be solved to find the nodal displacements.

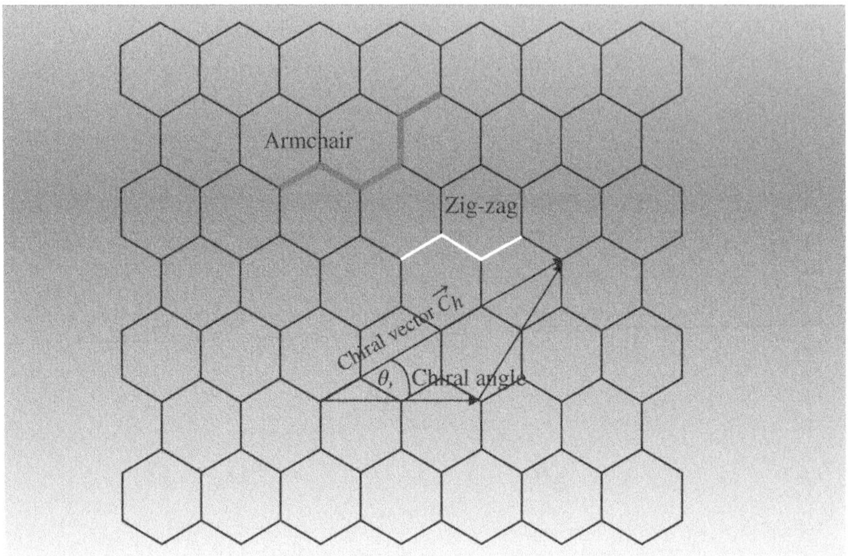

Figure 8.2 Structural characteristics of a CNT.

8.3 Carbon Nanotubes: Structural Mechanics Approach

A single-walled carbon nanotube (SWCNT) is obtained from a graphene sheet by rolling it into a tubular structure. A number of concentric CNTs arranged one inside the other results in the formation of a multi-walled carbon nanotube (MWCNT). The interlayer spacing between the concentric CNTs is kept as 0.34 nm. Graphene sheet is generally 1-atom thick and consists of a hexagonal array of carbon atoms arranged in a 2-D shape. Each carbon atom in a graphene sheet consists of three nearest neighboring carbon atoms. The structure of CNTs depends on their chirality or helicity. Chirality is defined by a chiral vector $\vec{C}_h$ and the chiral angle θ. Figure 8.2 shows the structural characteristics of a CNT. The chiral vector can be described as a roll-up vector and can be written as:

$$\vec{C}_h = n\,\vec{a}_1 + m\,\vec{a}_2 \tag{8.8}$$

where the integers (n,m) are the number of steps covered along the zig-zag carbon bonds and $\vec{a}_1$ and $\vec{a}_2$ are the unit vectors. Chirality represents the amount of twisting in the CNT. If $n = m$, the CNT is termed as armchair. If $m = 0$, then it represents a zig-zag CNT and if $n \neq m \neq 0$, it is termed as a chiral CNT. Thus, the armchair, zig-zag, and chiral CNTs can be represented as (n,n), $(n,0)$, and (n,m) respectively. The three types of CNTs are shown in Figure 8.3a–c. Chirality greatly influences the electronic properties of CNTs [8]. Armchair nanotubes are highly conducting, whereas zig-zag and chiral CNTs can be termed as semiconductors.

The CNTs can be treated from the structural mechanics point of view because there exist potential relations between the deformations of CNTs and the space-frame structures. When dealing with macroscopic entities made of real engineering materials, the sectional parameters and the material properties can be obtained from design data books and from calculations based on the sectional parameters. For CNTs, there is no definite information about the elastic and sectional properties of C—C bonds and the material properties. Thus, a link has to be established between the atomistic computational chemistry and the macroscopic structural

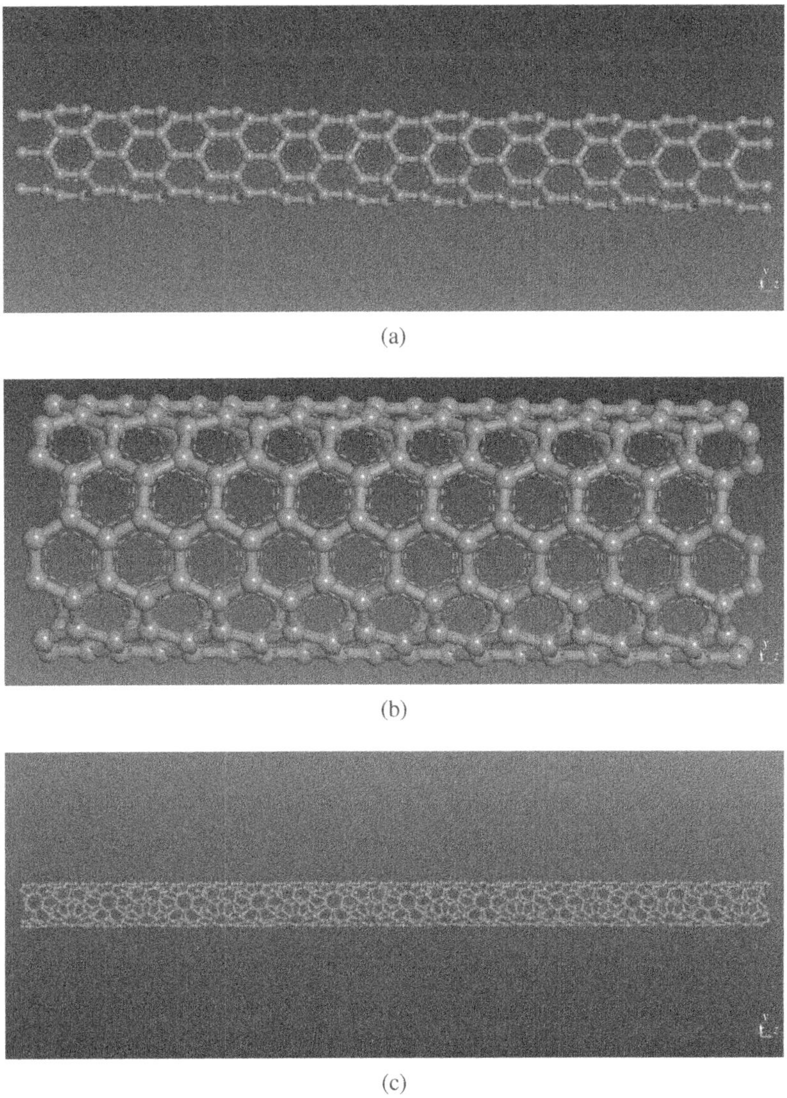

Figure 8.3 Atomistic models of (a) armchair, (b) zig-zag, and (c) chiral CNTs.

mechanics. The CNT can be considered to be made of a number of carbon atoms arranged in the form a large molecule. The atomic nuclei can be considered as material points, the motion of which is controlled by a force-field. The force-field is generated due to the interactions between the electron and the nucleus and also due to the nucleus–nucleus interactions [9]. The force-field can be expressed as steric potential energy. The steric potential energy is related to the relevant positions of the nuclei constituting the molecule. The general form of the steric potential energy, as given by Rappe et al. [10], is a sum of bonded (valence) interactions and nonbonded interactions as shown below.

$$U = \sum U_r + \sum U_\theta + \sum U_\varphi + \sum U_\omega + \sum U_{vdw} \tag{8.9}$$

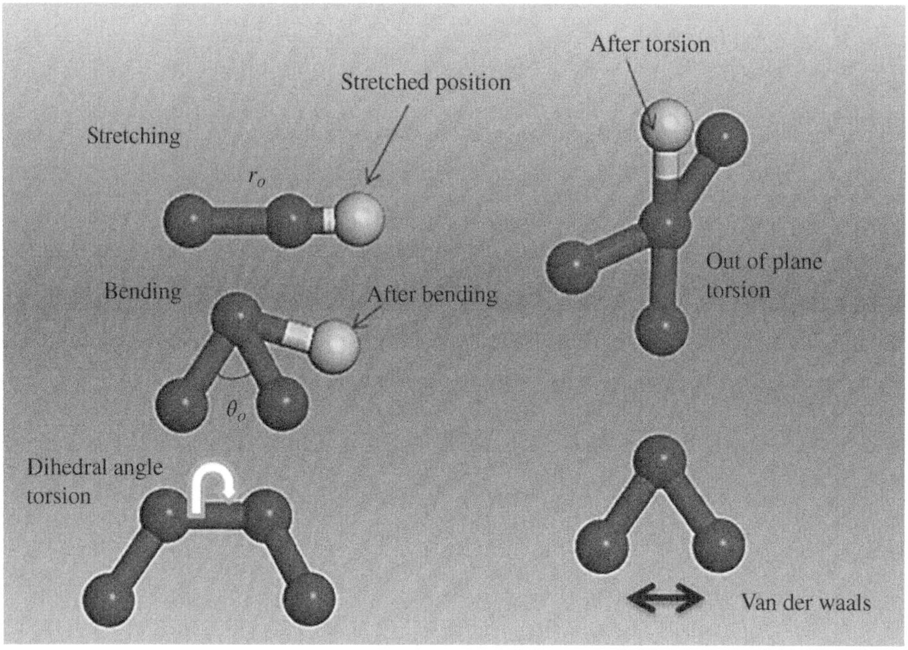

Figure 8.4 Interatomic interactions in molecular mechanics.

where

U_r = bond stretch interaction
U_θ = bond angle bending
U_φ = dihedral angle torsion
U_ω = out-of-plane torsion
U_{vdw} = nonbonded van der Waals interaction

The above interactions have been shown in Figure 8.4. These interactions can be represented by various terms depending on the material and loading conditions. Reasonable functional forms of these interactions have been given by Rappe et al. [10], Brenner [11], Mayo et al. [12], and Cornell et al. [13]. Generally, the major contribution to the potential energy function comes from the first four terms of Eq. (8.9). In case of small deformations, the harmonic approximation can be considered to be adequate for describing the potential energy (Gelin [14]). For simplicity, the simple harmonic function can be adopted, shown by Eqs. (8.10a) and (8.10b), and the dihedral angle torsion and improper torsion terms can be combined into a single term as given by Eq. (8.10c).

$$U_r = \frac{1}{2}k_r(r - r_o)^2 = \frac{1}{2}k_r(\Delta r)^2 \tag{8.10a}$$

$$U_\theta = \frac{1}{2}k_\theta(\theta - \theta_o)^2 = \frac{1}{2}k_\theta(\Delta\theta)^2 \tag{8.10b}$$

$$U_\tau = U_\varphi + U_\omega = \frac{1}{2}k_\tau(\Delta\varphi)^2 \tag{8.10c}$$

where

k_r = bond stretching force constant
k_θ = bond angle bending force constant
k_τ = torsional resistance
Δr = bond stretching increment
$\Delta\theta$ = bond angle change
$\Delta\varphi$ = angle change of bond twisting

8.4 Stiffness Parameters and Force Field Constants: Linkage

In a CNT, the carbon atoms are bonded together by covalent bonds in a hexagonal pattern. These bonds have their characteristic length and angles. The C—C bond in a CNT can be treated as a connecting element similar to a space frame structure. The carbon atoms can be treated as joints of these elements. For simplicity, assume that the sections of C—C bonds are uniformly round and identical, i.e. $I_x = I_y = I$, and only three stiffness parameters need to be determined viz. EA, EI, and GJ. Thus, when a CNT is subjected to external forces, the displacements are defined by the connecting elements, and the CNT can be treated as a space-frame structure. Using the principle of energy equivalence, the three stiffness parameters can be obtained. The deformation of the space frame structure results in changes in strain energy. Since the Eq. (8.10) represents individual interaction and no cross-interaction terms are there, only the strain energies associated with individual forces are considered. The strain energy of a uniform beam of length L subjected to a pure axial force N as shown in Figure 8.5 can be written as:

$$U_A = \frac{1}{2}\int_0^L \frac{N^2}{\mathrm{EA}}dL = \frac{1}{2}\frac{N^2L}{\mathrm{EA}} = \frac{1}{2}\frac{\mathrm{EA}}{L}(\Delta L)^2 \tag{8.11a}$$

where ΔL represent the axial stretching deformation. The strain energy of a uniform beam under pure bending moment M, as shown in Figure 8.5, can be written as:

$$U_M = \frac{1}{2}\int_0^L \frac{M^2}{\mathrm{EA}}dL = 2\frac{\mathrm{EI}}{L}(\alpha)^2 = \frac{1}{2}\frac{\mathrm{EI}}{L}(2\alpha)^2 \tag{8.11b}$$

where α represent the rotational angle at the ends of the beam. The strain energy of a uniform beam under pure torsion T, as shown in Figure 8.5, can be written as:

$$U_T = \frac{1}{2}\int_0^L \frac{T^2}{\mathrm{GJ}}dL = \frac{1}{2}\frac{T^2L}{\mathrm{GJ}} = \frac{1}{2}\frac{\mathrm{GJ}}{L}(\Delta\beta)^2 \tag{8.11c}$$

where $\Delta\beta$ represent the relative rotation between the ends of the beam.

Comparing Eqs. (8.10) and (8.11), it can be observed that both U_r and U_A give the stretching energy, both U_θ and U_M give the binding energy, and both U_τ and U_T represent the torsional energy. Thus, it can be assumed that the rotation angle 2α is equivalent to the change in bond angle $\Delta\theta$. Similarly, ΔL is equivalent to Δr and $\Delta\beta$ is equivalent to $\Delta\varphi$. Thus, comparing the Eqs. (8.10) and (8.11), a direct relation between the structural parameters (EA, EI, and GJ) and the molecular mechanics parameters (k_r, k_θ, and k_τ) can be established as shown below.

$$\frac{\mathrm{EA}}{L} = k_r, \quad \frac{\mathrm{EI}}{L} = k_\theta, \quad \frac{\mathrm{GJ}}{L} = k_\tau \tag{8.12}$$

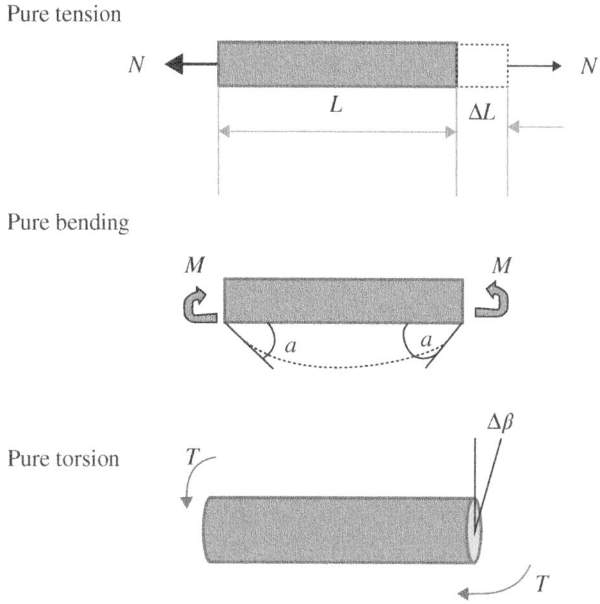

Figure 8.5 An element (C–C bond) under pure tension, bending and torsion.

Eq. (8.12) lays the foundation for the application of structural mechanics approach to the modeling of CNTs and similar structures. After determining the structural parameters, the procedure of stiffness matrix method is followed for the determination of deformation and stress–strain behavior of nanomaterials.

8.5 Young's Modulus of Graphene and CNT

As a case study, the Young's modulus of graphene sheet shown in Figure 8.6 has been calculated using the structural mechanics approach discussed in the previous section. The Young's modulus of any material is the ratio of normal stress to normal strain in a uniaxial tension test and is given as:

$$Y = \frac{\sigma}{\varepsilon} = \frac{\frac{F}{A_o}}{\frac{\Delta H}{H_o}} \tag{8.13}$$

where

F = total force acting on the atoms at one end of the sheet
$A_0 = Wt$ = Cross-sectional area of the sheet width (W) and thickness (t)
H_0 = initial length
ΔH = elongation

The thickness of the graphene sheet is taken as 0.34 nm which is equal to the interlayer spacing of graphite [15]. The force-field parameters can be taken from assisted model building with energy refinement (AMBER) force-field [13]. These parameters have been shown in Table 8.1. From Table 8.1, it can be seen that the results of Young's modulus of

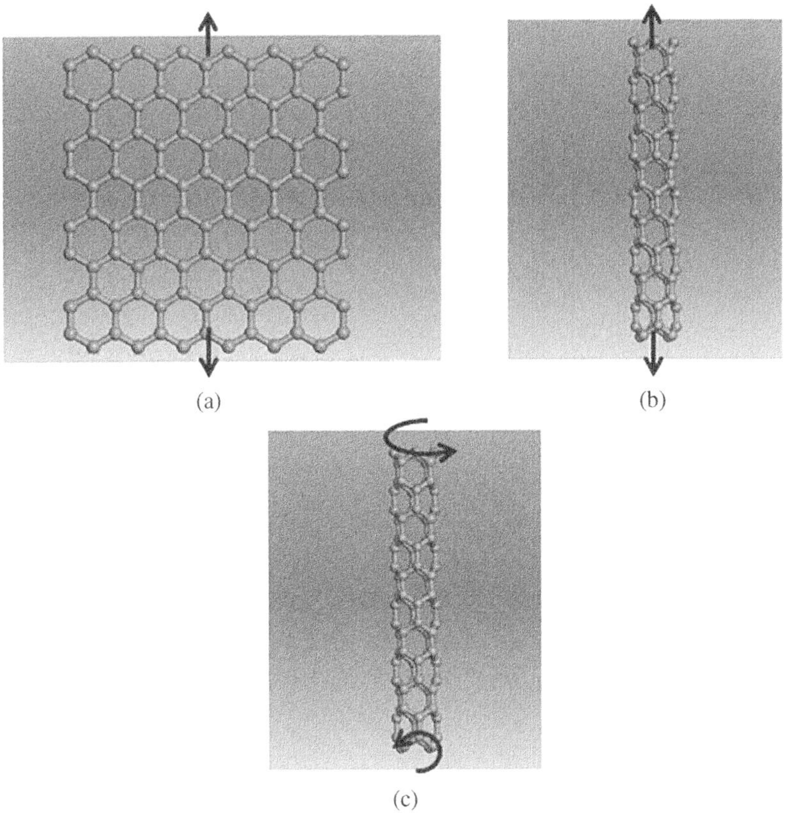

Figure 8.6 Atomistic models of graphene sheet and SWCNT.

graphene for both pairs of force-field constants are close to the value of 1.025 TPa given by Kelly [16]. Also, the model size has negligible influence on the Young's modulus of graphene.

Similarly, for modeling of a CNT, the values of force constants are chosen as $\frac{1}{2}k_r = 469\ \text{kcal}/(\text{mol}\ \text{Å}^2)$, $\frac{1}{2}k_\theta = 63\ \text{kcal}/(\text{mol rad}^2)$, and $\frac{1}{2}k_\tau = 20\ \text{kcal}/(\text{mol rad}^2)$. These values have been used by Cornell et al. [13] and Jorgensen and Severance [17]. The effect of k_τ on the Young's modulus of CNT is very weak. The CNT is subjected to a tensile force at one of its ends, while the other end is fixed. The only change in Eq. (8.13) is that the sectional area is now, $A_0 = \pi dt$, where d represents the tube diameter. Variation of Young's modulus (E_{11}) with radius for different types of SWCNTs has been shown in Figure 8.7. Yao and Lordi [18] explored the dependence of Young's modulus (Y) on diameter and helicity by using the universal force field developed by Rappe et al. [10]. It has been observed that Y decreases significantly with increasing tube diameter, e.g. Y for a (20,20) tube is 15% smaller than that for a (5,5) tube. The correlation between the variation of Y and the torsional energy with diameter and helicity was also noted. Yao and Lordi [18] also discussed the difference between such dependence of Y noted by them and near independence of Y on diameter and helicity as noted by Lu [19] on the basis of the absence of torsion energy or four-atom interaction energy term in the computations of Lu [19]. Thus, the results of Yao and Lordi [18] suggest that the diameter dependence observed in the present work is not surprising. All these observations suggest that either Y is constant with diameter or the variation of Y

Table 8.1 Variation of Young's modulus of graphene sheet with model size and force-field parameters k_r (kcal/[mol Å^2]) and K_θ (kcal/[mol rad^2]).

Model size		$\frac{1}{2}k_r, \frac{1}{2}k_\theta$	
Width	**Height**	**469, 63**	**447, 70**
Young's modulus (TPa)			
0.74	2.84	0.994	1.012
0.98	3.13	1.012	1.020
1.97	4.83	1.022	1.023
1.97	9.85	1.025	1.024
4.24	20.18	1.035	1.031

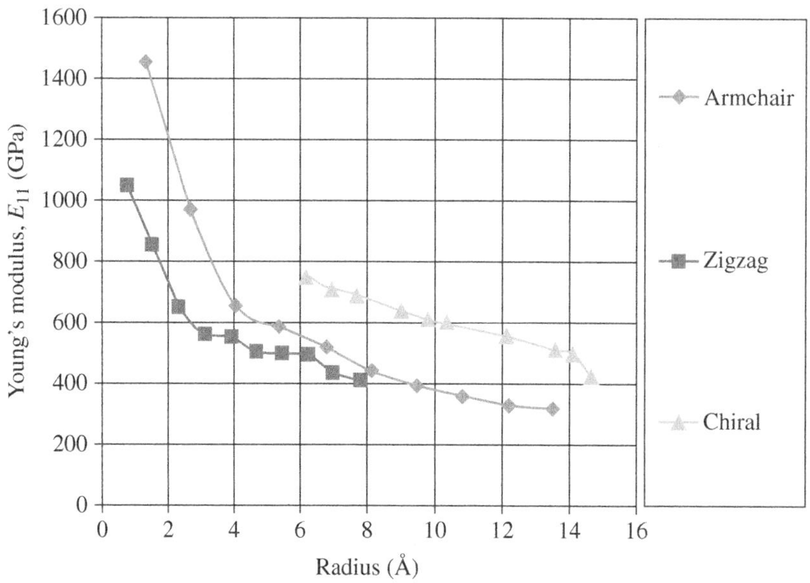

Figure 8.7 Variation of Young's modulus (E_{11}) with radius for different types of SWCNTs.

with diameter is small and it depends on some other factors such as four-atom interaction terms. Depending upon the strength of such terms at the configuration of the atoms of the tube, the computed results might show a decrease or increase or constancy of Y with diameter.

The carbon atoms in a CNT are in sp^2 configurations and connected to one another by three strong σ bonds. Due to the geometric orientation of the carbon–carbon bonds relative to the nanotube axis, armchair SWCNTs exhibit higher tensile strength and Young's moduli values compared to the zigzag SWCNTs. In general, the Young's moduli depend more on the radii than on the helicity. The Young's modulus of chiral SWCNTs is slightly lower than that of armchair and zigzag SWCNTs as observed from Figure 8.7. The results obtained for SWCNTs have been tabulated in Table 8.2. The average of Young's modulus, E_{11} of armchair SWCNTs

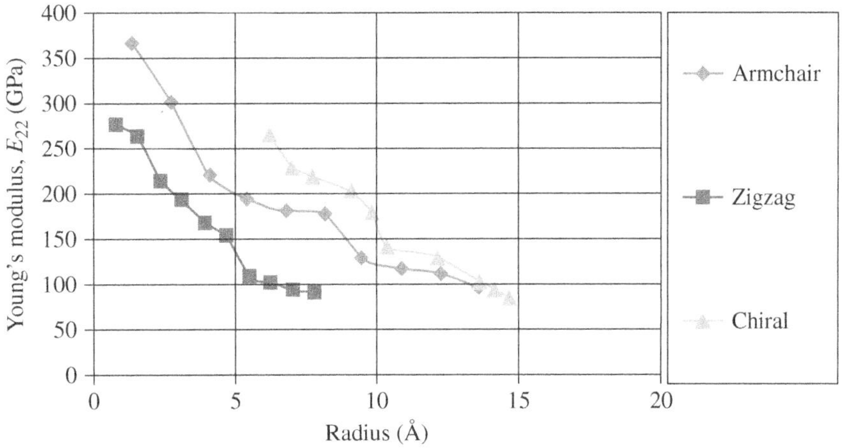

Figure 8.8 Variation of Young's modulus (E_{22}) with radius for different types of SWCNTs.

is 602 GPa. The average transverse Young's modulus, E_{22} of armchair SWCNTs is 190 GPa. Similarly, the average of Young's modulus, E_{11} of zigzag SWCNTs is 599 GPa, and the average transverse Young's modulus, E_{22} of zigzag SWCNTs is 166 GPa. The average of Young's modulus, E_{11} of chiral SWCNTs is 598 GPa, and the average transverse Young's modulus, E_{22} of chiral SWCNTs is 165 GPa.

Figure 8.8 shows the variation of Young's modulus (E_{22}) with radius for different types of SWCNTs. The transverse Young's modulus (E_{22}) also decreases with increase in radius of SWCNT. Also, the value of (E_{22}) is higher for armchair SWCNT in comparison to the zigzag and chiral SWCNTs. Table 8.3 shows the results obtained for Bulk modulus, shear modulus, and Poisson's ratio of SWCNTs for different tube radius. The average Bulk modulus of armchair SWCNTs is 94.2 GPa and of zigzag SWCNTs is 91.5 GPa. The average Bulk modulus of chiral SWCNTs is 134.82 GPa. The average shear modulus of armchair SWCNTs is 68.51 GPa and of zigzag SWCNTs is 72.36 GPa. The average shear modulus of chiral SWCNTs is 58.66 GPa, whereas the Poisson's ratio is the same (0.30) for both armchair and zigzag SWCNTs, the Poisson's ratio for chiral SWCNTs is 0.22.

The shear modulus of CNTs can be calculated by subjecting the CNT to a torsional moment at one end and constraining the other end. The shear modulus can thus be obtained using the following equation.

$$S = \frac{TL_o}{\theta J_o} \tag{8.14}$$

where

T = torque applied at the end of SWCNT
L_o = length of the SWCNT
θ = angle of torsion
J_o = cross-sectional polar inertia of the SWCNT

Figure 8.9 shows that the shear modulus (G) decreases with increase in radius of SWCNTs. It can be inferred from Figure 8.9 that shear modulus of armchair SWCNTs is greater than that of zigzag SWCNT. For diameters greater than 12 Å, the shear modulus is almost the same for all three types of SWCNTs. Also, the value of Poisson's ratio is the greatest for

Table 8.2 Young's modulus of SWCNTs for different tube radius.

(*n*,*m*)	Number of atoms	Radius (Å)	Length (Å)	Longitudinal Young's modulus, E_{11} (GPa)	Transverse Young's modulus, E_{22} (GPa)
Armchair					
(2,2)	48	1.35	14.75	1455	367
(4,4)	192	2.71	29.51	970	302
(6,6)	408	4.07	41.81	652	220
(8,8)	768	5.42	59.08	587	194
(10,10)	1120	6.78	68.86	519	181
(12,12)	1680	8.13	86.08	440	178
(14,14)	2240	9.49	98.38	391	129
(16,16)	2880	10.85	110.67	360	118
(18,18)	3744	12.2	127.89	330	112
(20,20)	4560	13.56	140.19	317	98
Average				602	190
Zigzag					
(2,0)	24	0.78	12.78	1050	275
(4,0)	80	1.56	21.30	850	263
(6,0)	168	2.35	29.82	650	214
(8,0)	288	3.13	38.34	560	193
(10,0)	400	3.92	42.60	550	167
(12,0)	624	4.7	55.38	500	153
(14,0)	840	5.48	63.9	498	108
(16,0)	1088	6.26	72.42	490	102
(18,0)	1368	7.04	80.94	430	94
(20,0)	1680	7.83	89.46	410	89
Average				599	166
Chiral					
(12,6)	1176	6.21	78.89	750	265
(14,6)	1896	6.96	113.59	711	230
(16,6)	2328	7.71	125.86	690	219
(18,8)	3192	9.03	147.38	639	204
(20,8)	3744	9.78	159.62	610	180
(20,10)	3360	10.36	135.25	598	142
(24,11)	3844	12.13	132.06	556	130
(30,8)	4816	13.58	147.81	512	102
(30,10)	6240	14.11	184.31	498	93
(30,12)	6552	14.66	186.22	424	85
Average				598	165

Table 8.3 Bulk modulus, shear modulus, and Poisson's ratio of SWCNTs for different tube radius.

(*n*,*m*)	Number of atoms	Radius (Å)	Length (Å)	Bulk modulus, *K* (GPa)	Shear modulus, G_{Voight}, (GPa)	Poisson's ratio, ν_{12}
Armchair						
(2,2)	48	1.35	14.75	160.46	113.62	0.34
(4,4)	192	2.71	29.51	106.32	89.01	0.36
(6,6)	408	4.07	41.81	84.56	84.70	0.34
(8,8)	768	5.42	59.08	75.59	61.20	0.29
(10,10)	1120	6.78	68.86	83.78	62.22	0.31
(12,12)	1680	8.13	86.08	89.9	58.07	0.32
(14,14)	2240	9.49	98.38	94.16	56.56	0.3
(16,16)	2880	10.85	110.67	89.93	56.94	0.29
(18,18)	3744	12.2	127.89	83.88	54.07	0.26
(20,20)	4560	13.56	140.19	73.43	48.78	0.27
Average				94.20	68.51	0.30
Zigzag						
(2,0)	24	0.78	12.78	126.82	110.51	0.34
(4,0)	80	1.56	21.30	119.92	97.5	0.38
(6,0)	168	2.35	29.82	90.34	73.31	0.35
(8,0)	288	3.13	38.34	80.46	70.21	0.33
(10,0)	400	3.92	42.60	75.48	67.84	0.31
(12,0)	624	4.7	55.38	74.22	69.02	0.28
(14,0)	840	5.48	63.9	74.96	60.28	0.29
(16,0)	1088	6.26	72.42	84.59	60.2	0.3
(18,0)	1368	7.04	80.94	90.87	58.63	0.26
(20,0)	1680	7.83	89.46	97.37	56.19	0.27
Average				91.50	72.36	0.30
Chiral						
(12,6)	1176	6.21	78.89	138.41	80.49	0.26
(14,6)	1896	6.96	113.59	128.98	75.12	0.28
(16,6)	2328	7.71	125.86	132.58	58.06	0.27
(18,8)	3192	9.03	147.38	138.68	56.33	0.22
(20,8)	3744	9.78	159.62	141.05	60.87	0.23
(20,10)	3360	10.36	135.25	147.49	54.78	0.19
(24,11)	3844	12.13	132.06	137.19	52.25	0.18
(30,8)	4816	13.58	147.81	130.66	51.43	0.19
(30,10)	6240	14.11	184.31	125.8	47.62	0.17
(30,12)	6552	14.66	186.22	127.44	49.7	0.173
Average				134.82	58.66	0.22

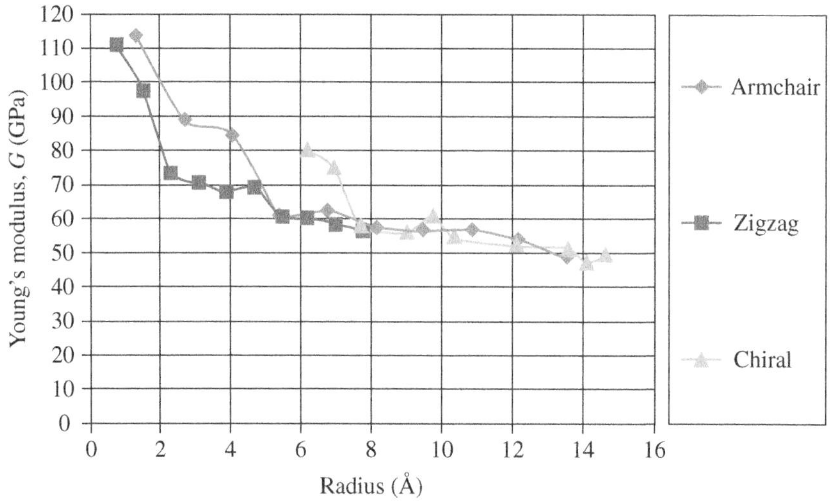

Figure 8.9 Variation of Shear modulus (*G*) with radius for different types of SWCNTs.

armchair SWCNT and the Poisson's ratio decreases with increase in diameter of SWCNTs. For diameters less than 12 Å, the average bulk modulus (*K*) of armchair SWCNTs is greater than that of zigzag nanotubes. For diameters greater than 12 Å, the average bulk modulus (*K*) of zigzag SWCNTs is greater than that of armchair nanotubes. Also, it can be observed from Table 8.3 that for diameters greater than 12 Å, the average bulk modulus (*K*) of chiral SWCNTs is the greatest.

8.5.1 Modeling of Polymer Matrix

The polymer matrix can be treated as a continuum, and finite element method (FEM) can be applied to it. Meshing of the matrix can be achieved using two types of three dimensional finite elements viz. the three-dimensional 20-nodes isoparametric cubic element and the 15 nodes isoparametric cubic element. These two elements have been shown in Figure 8.10. The 20-nodes isoparametric cubic elements are used in the circumferential region around the SWCNT. The 15 node elements are used in the regions directly above or below the SWCNT as shown in Figure 8.11.

8.6 Modeling of CNT/Polymer Interface

The interface can be modeled as a truss rod. The truss rod represents the van der Waals interactions at the interface. This model was first developed for simulating the van der Waals forces between the layers of an MWCNT by Li and Chou [20]. According to this model, the van der Waals force between the interacting atoms is given by the Lennard–Jones (6–12) potential. This potential is given as:

$$F(r) = -\frac{dU(r)}{dr} = 24\frac{\varepsilon}{\sigma}\left[2\left(\frac{\sigma}{r}\right)^{13} - \left(\frac{\sigma}{r}\right)^{7}\right] \tag{8.15}$$

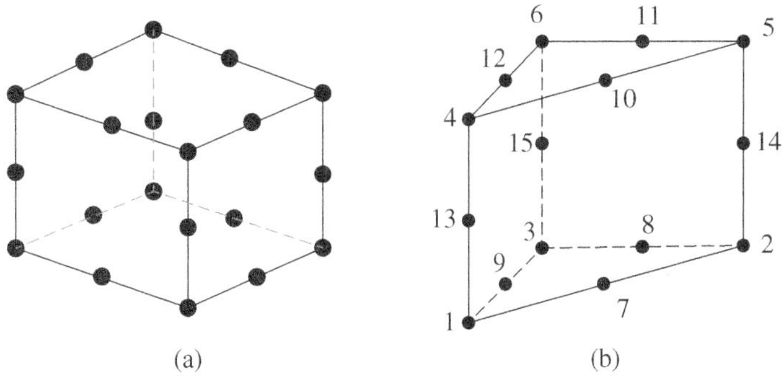

Figure 8.10 Elements for modeling of polymer: (a) 20 node cubic element (b) 15 node wedge element.

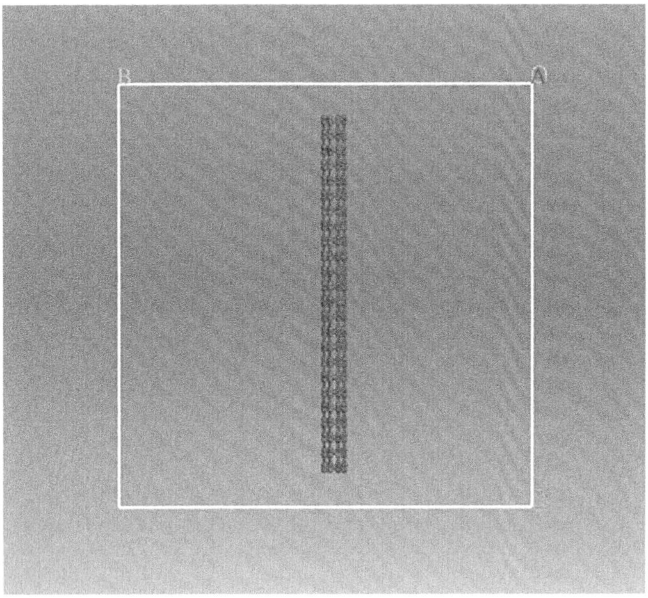

Figure 8.11 Computational model of CNT/polymer composite.

where

r = interatomic distance
ε and σ = Lennard–Jones parameters

For carbon atoms, the Lennard–Jones parameters are $\varepsilon = 0.0556$ kcal/mol and $\sigma = 3.4$ Å [21].

For simplicity in computations, only the van der Waals interactions between the CNT and the surface of polymer immediately adjacent to the CNT are considered. This assumption may give an underestimation of the load transfer capability of the CNT/polymer interface. Two types of interface can be considered: (i) van der Waals interface and (ii) perfect interface. In van der Waals interface, the center of the atoms in SWCNT is assumed to be located in the midsection of the tube thickness, which can be assumed as 0.34 nm, as discussed earlier. The inner surface of the polymer matrix is assumed to be located at the same position as

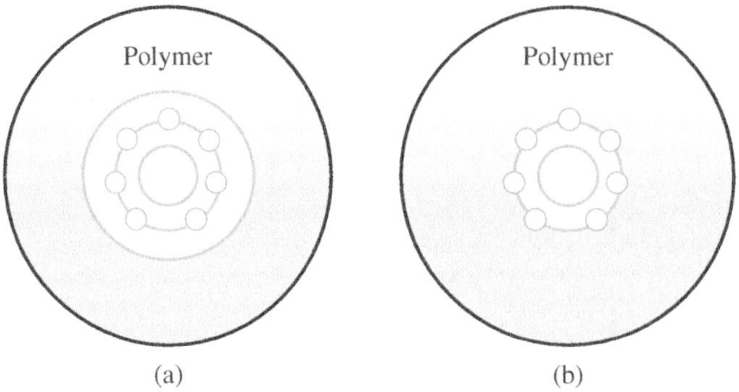

Figure 8.12 Two types of interface between CNT and polymer: (a) van der Waals interface and (b) perfect interface.

the outer surface of the CNT. The activation of a truss rod is based on the distance between a node in the finite element and an atom in the CNT. If this distance is less than 2.5 σ, the truss rod is activated. In perfect interface model, the outer surface of the CNT is assumed to be coinciding with the inner surface of the polymer. The center of atoms in the CNT is now assumed to be located on the outer surface. Both these models of the CNT/polymer interface have been shown in Figure 8.12.

Assuming the polymer as epoxy having Young's modulus 2.41 GPa and Poisson's ratio of 0.35, the Young's modulus of CNT/polymer composite can be obtained. For finding the Young's modulus, an isostrain of 0.001 can be applied. The outer radius of CNT can be taken as 4.71 Å, whereas the outer radius of the polymer matrix can be taken as approximately five times the outer radius of the CNT. The sectional areas of the CNT and the polymer matrix can be calculated using the following equations:

$$A_{\mathrm{nt}} = \pi[(R_{\mathrm{nt}}{}^2 - (R_{\mathrm{nt}} - 3.4)^2] \tag{8.16}$$

$$A_{\mathrm{m}} = \pi[(R_{\mathrm{m}}{}^2 - \mathrm{R}_{\mathrm{nt}}{}^2] \tag{8.17}$$

where, R_{nt} and R_m denote the outer radius of the CNT and polymer, respectively. The effective Young's modulus of the composite is calculated from the ratio of the total force in the composite to the product of sectional area and the applied strain. For the applied strain of 0.001, the effective Young's modulus of the composite was found to be 36 GPa for CNT volume fraction of 0.03. For the same loading conditions, the shear stresses due to perfect interface were found to be much larger in comparison to the shear stresses due to van der Waals interface.

8.7 Elastic Buckling of CNT/Polymer Composite

Buckling is one of the prominent failure modes of a structural component under compressive loading. Behavior of a CNT/polymer composite can be understood more clearly by addressing the issue of elastic buckling of these composites. Using the eigenvalue analysis technique, the buckling force of any structural component can be determined. This requires the consideration of a geometrically nonlinear problem. Once the total elastic stiffness matrix $\boldsymbol{K}_{\mathrm{E}}$ and the total geometric stiffness matrix $\boldsymbol{K}_{\mathrm{G}}$ are obtained using the conventional structural mechanics approach, the eigenvalue equations for finding the load factor can be written as:

$$(\boldsymbol{K}_{\mathrm{E}} + \lambda \boldsymbol{K}_{\mathrm{G}})\boldsymbol{U} = 0 \tag{8.18}$$

where

λ = loading factor
$\boldsymbol{U}$ = displacement vector

The lowest value of $\lambda_{\min}$ is obtained from Eq. (8.18). The buckling force can be calculated as:

$$\boldsymbol{P} = \lambda_{\min} \boldsymbol{P}^{*} \tag{8.19}$$

where $\boldsymbol{P}^{*}$ denotes the reference load.

The element stiffness matrix ($\boldsymbol{K}_{\mathrm{e}}$) for the CNT modeled using molecular structural mechanics approach can be written as:

$$\mathbf{K}^{\mathrm{e}} = \begin{bmatrix} \boldsymbol{K}_{ii}{}^{\mathrm{e}} & \boldsymbol{K}_{ij}{}^{\mathrm{e}} \\ \boldsymbol{K}_{ij}^{\mathrm{eT}} & \boldsymbol{K}_{jj}{}^{\mathrm{e}} \end{bmatrix} \tag{8.20}$$

where $\boldsymbol{K}_{ii}{}^{\mathrm{e}}$, $\boldsymbol{K}_{ij}{}^{\mathrm{e}}$, and $\boldsymbol{K}_{jj}{}^{\mathrm{e}}$ are same as Eqs. (8.4–8.7) and $\boldsymbol{K}_{ij}{}^{\mathrm{eT}}$ is the transpose of $\boldsymbol{K}_{ij}{}^{\mathrm{e}}$. The element geometric stiffness matrix $\mathbf{K}^{\mathrm{g}}$ is given as [22]:

$$\mathbf{K}^{\mathrm{g}} = \begin{bmatrix} \boldsymbol{K}_{ii}{}^{\mathrm{g}} & \boldsymbol{K}_{ij}{}^{\mathrm{g}} \\ \boldsymbol{K}_{ij}^{\mathrm{gT}} & \boldsymbol{K}_{jj}{}^{\mathrm{g}} \end{bmatrix} \tag{8.21}$$

where $\boldsymbol{K}_{ii}{}^{\mathrm{g}}$, $\boldsymbol{K}_{ij}{}^{\mathrm{g}}$, and $\boldsymbol{K}_{jj}{}^{\mathrm{g}}$ are given by Yang and Kuo as [23]:

$$\boldsymbol{K}_{ii}{}^{\mathrm{g}} = \begin{bmatrix} a & 0 & 0 & 0 & -d & -e \\ 0 & b & 0 & d & g & k \\ 0 & 0 & c & e & h & g \\ 0 & d & e & f & i & l \\ -d & g & h & i & j & 0 \\ -e & k & g & l & 0 & m \end{bmatrix} \tag{8.22}$$

$$\boldsymbol{K}_{ij}{}^{\mathrm{g}} = \begin{bmatrix} -a & 0 & 0 & 0 & -n & -o \\ 0 & -b & 0 & n & -g & k \\ 0 & 0 & -c & o & -h & -g \\ 0 & -d & -e & -f & -i & -l \\ d & -g & h & -i & p & -q \\ e & -k & -g & -l & q & r \end{bmatrix} \tag{8.23}$$

$$\boldsymbol{K}_{jj}{}^{\mathrm{g}} = \begin{bmatrix} a & 0 & 0 & 0 & n & o \\ 0 & b & 0 & -n & g & -k \\ 0 & 0 & c & -o & h & g \\ 0 & -n & -o & f & i & l \\ n & g & h & i & j & 0 \\ o & -k & g & l & 0 & m \end{bmatrix} \tag{8.24}$$

where

$$a = \frac{F_{xb}}{L}, \quad b = c = \frac{6F_{xb}}{5L} + \frac{12IF_{xb}}{AL^{3}}, \quad d = \frac{M_{ya}}{L}, \quad e = \frac{M_{za}}{L}, \quad f = \frac{JF_{xb}}{AL}$$

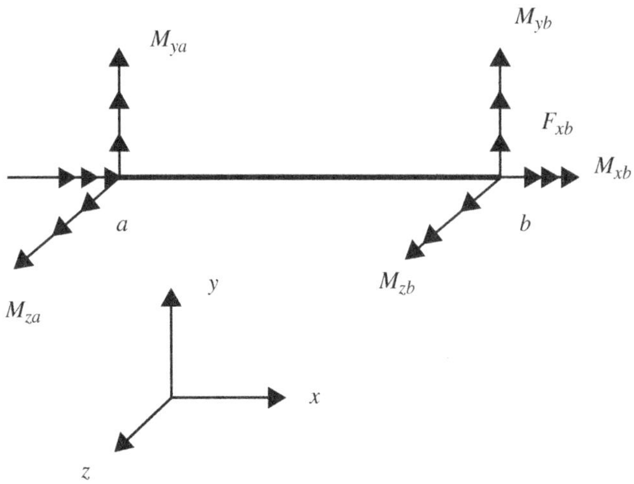

Figure 8.13 A beam element showing the internal forces.

$$g = \frac{M_{xb}}{L},\quad h = k = \frac{F_{xb}}{10} + \frac{6IF_{xb}}{AL^3},\quad i = \frac{M_{za} + M_{zb}}{L},\quad j = m = \frac{2F_{xb}L}{15} + \frac{4F_{xb}I}{AL}$$

$$l = -\frac{M_{ya} + M_{yb}}{6},\quad n = \frac{M_{yb}}{L},\quad o = \frac{M_{zb}}{L},\quad p = r = -\frac{F_{xb}L}{30} + \frac{2F_{xb}I}{AL},\quad q = -\frac{M_{xb}}{L}$$

F_{xb}, M_{xb}, M_{ya}, M_{yb}, M_{za}, and M_{zb} are the internal forces in a beam element shown in Figure 8.13.

8.8 Conclusions

A structural mechanics approach has been developed for modeling CNTs. A simple linkage between structural mechanics and molecular mechanics is established. In this approach, the computational strategy is essentially classical structural mechanics, but the theoretical concept stems from modern computational chemistry, and the modeling is kept at the atomistic scale. The computational results for elastic properties of CNTs are comparable to those obtained from other modeling techniques. The major advantages of the method are the simplicity of the concept and the improved computational efficiency for analyzing the deformation of CNTs. It is expected that the present methodology will be further developed to model other covalent-bonded materials and facilitate analysis of the deformation of nanotube-based composites.

References

1 Iijima, S., Brabec, C., Maiti, A., and Bernholc, J. (1996). Structural flexibility of carbon nanotubes. *Journal of Chemical Physics* 104: 2089–2092.

2 Yakobson, B.I., Campbell, M.P., Brabec, C.J., and Bernholc, J. (1997). High strain rate fracture and C-chain unraveling in carbon nanotubes. *Computational Materials Science* 8: 341–348.

3 Hernandez, E., Goze, C., Bernier, P., and Rubio, A. (1998). Elastic properties of C and $B_xC_yN_z$ composite nanotubes. *Physical Review Letters* 80: 4502–4505.
4 Sanchez-Portal, D. (1999). Ab initio structural, elastic, and vibrational properties of carbon nanotubes. *Physical Review B* 59: 12678–12688.
5 Tersoff, J. (1992). Energies of fullerenes. *Physical Review B* 46: 15546–15549.
6 Yakobson, B.I., Brabec, C.J., and Bernholc, J. (1996). Nanomechanics of carbon tubes: instabilities beyond linear range. *Physical Review Letters* 76: 2511–2514.
7 Weaver, W. Jr. and Gere, J.M. (1990). *Matrix Analysis of Framed Structures*, 3rde. New York: Van Nostrand Reinhold.
8 Popov, V.N., Van Doren, V.E., and Balkanski, M. (2000). Elastic properties of single-walled carbon nanotubes. *Physical Review B* 61: 3078–3084.
9 Machida, K. (1999). *Principles of Molecular Mechanics*. Tokyo: Kodansha and John Wiley & Sons Co-publication.
10 Rappe, A.K., Casewit, C.J., and Colwell, K.S. (1992). UFF, A full periodic-table force-field for molecular mechanics and molecular dynamics simulations. *Journal of the American Chemical Society* 114: 10024–10035.
11 Brenner, D.W. (1990). Empirical potential for hydrocarbons for use in simulating the chemical vapor deposition of diamond films. *Physical Review B* 42: 9458.
12 Mayo, S.L., Olafson, B.D., and Goddard, W.A. (1990). Dreiding-a generic force-field for molecular simulations. *Journal of Physical Chemistry* 94: 8897–8909.
13 Cornell, W.D., Cieplak, P., and Bayly, C.I. (1995). A second generation force-field for the simulation of proteins, nucleic-acids, and organic-molecules. *Journal of the American Chemical Society* 117: 5179–5197.
14 Gelin, B.R. (1994). *Molecular Modeling of Polymer Structures and Properties*. Cincinnati: Hanser/Gardner Publishers.
15 Dresselhaus, M.S., Dresselhaus, G., and Saito, R. (1995). Physics of carbon nanotubes. *Carbon* 33: 883.
16 Kelly, B.T. (1981). *Physics of Graphite*. London: Applied Science.
17 Jorgensen, W.L. and Severance, D.L. (1990). Aromatic interactions-free-energy profiles for the benzene dimer in water, chloroform, and liquid benzene. *Journal of the American Chemical Society* 112: 4768–4774.
18 Yao, N. and Lordi, V. (1998). Young's modulus of single-walled carbon nanotubes. *Applied Physics* 84: 1939–1943.
19 Lu, J.P. (1997). Elastic properties of carbon nanotubes and nanoropes. *Physical Review Letters* 79: 1297–1300.
20 Li, C.Y. and Chou, T.W. (2003). Elastic moduli of multi-walled carbon nanotubes and the effect of van der Waals forces. *Composites Science and Technology* 63: 1517–1524.
21 Battezzatti, L., Pisani, C., and Ricca, F. (1975). Equilibrium conformation and surface motion of hydrocarbon molecules physisorbed on graphite. *Journal of the Chemical Society* 71: 1629–1639.
22 Li, C.Y. and Chou, T.W. (2004). Modeling of elastic buckling of carbon nanotubes by molecular structural mechanics approach. *Mechanics of Materials* 36: 1047–1055.
23 Yang, Y.B. and Kuo, S.R. (1994). *Theory and Analysis of Nonlinear Framed Structures*. New York: Prentice-Hall.

9

Conclusions and Future Scope

Following conclusions have been drawn on the basis of research conducted and presented here:

(i) FEM/Strain energy method can be used as an alternative tool for predicting the elastic and damping properties of VGCF/PP nanocomposites for longitudinal, transverse, and shear loading by considering geometric parameters.
(ii) Elastic modulus and loss factor predicted by FEM/Strain energy method correlate well with those obtained from experimental measurements.
(iii) There is considerable improvement in longitudinal modulus of the VGCF/PP nano-composite for addition of small amount of nanofiber.
(iv) The longitudinal loss factor decreases 40.54% after addition of 0.231% volume fraction of fiber and 50.76% decrement at 0.464% fiber volume fraction.
(v) At low fiber volume fractions, the matrix stiffness strongly influences the composite stiffness.
(vi) There is better correlation of the loss factor predictions made by Eshelby's model and Hashin–Shtrikman model.
(vii) Mori–Tanaka model may be considered as the best choice for estimating the properties of aligned short fiber composites in comparison to other analytical models such as Cox model, Eshelby's model, Chow model, Lielens model, and modified Halpin–Tsai model.
(viii) Aspect ratio significantly influences the longitudinal storage modulus E'_{11} but has little effect on transverse storage modulus E'_{22}, transverse shear storage modulus G'_{23}, and longitudinal shear storage modulus G'_{12}.

Molecular modeling combined with continuum modeling can be used for predicting the elastic and dynamic properties of nano-composites. Finite element modeling of wavy fibers can also be done on simulation software. Sometimes the nano-reinforcement is not homogeneously dispersed, forming clusters with a higher density of single-walled nanotube (SWNT). To predict the elastic and damping properties of these types of materials is a difficult task. The elastic properties of carbon nano-tubes can be determined using various software such as Materials Studio, Digimat, etc. These software can simulate the nanotubes of various types such as zig-zag, armchair, and chiral structures. This requires a command over programming in Linux/Java. To study the properties at such a level remains a challenging task. Though, much data is available about the longitudinal storage and loss factor, very little research has been carried out for transverse and shear moduli and the corresponding loss factors. This is another area where the research can be carried out.

Mechanics of Particle- and Fiber-Reinforced Polymer Nanocomposites: From Nanoscale to Continuum Simulations, First Edition. Sumit Sharma.

Molecular interactions play a key role in the enhancement mechanism of polymer nanocomposite materials. Past few years have been devoted by academic and industrial researchers to understand physical phenomena and origin of properties improvement in this new class of materials. Micromechanical approaches seem to be limited in the prediction of polymer nanocomposite properties, since the active molecular interactions between nanofillers and polymer matrix as well as those between nanofillers themselves are far from being explicitly taken into consideration. At the nanoscale, analytical models are difficult to establish or too complicated to solve, and tests are extremely difficult and expensive to conduct. Modeling and simulations of nanocomposites, on the other hand, can be achieved readily and cost effectively on even a desktop computer. Characterizing the mechanical properties of carbon nanotube (CNT)-based composites is just one of the many important and urgent tasks that simulations can accomplish.

It has been observed that the properties of nanocomposites depend on many parameters relating structural configurations of their constituent materials at the molecular level. CNT/CNF (carbon nanofiber) concentrations and aspect ratio (l/d) are critical controlling parameters for achieving desirable mechanical properties. As a result, developing reliable computational tools are extremely important in order to understand the interactions of various molecular entities of nanocomposites.

Molecular dynamics (MD) simulation is a powerful tool for numerical experiments of nanocomposites. One of the aims of atomistic simulation is to reproduce experiment to elucidate the invisible microscopic details and further explain the experimental results. The fundamental concept behind molecular dynamics is to apply Newton's equations of motion in solving multi-body interacting system (i.e. polymers, crystals, nanocomposites, etc.). The molecular interactions are the vital parts in MD simulation. This is defined by an appropriate force-field. The key components of a force-field are bonded interactions and nonbonded interactions such as van der Waals force. This makes molecular dynamics simulation to be a primary tool in assessing mechanical properties of polymer nanocomposites.

Index

Printed and bound by CPI Group (UK) Ltd, Croydon, CR0 4YY